BEYOND THE CLASSROOM AQUARIUM

A CURRICULAR SEQUENCE OF 14 PROJECT-BASED, STEM-ORIENTED AQUATIC-ECOSYSTEM THEMED LEARNING ACTIVITIES YOUR STUDENTS WILL LOVE

Robert M. Jakus

"Give students something to do,
Not something to learn;
And the doing is of such a nature as to demand thinking;
Learning naturally results!"
- John Dewey

"We will only save what we love.
We will only love what we understand.
We will only understand what we are taught."
- Chief Seattle

Published by

Robert M. Jakus (* p. 347)

2014

ISBN 13: 978-0-615-99606-6 (Paperback edition)

Cover design and photograph by Robert M. Jakus

Grateful acknowledgement is made for permission to reprint materials published previously by the following individuals or governmental agencies:

Excerpts from the Marine Biology Coloring Book; Copyright ©2000 by Niesen, T.M.; reprinted with permission of Coloring Concepts, Inc.; Collins Reference
Excerpts from the Marine Aquarium Reference: Systems and Invertebrates; Copyright ©1992 by Moe Jr., Martin A.; used by permission of Mr. Martin A. Moe Jr.
An excerpt from: Tales of the South Pacific; Copyright ©1947 by Michener, J. A.; used by permission of Simon & Schuster, NY
Excerpts from Coral Restoration: Giving Mother Nature a Kick Start; copyright ©2009 by Alex Brylske; used by permission of Alex Brylske
Excerpts from The Reef's Tale: Smithsonian Transplants Coral System from Tropics to Tank; copyright ©1980; Walton, Susan; used by permission of Bioscience
Excerpts from Culturing Lumbriculus variegatus by Drewes, Charles; http://www.eeob.iastate.edu/faculty/DrewesC/htdocs/LVCULT.htm; used by permission stated on website
Excerpts from Lights, Lenses and Lasers; Copyright ©1987 by Berger, Melvin used by permission of G.P. Putnam's Sons, New York
Excerpts from Practical Research: Planning and Design; Copyright ©1980 and 2010 by Leedy, Paul D.; Jeanne Ellis Ormrod; Upper Saddle River, N.J.: Merrill Prentice Hall used by permission of Prentice Hall
Modified with Permission of the Chicago Public Schools Science Fair Inc. Form (2014); Form requesting endorsement to use non-human Vertebrates in class "labs/projects"
Excerpts from Marine Aquaculture: Raising Saltwater Fish in Your Classroom; Copyright ©2001 by Moran, Brandy; Goudy, Clifford; and Hankin, Alan; used by permission of MITSG 01-13 Activity One used by permission of Ms. Brandy Moran
Excerpts from Learning Ocean Science through Ocean Exploration: A Curriculum for Grades 6-12; 3rd Edition; Copyright ©2003 by Chase, Valerie (editor); used by permission of NOAA; http://oceanexplorer.noaa.gov used by claim of public domain nature of document
Excerpts from Intensive Cultivation of a Calanoid Copepod - Gladioferens imparipes: A Guide to Procedures; Copyright ©2001 by Rippingale, R.J; Payne, M.F used by claim of public domain nature of the document
Excerpts from Bybee, Rodger; Scientific and Engineering Practices in K–12 Classrooms: Understanding a Framework for K–12 Science Education; December 2011; NSTA's Journals used by permission of NSTA Journals
Excerpts from Under the Microscope; Copyright ©1982 by Curry, Alan; reprinted with permission of Van Nostrand Reinhold Co. Inc.

Internet
Excerpts from Fragmentation of a Branching Coral (Type I), e.g. Acropora sp., Copyright ©2009 (September) by Santos, Elsa; Baylina, Núria; Stylophora pistillata; Pocillopora, damicornis; in Osinga, Ronald and Janse, Max (eds.); Fragmentation; CORALZOO used under claim of public domain nature of the document
Excerpts from Fragmenting Branching Corals; Copyright ©2009 (September) by Schutter, Miriam; Shafir, Shai; in Osinga, Ronald and Janse, Max (eds.); Fragmentation; CORALZOO used under claim of public domain nature of the document
Excerpts from Fragmentation of Massive Corals: Favia Sp.; Lobophyllia Sp.; Copyright ©2009 (September) by Westerd, Leon; Janse, Max; in Osinga, Ronald and Janse, Max (eds.); Fragmentation; CORALZOO used under claim of public domain nature of the document
Excerpts from How Success Works; 2012, by Marten, Gerry; Marten, Julie; reprinted with permission of Mr. Gerry Marten
Excerpts from - NASA Mission Geology: Geography (What's up with Earth's Water Resources?): http://er.jsc.nasa.gov/seh/Mission_Geography/9-12/Module_1/III-1-1.pdf used under claim of public domain nature of the document
Excerpts from Buchheim, Jason; Coral Reef Bleaching; 1998; Odyssey Expeditions- Marine Biology Learning Center Publications; http://www.marinebiology.org/coralbleaching.htm used by permission granted on website
Excerpts from Buchheim, Jason; Coral Reefs: Oceanography: Water, Seawater and Ocean Circulation and Dynamics; 1998, Odyssey Expeditions- Marine Biology Learning Center Publications; http://www.marinebiology.org/oceanography.htm; accessed 3/2/99 used by permission granted on website
Excerpts from Buchheim, Jason; Coral Reefs: Living Corals; 1998, Odyssey Expeditions- Marine Biology Learning Center Publications; http://www.marinebiology.org/coralreefs.htm used by permission granted on website
Excerpts from Buchheim, Jason; 1998, Odyssey Expeditions- Marine Biology Learning Center Publications; http://www.marinebiology.org/fish.htm; used by permission granted on website
Excerpts from Curious About Water ? Copyright ©1998 by Williamson, Robert; Smoak, Ellen; used by permission of NC A&T State University – Cooperative Extension used under claim of public domain nature of the document
Excerpts from - Brooks, David W.; Electrolysis of Water: Becker Demos; http://chemmovies.unl.edu/chemistry/dochem/DoChem044.html; accessed Dec. 2012; (NSF developed materials – Public Domain)
Excerpts from DIY Reef Type Fluorescent Light Hood Plans for Saltwater Aquariums; Copyright ©1997 by Carner, Don; 1997; http://saltaquarium.about.com/od/diyhoodcanopylighting/ss/sbsdiydonhood_4.htm; accessed July 2012 (used by Permission of Don Carner)
Excerpts from Simulated Laminar Flow Leapfrog Fountain; Copyright ©1997 by Schauer, Jerry L.;. http://www.hiddenmickeys.org/Imagineering/LeapFrog.html(usedby permission of hidden mickey's website copyright holder Tom Shaw
Excerpts from History of Hydroponics by Gary V. Deutschmann (1998) used by permission granted by Mr. Gary V. Deutschmann

Image Credits listed on back page of book – Cover image of Peppermint Shrimp in a Classroom Aquarium by Robert M. Jakus

"SailDog" (1973) logo on back cover by permission of Jim Cummings

ISBN	978-0-615-99606-6
Title	*Beyond the Classroom Aquarium: A Curricular Sequence of 14 Project Based, STEM Oriented Aquatic Ecosystem Themed Learning Activities Your Students will Love*
# of pages	339
Format	Paperback
Subject	STEM Curriculum
Author	Robert M. Jakus - robertjakus@yahoo.com; http://www.robertjakus.wordpress.com
Copyright year	2014

Beyond the Classroom Aquarium:
A Curricular Sequence of 14 Project-Based, STEM-Oriented Aquatic Ecosystem Themed Learning Activities Your Students Will Love

Table of Contents

Table of Contents ..... **iii**

Preface vi

Acknowledgements vii

CHAPTER I: Your Classroom as an Aquatic Ecosys-STEM Learning Lab: Setting the Stage for STEM and Taking the Plunge – Introductory Considerations **1**

Connecting Classroom Aquatic Ecosys-STEM Projects to Scientific Inquiry 7

Connecting Technological Applications to Classroom Aquatic Ecosys-STEM Projects 10

Connecting Aquatic Ecosys-STEM Projects to Engineering and Design 12

Connecting Your Student's Individualized Math Curriculum with Aquatic Ecosys-STEM Projects: Logistics of Math Assessment and Daily Work in the Math Lab 14

Aquatic Ecosys-STEM Projects: Budgetary Considerations 18

A School Year Calendar 19

A Word About Beyond the Classroom Aquarium Field Trips 19

Field Trip Feedback Form 21

Budget Sheet Guide (Chronologically Itemized Project Cost Assessment [Blank]) 22

CHAPTER II: Student Awareness of Universal Aquatic Ecosystem Parameters: Introductory System Design Applications (Projects 1, 2, and 3) **23**

PROJECT ONE Introductory Activities: Marine Biology Art, Vocabulary, and Beyond 25

Coloring Instructions 27

Tides *28*

Photic Zone *30*

Types of Coral Reef *32*

Marine Flowering Plants *34*

Crustaceans – Crab Life Cycle *36*

Challenge your Students to Expand their Powers of Observation and Inquiry: Introduction to Operational, Instructor-Established, Aquatic Ecosys-STEMs 38

"In House" Classroom Aquatic Ecosystem Model Number One (An Overview) *39*

"In House" Classroom Aquatic Ecosystem Model Number Two (An Overview) *40*

"In House" Classroom Aquatic Ecosystem Model Number Three (An Overview) *41*

"In House" Classroom Aquatic Ecosystem Model Number Four (An Overview) *42*

A Classroom-Based Aquatic Ecosystem Observation Activity 43

Preliminary Steps – Toward Achieving Student-Designed Aquatic Ecosystems in the Classroom 46

Guidelines for Deciding Which Organisms to Keep in Student Designed Aquatic Ecosystems: A Library Assignment 46

Fish for Aquaculture *47*

Requirements for survival *47*

Fish Descriptions *48*

Aquatic Ecosystem Livestock Selection 50

Aquaponic Species Selection and Lighting System Proposal Form *50*

Coral Species Selection and Lighting System Proposal Form *51*

Broodstock Species Selection and Lighting System Proposal Form *51*

Species Specific Lifecycle Requirements Form *52*

Classroom Discussion of System and Organism Data – An Overview *53*

Project One Enriching Assessment Activities 54

Are Your Fish Really Suffering From Disease? - A guide to non-disease causes of illness * 58

Getting Ready for Project Nine – "Research-ability" Potential Topics Form (Foreshadowing) 59

PROJECT TWO Beyond the Sun - The Nature and Function of Light in Simulated Aquatic Ecosystems 60

Light, Vision and Color – Project Two Opening Activity 60

Color in the Sea *63*

Introduction to Important Lighting Related Concepts, Skills, and Aquatic Ecosystem Applications 64

Specific Lighting System Design and Assembly – An Example ... 74
General Protocols for Lighting System Design and Construction of Three Primary Aquatic Ecosystems ... 75
The Aquaponics Frame and Lighting System (Field Trip #1 Applications) ... *76*
Coral Propagation Frame and Lighting System (Field Trip #1 Applications) ... *80*
The Broodstock/Larval Rearing System Frame and Lighting System (Field Trip #1 Applications) ... *83*
Summary of Lighting System and Frame Design and Assembly ... 86
Project Two Enriching Assessment Activities ... 87
PROJECT THREE ...Beyond the Shore - Water Containment & Control in Simulated Aquatic Ecosystems ... 91
Electrolysis of Water - Project Three Opening Activity ... 93
Introduction to Important Water Related Concepts, Skills, and Aquatic Ecosystem Applications ... 95
Water Management and Control Equipment ... 100
Water Processing, Filtration (Prospective Considerations for System Design and Research) ... 102
The Aquaponics System Water Containment and Flow Design (Field Trip #2 Applications) ... 105
Dissolved Oxygen Measurement in Classroom Aquatic Ecosystems ... *108*
Calculating Classroom Aquatic Ecosystem Bio-Filter Surface Area Requirements ... *110*
The Coral Propagation System Water Containment and Flow Design (Field Trip #2 Applications) ... 112
Measuring Classroom Aquatic Ecosystem Salinity by Chloride Titration ... *114*
Ammonia, Nitrite and Nitrate Waste Product Monitoring and Control in Classroom Aquatic Ecosystems ... *115*
The Broodstock and Larval Rearing System Water Containment and Flow Design (Field Trip #2 Applications) ... 117
pH in Classroom Aquatic Ecosystems ... *120*
Project Three Enriching Assessment Activities ... 122
CHAPTER III: Students Design, Assemble, & Maintain Primary Aquatic Ecosys-STEMs (Projects 4 - 6) ... 126
PROJECT FOUR Aquaponics and Beyond ... 130
Seed Germination - Project Four Opening Activity ... 130
Materials, Equipment, and Instructions for Germinating Seeds ... 131
Introductory and Background Aquaponics Information ... 132
Completing the Aquaponics System ... 137
Photosynthesis and Respiration in Classroom Aquatic Ecosystems ... *139*
Coordinated Sharing of Aquaponic System Development and Operation Information with Classmates ... 141
Project Four Enriching Assessment Activities ... 142
PROJECT FIVE Coral Propagation and Beyond: Biology & Natural History of Corals (Soft & Hard) ... 146
Coral Frag Substrate Production - Project Five Opening Activity ... 146
Introduction to Coral Related Topics ... 147
Completing the Coral Propagation System ... 160
Fragging Protocols ... *161*
Coordinated Sharing of Coral Propagation System Development and Operation Information with Classmates ... 165
Project Five Enriching Assessment Activities ... 166
PROJECT SIX Aquaculture and Beyond: Gravity-Flow Habitats for Three Sets of Broodstock Fish or Other Compatible Aquatic Organisms ... 169
How to Culture Lumbriculus, variegatus - Project Six Opening Activity – (Freshwater) ... *169*
Design a System for Hatching and Rearing Artemia, salina – Project Six Opening Activity – (Saltwater) ... *171*
Introduction to Aquaculture Related Topics ... 172
Introductory Information about Fish ... 175
Completing the Broodstock and Larval Rearing System ... 185
Coordinated Sharing of Broodstock and Larval Rearing System Development and Operation Information with Classmates ... 186
Project Six Enriching Assessment Activities ... 187
CHAPTER IV: Students Design, Assemble & Maintain Auxiliary Aquatic Ecosys-STEM Components (Projects 7-8) 191
PROJECT SEVEN Design, Construction, and Ongoing Maintenance of an Algae-Rotifer Production System . . . and Beyond ... 193
Laboratory Production of Guillard's f/2 Formula – Project Seven Opening Activity ... 193
Algae and Rotifer Biology and Production – Auxiliary System Designs ... *195*
Computer Automated Plankton Culture and Collection System ... *200*
Project Seven Enriching Assessment Activities ... 203
PROJECT EIGHT ...Beyond Microscope Design and Construction: Microscopic Aquatic Organisms ... 206
Introductory Optics Activity ... 207
Design and Construction of a Homemade Microscope ... 210
Assembly of the Mechanical Stage ... *216*
Microscope Activities with Aquatic Organisms ... 217
Using a Compound Microscope and Hemocytometer to Monitor Classroom Cultures of Unicellular Algae ... *218*
Project Eight Enriching Assessment Activities ... 220

CHAPTER V: Students Design and Execute an Aquatic Ecosystem Research Project (Project 9) 225
PROJECT NINE ...Beyond the Science Fair - Design & Execute an Aquatic Ecosystem Research Project 227
Introductory Research Related Activities – Project Nine 227
Introductory Research Reviewing Activity – Field Trip Six 231
Field Trip Seven – An Exploratory Data Collection Experience 234
Research Project Selection, Design, and Approval...237
Project Nine Enriching Assessment Activities 246
CHAPTER VI: Application of Aquatic Ecosys-STEM Knowledge and Skills to Local, Regional, and Global Issues (Projects 10 – 13)..249
PROJECT TEN Beyond Larval Rearing – Optimal Aquatic Ecosystem Conditions for Juvenile Grow-Out 251
Design and Implement Classroom Relevant Juvenile Growout Systems – Project Ten Opening Activity...........................251
Classroom Fertilization and Larval Rearing of Sea Urchins (Echinoderms) 253
Student Design & Assembly of Bivalve Mollusc, Cnidarian or Crustacean Larval Growout Systems in the Classroom 256
Project Ten Enriching Assessment Activities.. 258
PROJECT ELEVEN Human Ecology – Feedback Diagram Analysis and Beyond: Design an Ecologically Sustainable Service Learning Program Proposal 261
Chesapeake Bay Educator's Guide: Project Eleven Opening Activity 261
How do Natural Processes and Human Actions Affect the Chesapeake Bay as a Water Resource? 272
Critical Issues for Ecosystem Management and Restoration 273
Using Remote Sensing as a Research Tool: Three Examples from NASA 275
How Success Works... 277
Feedback Diagrams: Apo Island (Student Worksheet) - Diagram for students to fill in..281
Class Preparation for Design/Delivery of an Ecologically Sustainable Community Service Program.. 282
Ecologically Sustainable Community Service Learning Program – Design Process (Examples & Forms) 283
Project Eleven Enriching Assessment Activities... 288
PROJECT TWELVE Community Involvement and Beyond: Implementation of an Ecologically Sustainable Service Learning Program..290
Design a Large Scale Aquatic Ecosystem Trade Show Booth - Diorama – Project Twelve Opening Activity................................... 290
Exploration of Future Aquatic Ecosystem Opportunities: Businesses and Careers.. 292
Project Twelve Enriching Assessment Activities.. 293
PROJECT THIRTEEN ...And Beyond! A Prospective Vision of Future Design Projects 296
Design and Construct a Basic Laminar Flow Fountain - Project Thirteen Opening Activity.. 296
Suggestions for Final Independent Project Designs 299
Laminar Flow Fountain Design Activity – Part Two (Supplemental Fountain Components)...301
Mobile Video/Acoustic System (MVA) Design Activity 302
Submersible ROV (Remotely Operated Vehicle) Design Activity..303
Renewable Sustainable Energy Design Activity for Electronic Aquaponics/Greenhouse Applications 304
Calcium Reactor Design Activity...305
A Futuristic Experimental Aquarium System Design Activity...306
Project Thirteen Enriching Assessment Activities 307
CHAPTER VII: A Student Produced *Beyond the Classroom Aquarium* Expo & Conference (Project 14) 309
PROJECT FOURTEEN Back From Beyond: Student-Organized Aquatic Ecosystem Conference Summative Assessment..310
Conference Staging and Run-Through Activity 310
STUDENT ASSESSMENT CHECKLIST for Conference Preparation.. 311
Summative Assessment Grids and Timelines..312
Bibliography 319
Index...334
Pre and Post Assessment 336
Image Credits 337
Connections with Next Generation Science Standards Matrix - by Project 338
About the Author – Robert M. Jakus..339

Preface

Beyond the Classroom Aquarium is a curricular sequence of project based learning activities designed to guide educators and students who use it through a wide range of real-world adventures. Many educators appreciate having such a project-based, STEM-oriented, aquatic ecosystem themed guide that they can modify to best suit the needs of their own students. The bibliography provides resources educators may wish to access for additional information. The sequence and structure of the projects, including end of project activities, provide for ongoing assessment of developing concepts and skills. The scalable aquatic ecosystem themed project sequence maximizes student experiences with multidisciplinary STEM applications. And, each student project has the capacity to be as real-world as their teacher will allow.

I first thought of writing this book as I sought to locate a resource for educators that encouraged a unified compilation of project based, STEM-Oriented, and aquatic ecosystem themed educational resources and activities. I realized I wanted to take on the task of writing this book only after locating and reviewing thousands of pages of documents and speaking to many professional development providers.

Educators who use this book with their students will derive more than a generic sense of "daily adventure." The authentic, real-world flavored, multidisciplinary nature of each project is capable of generating explosive incentives for students to learn and motivation for students to work (alone and in groups). The potential impact of these projects upon a class of students includes appreciation of many aspects of real scientific endeavor, from initial inquiry to personal experiences with numerous technologies and work performed by skilled tradespersons or craft workers, involvement in system design, critiquing and problem solving, and application of an extensive array of mathematical applications. Beyond this, work on these projects is entertaining for many participants.

In addition to keeping aquariums I have, as an educator, pursued numerous professional development experiences to supplement my own curiosity and knowledge about classroom aquatic ecosystems. It has taken me twenty years to develop the sequence of projects presently offered in this book. To this day, I find that I often gain knowledge and develop skills as I apply the principles of inquiry, exploration, and discovery as modeled by the learning activities in this book. I suppose others might consider me a" lifelong learner" and this reminds me to mention that I hope teachers who use the projects in this book become comfortable asking for additional help and use every project related opportunity to model this assistance seeking behavior for their students.

It is also important to reemphasize benefits that educators using this book may derive by personally modifying/adapting activities so as to make each activity/project most useful for themselves and each group of students with which they work. Design of *Beyond the Classroom Aquarium* field trips are intended to encourage educators to reach out to the community for help implementing any number of mechanical/technological details of ecosystem design and development that require time, skill sets, or money/materials that may extend beyond their personal comfort zones. Field trips in this book are also designed to lend a feel of authenticity and credibility to learning activities. These field trip events also provide flex time educators may need in order to keep the sequence of projects on track; to assess a group's progress in moving through the sequence of projects. As mentioned in the book, some of the field trips are optional. Educators should make use of projects and activities in the book with which they believe they can achieve success. In my vision, *Beyond the Classroom Aquarium* is a living document with which each educator safely merges his/her own unique spirit of adventure and learning.

Acknowledgements

In reverse chronological order I first wish to acknowledge the intrepid work Tara Gnagy performed by editing the final manuscript. I also wish to express my appreciation for like-minded colleagues and peers who attended conference sessions I presented and, in subsequent conversations, provided me with insightful feedback and encouragement like Sonya Wood Mahler, Scott Simon, and Admiral Richard Seesholtz to name but a few. A number of classmates and trainers with whom I attended professional development workshops and classes also supported my vision for this book including Beth Day Miller (and a cohort of dedicated educators that worked tirelessly to establish Ocean and Great Lakes literacy standards), "Doc" Herrnkind, Jeff Dutrow, Darlene Richardson and Martha Cordell (from FSU summer '03), David Nadeau at MarineLab and the education staffs at the Mote Laboratory and Shedd Aquarium. I am grateful for the acknowledgement of the expanding audience committee of the National Marine Educators Association, under the direction of Mellie Lewis and Tammy Lunsford for granting me a stipend to share my work with NMEA members. I am grateful to NMEA and the IEEE/MTS organizations for inviting me to share various aspects and evolving versions of this work at conferences; particularly Mare Timmons, Rita Hall, Padgett Kelly, Becky Cox, Bob Rocha and again Tammy Lunsford. I am also grateful for understanding and supportive school administrators Gene Head, Cabrina Williams-Leneau, Joan Brixey, Nicole Fishman who found the classroom aquariums to be of some value to our students. A number of school teacher colleagues have also provided thoughtful feedback and support about this project including Chuck Kreml, Michelle Parish, Mary Mlinar, Cressie Reese, Mike Devins, Nelson Campos, Mary Colwell, Tim Gaus, Ms. Petersen, Laura Abundes and Ms. Lin. I also received great encouragement from professor Tom Anderson at the University of Illinois (and the Teaching Enhancement Activities in Math & Science Instructional Design Workshop TEAMS) – Urbana who nominated me for a Golden Apple teaching award and his colleagues at the University Center of Lake County, IL Hillary Schnadt and Lynne Reynolds for encouraging me to develop this book. Lynne Reynolds even reviewed a very early version of it.

Many students were captivated by the aquatic specimens that lived in our classrooms throughout the years. Most of the organisms the students got to observe and work with were available due to the generous donations of local business persons such at Matt Jensen (Sho-Tank) aquariums and Jim Whitney (SeaQuest Aquariums). All Glass Aquariums also supported my effort to share marine ecosystems with my students. Joe Lichtenbert even provided my class with an aging veteran pair of spawning Tomato Clowns (holy smokes). Rockwell Automation contributed to the technical flavor of this book by donating a programmable logic controller (Micro 830) so that I might gain sufficient personal experience with it to be able to explain how secondary students might be guided to design and create computer automated aquatic ecosystems (and components). Also, great thanks to generous craft persons such as Dave Harrington who helped me learn to work with saws (and routers and beyond) and for the patience and support I received from employees of Home Depot, Menards and Lowes (and many other local neighborhood hardware stores).

I am indebted to many educators and writers who have worked over the past decades to develop engaging aquatic ecosystem themed activities for students. I deeply appreciate the work of organizations that support development of learning activities that include the ethical use of living creatures in the classroom. Programs of this type (to name a few) include: Salmon in the Classroom, Lobsters in the Classroom and, Conch in the Classroom. You will find the work of some of these individuals and groups listed in the bibliography of this book. Given the sequence and structure of projects in this book any student may find an idea in resources listed in the bibliography that illuminate his/her curiosity and desire to explore, discover, and learn. I am also grateful to the organizations that offer aquatic ecosystem themed professional development programs for educators and the organizations that help fund them.

Thank you is also owed to the many writers featured in this book who have granted me permission to share their ideas with fellow educators on a "fair use" basis. Among these writers are included: Martin Moe, Gerry Marten, and Brandy Moran, And similar appreciations must also be delivered to the many governmental agencies and NGO's that have also produced materials that are shared herein as common property of taxpayers and citizens. Such organization include: NASA, NOAA, NSF, DOC, DOD, CORALZOO, the Fisheries Research & Development Corporation, North Carolina A&T University Agriculture Extension and, Odyssey Expeditions. And of course to dear Sally and cousin Larry whose innocent "gift," those many years ago, nudged me into this state of Beyond-ness.

Finally, I am grateful for Kathy's ever-present encouragement to develop and express my own voice... I can never give back to you all the hours, mornings, evenings (after school), weekends (at the school green house or shopping or tending aquaria), weeks (at professional development classes) and months (at more professional development classes). And the constant reading and writing and planning. Thank you for lending a hand (and more encouragement), when a livestock delivery was arriving or a piece of wood or PVC pipe needed trimming. Thanks for going beyond with me in so many ways. And thanks for letting me share it with you. It has been a much nicer journey because of you.

Dedicated to

The youthful awakening of love for Earth's natural aquatic ecosystems

And the desire

To preserve her native aquatic habitats and species

For the beauty and joy they give to the world

With hope for the human race and all her children

In Honor of

My Beautiful and Supportive Wife Katherine – I Love You

Other Inspirational Individuals – Past, Present and Future

Family Members, Friends, Classmates, Students, Colleagues, Educators, and Folks

Who Have Touched My Heart (and Soul) along the Way

Bless Your Hearts - Each One

In Grateful Memory of

Thoughtful, caring, and loving parents: Mindo and Leona Jakus

Amazing educators, colleagues, and mentors:

Michael Lynch ("***Kindness Spoken Here***" *Triton College*) - mathematics
Alex Lane (*Triton College* "***the Big T.***") - Anatomy & Physiology and
Barbara McCormick (*Chicago City Colleges*) – Invertebrate Zoology

That this work may serve to carry forth their learned spirit and kindness to all

CHAPTER I: Your Classroom as an Aquatic Ecosys-STEM Learning Lab: Setting the Stage for STEM and Taking the Plunge – Introductory Considerations

The innovative tapestry of aquatic ecosystem projects in *Beyond the Classroom Aquarium* is designed to enhance student learning opportunities, motivate academic excellence, and guide learners of every kind to adopt a more sustainable sense of global stewardship. Over the years, numerous colleagues, peers, and friends have asked questions of the "What and how have your students done this?" type. Information in *Beyond the Classroom Aquarium* is intended to answer such questions more fully than is possible in a teacher's meeting or during brief professional development workshops. *Beyond the Classroom Aquarium* contains all the information educators need to provide similar experiences for their own students.

In general, *Beyond the Classroom Aquarium* may also serve as a model for educators who seek to unify, for the benefit of students everywhere, a range of disparately interrelated activities under one primary theme. This book is my humble contribution to a growing body of work through which teachers and other leaders of the nation's educational reform movement may glean as they search for educational ideas, visions, and voices worthy of our children. Readers may also discover a light sprinkling of casual narrative insights and reflections from my own teaching experience of 35 years. Please find and use the parts of the book that best meet your own purposes. I hope the resources offered here are of use as you develop classroom activities, consider educational reform ideas, or find the ideas it contains a worthwhile source of entertainment.

My instructional experience with these projects did not follow the linear path that is described here. The extensive collection of aquatic ecosystem-related materials and curricular products upon which the projects in *Beyond the Classroom Aquarium* are based is ever expanding and has been updated often. Some of the resources in this collection present activities that were designed for traditional period-long school settings. Other activities are derived from a range of non-formal educational programs such as intensive immersion seminars, workshops, and educational activities offered by facilities like museums, aquariums, hatcheries, water treatment facilities, research laboratories, camps, governmental agencies, and park and national/local forest and wetlands. At times, simply choosing which one of many similar resources to illustrate project ideas and strategies took me in unexpected directions. I have continued to learn new things, develop new skills, and update the contents of this book, especially related to computer-automation and programmable logic controllers, up to the time it was published. Educators interested in using aquatic ecosystems as a medium for integration of instruction in fields of science, technology, engineering, mathematics, (STEM) and beyond will save a lot of time and money by using the information in *Beyond the Classroom Aquarium* to guide their instructional efforts.

In *Beyond the Classroom Aquarium*, I am excited to share descriptions of activities that guide student participation in a structured sequence of 14 intensely ambitious STEM-oriented aquatic-ecosystem projects. While beginning with modest classroom aquariums, the projects/activities in this book take aquariums in the classroom to new levels by helping teachers guide students to utilize tools of math and technology to design, assemble, maintain, and investigate three unique and fully operational aquatic ecosystems (i.e. aquaponics, coral propagation, and broodstock and larval rearing system) and beyond. The 14 projects in *Beyond the Classroom Aquarium* guide students through a structured sequence of activities in which they have opportunities to work on a range of activities: from those that may involve the simplest aquatic ecosystems and humblest single-celled organisms, to those that involve highly complex organisms and aquatic ecosystems, utilizing many modern technologies and extending from the classroom to the community. Over the years, numerous educators have used the majority of projects described in this book as stand-alone educational activities. In *Beyond the Classroom Aquarium*, I have sought to seamlessly merge many unique stand-alone activities into an integrated curricular plan. In general, projects have been placed into the following six sections and follow one another in this sequence:

- Introductory Activities – STEM Connections – Universal Aquatic Ecosystem Parameters
- Design, Assembly, and Maintenance of Three Primary Systems
- Design, Assembly, and Maintenance of Auxiliary System Component Projects
- Opportunities for Research
- General Applications of Aquatic Ecosystem Knowledge and Skills – Beyond the Classroom
- Summative Assessment: *A Beyond the Classroom Aquarium* Conference – student run

Many teachers will be most comfortable using *Beyond the Classroom Aquarium* projects to supplement an existing STEM curriculum. Other educators may be more comfortable structuring most student assignments though the projects. Per Ron Berger's description in: *An Ethic of Excellence* (2003, p. 65), many educators may prefer to use projects as a "primary framework through which skills and understandings are learned. They are not extensions of the curriculum or extras when the required work is done. They are themselves, the core of the curriculum." *Beyond the Classroom Aquarium* is flexible and students will benefit most when educational purpose, educator interests and abilities, sources of technical support, and setting are aligned.

One of the first questions many educators ask about the projects in *Beyond the Classroom Aquarium* concerns how these activities align with recently adopted educational standards. In order to help educators with assessment and alignment, the end of each project contains an assessment sheet that emphasizes individualized assessment of progress with common core math standards (with emphasis on five easily overlooked process standards), suggestions for additional assessment activities, prompts for students to respond to in their daily aquatic ecosystem journal and when appropriate the Field Trip Feedback Form (see page 21) may also be used. In addition, a NGSS matrix is provided on page 338 to help educators determine how they would like to use the projects in this book to expose their students to learning activities related to current science standards. In creating the assessment activities at the end of each project description, assessment criteria reported by Hung (2008), Barron and Darling-Hammond (2008), and Patton & Robin (2012) for use in the context of project based learning activities, as presented in the following table, were consulted.

Hung (2008)	Barron and Darling-Hammond (2008)	Patton & Robin (2012)
necessary knowledge acquisition (for example, "need to know" or content objectives)	use of evidence	Does student performance in the project meet or exceed criteria set at the start of the project?
depth of study	accuracy of information	Has the student developed the skills required for the execution of the project?
effectiveness and efficiency of research methods	evaluation of competing views	Has the student learned the curriculum content required for this project?
logical and effective reasoning	development of a clear argument	
conceptual integration of knowledge	attention to writing conventions	
effective problem-solving strategies	collaboration	

Rivers (2011) online quotation of language contained in the Common Core Standards Initiative (CCSI, 2010) may provide the encouragement teachers need to more fully appreciate the opportunities for instructional flexibility these standards offer:

> Standards do not tell teachers how to teach, they help teachers figure out the knowledge and skills their students should have so that teachers can build the best lessons and environments for their classrooms. Standards also help students and parents by setting clear and realistic goals for success. Standards are a first step – a key building block – in providing our young people with a high-quality education that will prepare them for success in college and work. Of

course, standards are not the only thing that is needed for our children's success, but they provide an accessible roadmap for our teachers, parents, and students. (p. 3)

In addition to pencil and paper assessments that are part of most standard curricula, *Beyond the Classroom Aquarium* is designed to enable educators to readily assess student learning, knowledge, appreciation and skill development based upon evidence gained directly from student presentations of their work. In *Beyond the Classroom Aquarium*, an assessment rubric and other activity-based suggestions for assessment are provided for use; both during a project's development and after project completion. At any point in a project, students may be asked to publically present evidence of their STEM abilities (and beyond). Berger (2003) explains the selection of formats in which public displays of student work may occur include: exhibitions, assembly presentations, class displays, and portfolios. For an entire class, assessment via public display may take a couple of days. When conducted in class amongst peers, Berger explains that it is important to remind fellow students to follow three rules related to critique protocols and a simple set of guidelines. His three protocol rules and list of simple guidelines, adapted from Seidel's Collaborative Assessment Protocol (1997) and Allen and MacDonald's Tuning Protocol (2003), accompany the suggested assessment activities that are provided in the conclusion of each project description.

I believe the goals and objectives related to aquatic ecosystem project work presented in *Beyond the Classroom Aquarium* transcend classroom boundaries and reflect student learning and skill as measured against real-world exigencies using an approach similar to that which Jamie Escalante used as he guided his students to pass college board AP calculus exams. A student needs a program that "imposes a valid standard, some outside measure of how well he/she is doing" (Mathews, 1991). As students work with their living systems, they find that each project is instilled with direct connections to life, joy, cooperation, interdependence, existence, and the recognition that our abilities to use our intelligence and skills can be designed to have globally profound and positive impacts. Each project propels students to transform their daily ecosystem experience into a personal encounter with excellence. Students working on these projects quickly realize that these aquatic ecosystem projects are authentic. If this sequence of projects is followed in the spirit for which they were written, most classrooms will contain a minimum of 15 simultaneously running aquatic ecosystems and a few additional outdoor water-related projects by the end of the school year.

As student-developed aquatic ecosystems mature and as system productivities increase I often notice subtle transformations in the nature of student motivation to work/learn and in the spontaneous often passionate expressions of appreciation towards nature and life they express as they share their understanding, wonder, and awe regarding the growth in their aquatic ecosystems. Chapter VI in *Beyond the Classroom Aquarium* is designed to help teachers guide their students through a sequence of projects that involves them more deeply in the process of communicating aquatic ecosystem experiences and knowledge and interacting with others (within the school community and beyond), including opportunities to engage in meaningful community-service work. By working on these projects with their peers, students are participating in a model of education that is socially and emotionally satisfying. Given the nature of the projects, it becomes natural (with the help of some gentle guidance) for students to sustain a productive spirit of enthusiasm about all aspects of their project work, including keeping track of materials and equipment and in providing for the proper upkeep and storage of items and tools used regularly. I also often notice an increased sense of competence and self-assurance in many students once systems in the classroom are operational and students are comfortable with the daily routines/structure required to monitor & maintain their systems and once they have had time to reflect on their classroom aquatic ecosystem work.

My vision is that educators use *Beyond the Classroom Aquarium* to help establish classrooms and maintain educational practices that characterize "what an ethic of excellence can look like in a school and to share strategies for building and sustaining a culture predicated on this ethic" (Berger, 2003). In this sense, *Beyond the Classroom Aquarium* is not really about classroom aquariums, or STEM-oriented aquatic ecosystem projects, "or gifted kids, or clever teaching, or curriculum that should be marketed." To apply phrases from Berger's description of his own work to this description of *Beyond the Classroom Aquarium*, it is more about a pervasive spirit, "an ethic in the air . . . ." It is about the culture of schools and classrooms that encourage kids "to work together, and to care deeply about the quality of what they do. It is the ethic that a school culture instills." It answers the question: what can we do in our schools and communities to bring out the best in kids? It suggests ways to use a structured sequence of 14 aquatic ecosystem-oriented projects as a spark to catalyze "a passion for beautiful student work and developing conditions for making that possible." *Beyond the Classroom Aquarium* is as much about the "Beyond" as it is about aquariums and aquatic ecosystems.

As your students work on the projects in *Beyond the Classroom Aquarium*, their work will provide plentiful evidence that each aquatic ecosystem project progressively complements and reinforces ideas and skills developed in previous projects. By using the projects as the core of a curriculum, students can be guided to discover an unlimited variety of opportunities to demonstrate knowledge and mastery of common core standards, goals, and objectives in many STEM-related areas. The introduction to each of the seven chapters includes a general statement of standards that are addressed by projects within that chapter. The projects are grounded in common core educational standards and in addition, have a strong basis in recently developed Ocean and Great Lakes Literacy standards (Ocean Literacy Network, 2011).

The projects in *Beyond the Classroom Aquarium* have stretched my knowledge and skill levels beyond my highest expectations. I do not know many educators who can independently perform the myriad preparations and set-ups that the aquatic ecosystem projects described here require. Besides expanding my teaching abilities, my work with students on aquatic ecosystems has led me to become a more proficient carpenter, plumber, and electrician. I am now known in many local hardware stores. And, the helpful hardware personnel do not immediately laugh when I try to describe the purpose to which I want to put some esoteric piece of hardware. In fact, many of them have become strong allies in the effort to bring aquatic ecosystems into the lives of my students. I have asked many expert craftsmen for assistance over the years. I suggest that your students, and yourself as well, will often benefit from recruiting assistance from student's parents and local craftspeople. Without them, it will be more difficult to complete all of these projects in one school year – especially on a limited budget.

Beyond the Classroom Aquarium includes a model school-year calendar displaying a schedule of suggested field trips. For the purpose of creating and maintaining aquatic ecosystems, I believe field trips are essential for exposing students to operational models and standard protocols beyond the classroom, to which they may refer for guidance as they work on their aquatic ecosystem projects. Within weeks of the start of the school year, before embarking on project two, I suggest the organization of the important first aquatic ecosystem field trip. Logistically, the AM portion of the trip involves providing one of three groups of students with a tour that exposes them to an operational aquaponics system such as Growing Power. The second of the three student groups is given a behind-the-scenes tour in an operational coral propagation operation such as may be found in a local aquarium store or public aquarium. The third of the three student groups attends a tour of a productive aquaculture or fish hatchery facility. The logistics of the PM portion of the trip, including lunch and travel if necessary, deliver the three student groups to a carpentry and electrical workshop site where each group is introduced to a carpentry and electrical tradesperson team. During this two-hour hands-on session, students are introduced to steps to follow to design and assemble their group's respective aquatic ecosystem frame and lighting system components. The team of craft persons emphasizes safety while working with machinery, materials, and aquatic ecosystem equipment as they demonstrate the process of framing and lighting an aquatic ecosystem. Teachers may arrange in advance that students deliver resulting framing and electrical work to class.

The calendar also suggests a number of Saturdays during which community oriented public exhibitions of student projects may be scheduled. These events are for the purpose of sharing project progress and displaying skill and knowledge development in the students. The exhibition events are scheduled to follow extensive, end of grading period in-depth critiquing assessments during which teachers and fellow students examine the work of each student and spend a good deal of time critiquing it thoroughly. Students are most well prepared to display their work publically directly after the critiquing and assessment periods have ended.

Teachers may use the calendar as a useful project pacing guide. It also reminds teachers to be patient since, by nature, many *Beyond the Classroom Aquarium* learning experiences cannot be commanded to occur at the snap of our fingers. Such events naturally occur as organisms and systems become acclimated. Although it may be possible to race through these 14 projects in a 20 week semester, please resist this temptation. I guarantee students who work with the systems, organisms, and other related topics in this curriculum over an entire 40 week school year will experience many unique and more significant learning opportunities than those who are rushed through the program in 20 weeks.

Of both academic and ethical importance, student teams adopt laboratory policies that forbid any educational lab research animal to be kept outside an ecosystem (or its model) for which it evolved. Neither cruelty nor neglect toward living creatures, or the environment, is sanctioned in these projects. As students place living organisms into their aquatic ecosystems, they must formally agree to treat their organisms and the environment ethically by signing a bio-ethics Research Form that binds them to this agreement. A copy of this bio-ethics Research Form is provided to help teachers guide students to behave ethically toward species in their systems.

By providing students with an operational aquatic ecosystem lab environment, operational from the beginning of the school year; by using art to help introduce major natural aquatic ecosystems and vocabulary; and by introducing students to numerous aquatic organisms, students are immediately exposed to the cycle that recurs throughout this sequence of projects as advanced systems and more complex organism dynamics are studied. The introductory projects challenge students to expand their powers of observation, inquiry, investigation, data collection, and analysis. The early projects also broaden student experiences in preparation for their documentation of the experiences and procedures in which they and their team members will participate. Students are also challenged to produce some incredible applications of technological equipment from chapter II and beyond.

The systems and activities in *Beyond the Classroom Aquarium* are designed to provide large and safe areas for out-of-seat ecosystem work and access. I hope the project descriptions in this book also encourage you make full use of safety technologies related to movement and positioning of bulky and heavy aquatic ecosystems and their components. In addition, I believe student work is much more efficient when a sizable area for cleanup (with a large sink and hoses) is accessible. A large area for storage of equipment and materials is also desirable. I cannot overemphasize the importance of vigilance with regard to safety, injury prevention, and ethical treatment of living organisms. Equipment and materials can be harmful. Ideally, students will have opportunities to work safely with electricity, saws, glass, drills, chemicals, paints, epoxies, and more (in potentially wet environments). Not one project is worthy of use if any student participating in it becomes injured. Parental awareness and consent for student participation and student agreement to practice all safety precautions, at all times, is necessary.

Teachers and administrators who implement the structured sequence of aquatic ecosystem projects presented in *Beyond the Classroom Aquarium* are often familiar with common obstacles in educational policy related to adoption of new curricula described in documents such as *An Ocean Blueprint for the 21st Century* (U.S. Commission on Ocean Policy, 2004):

> While federal programs provide many opportunities for ocean-related education, education is primarily a state responsibility, with direct control exerted at the local level. Therefore, the interaction and involvement of education administrators at the state, district, and individual school levels will be fundamental to the success of any effort to use ocean-based examples to enhance student achievement. . . . The value of

> ocean-based learning must be recognized within local school districts to create a demand for ocean-related education products. (pp. 124-125 and 132)

At times, educational decisions and policies are adopted without recognizing a natural (commonly overlooked) fact: that much of the power and value in education for each student, every day, rests in the hands of individual teachers and their honest/informed work with students. The aquatic ecosystem projects and style of presentation in *Beyond the Classroom Aquarium* are intended to make the curricular ideas accessible to and modifiable by the many caring, free-thinking, knowledgeable and capable educators in classrooms across the nation, educators upon whom the foundation of student knowledge, skill, and a fabulous future of fulfilled opportunities are based.

From 1938, when Dewey wrote on the practice of a local and flexible, student-centered curricular paradigm (Dewey, 1998 pp. 5-6) to when Neill first described his Summerhill philosophy (Neill, 1960) the rate at which educators are adopting similar educational paradigms has accelerated. According to the International Democratic Education Network website, "there are 200 educational facilities in over 30 countries dedicated to the vision of maximized student participation in school government and freedom to learn. More than 40,000 students are enrolled in schools of this kind" (International Democratic Education Network, 2012). Zhao (2012) indicates that aside from populist models of student-centered educational practitioners such as Berger (2003), Esquith (2010), Escalante (Mathews, 1988), and Clark (1989), the most prominent practitioners of a student-centered educational programs include Rubenstein (2008), Sudbury Valley (Greenberg, Sadofsky, & Lempka, 2005), Jefferson County Open School (Posner, 2009), and the Studio School Trust (2010).

Today's educators are critically challenged to nurture all students to acquire entrepreneurial skills that enable students to turn ideas into action and to develop a critical mix of success-oriented attitudes of initiative, intelligent risk-taking, collaboration, and opportunity recognition. A growing number of educators today are finding that this set of knowledge and skills transcend common core standards, no child left behind legislation, and the intentions of powerful national and international agencies to reform and unify educational policy and practice (Loveless, 2012 p. 5). In *World Class Learners*, Zhao (2012) explains that the student-centered paradigm:

> Does not assume all children are the same; therefore it does not impose artificial standards or age-based, grade-level expectations. It helps children move forward from where they are. Furthermore, it does not believe children are simply empty vessels ready to be filled with knowledge, but rather, it assumes that is child is purposeful agent who actively interacts with the outside world. (p. 152)

As increasing numbers of locally empowered educators work to develop student-oriented and project-based curricula that foster world class entrepreneurial mindsets in their students, it is my mission to see that they will have resources such as *Beyond the Classroom Aquarium* to guide them in their efforts. In the words of Zhao (2011) "Education should always start from the child. . . . It begins with the children: what they are interested in, what excites them, what they are capable of, and how they learn...."

In *Beyond the Classroom Aquarium*, teachers carefully guide students to participate in the first eight aquatic ecosystem projects so that by project nine students may exercise newly gained skills and knowledge to design a research project that explores an aquatic ecosystem-related hypothesis (see page 59). As students become personally invested in the systems and the organisms in them (in projects one through eight), many find they possess the resources and knowledge to perform such an investigation in a scientifically productive manner. In the process of designing and conducting an aquatic ecosystem research project, students are guided to apply statistical techniques and vocabulary that are an important part of the scientific process. In many cases, students find the aquatic ecosystem materials and equipment they have been developing make it easy for them to replicate a variety of biochemical or microbiological studies. In project nine, students formulate conclusions that rely upon analysis of experimental data they collect, following techniques similar to those they were guided to observe as they reviewed "literature" in preparation for their research. In addition to providing direct guidance in completing a project, introductory background information and guided discussion/decision making activities (embedded within each project description) are also intended to serve as precursors to a literature review to establish an anticipatory framework for project nine.

By project ten, many students are prepared to accept suggestions as to how recently acquired research results might direct them to modify an aquatic ecosystem decision or action they currently practice. In projects ten through fourteen, students are guided to develop local applications of discoveries and conclusions from aquatic ecosystem studies whether performed by students in the classroom or by researchers in laboratories half-way around the world. Students are guided to successfully incorporate conclusions from other aquatic ecosystem studies to help them further explore a range of environmentally significant aquatic ecosystem-related topics. Finally, in project fourteen students publically document the STEM-related aquatic ecosystem experiences and discoveries they encountered personally, throughout the year, while working on projects one through thirteen.

Student presentations and critiques constitute an important component of assessment in many projects. Teachers may also use the eight categories in the rubric at the conclusion of each project [as presented in the publication: A Framework for K–12 Science Education: Practices, Crosscutting Concepts, and Core Ideas (Quinn et.al, 2011) and as described by Bybee (2011, pp. 2 - 4)] to assess student knowledge and skill in science.

- *Asking questions and defining problems*—a basic practice of the scientist is the ability to formulate empirically answerable questions about phenomena to establish what is already known and to determine what questions have yet to be satisfactorily answered.
- *Developing and using models*—science often involves the construction and use of models and simulations to help develop explanations about natural phenomena. Models make it possible to go beyond observables and simulate a world not yet seen. Models enable predictions to be made in order to test hypothetical explanations.
- *Planning and carrying out investigations*—a major practice of scientists is planning and carrying out systematic investigations that require clarifying what counts as data and in experiments identifying variables.
- *Analyzing and interpreting data*—scientific investigations produce data that must be analyzed in order to derive meaning. Scientists use a range of tools—including tabulation, graphical interpretation, visualization, and statistical analysis—to identify the significant features and patterns in the data. Modern technology makes the collection of large data sets much easier, providing secondary sources for analysis.

- *Using mathematics and computational thinking*—in science, mathematics and computation are fundamental tools for representing physical variables and their relationships. They are used for a range of tasks such as constructing simulations, statistically analyzing data, and recognizing, expressing, and applying quantitative relationships. Mathematical and computational approaches enable prediction of the behavior of physical systems along with the testing of such predictions. Moreover, statistical techniques are also invaluable for identifying significant patterns and establishing correlational relationships.
- *Constructing explanations and designing solutions*—the goal of science is the construction of theories that provide explanatory accounts of the material world. A theory becomes accepted when it has multiple independent lines of empirical evidence, a breadth of phenomena it accounts for, and explanatory coherence and parsimony.
- *Engaging in argument from evidence*—in science, reasoning and argument are essential for clarifying strengths and weaknesses of a line of evidence and for identifying the best explanation for a natural phenomenon. Scientists must defend their explanations, formulate evidence based on a solid foundation of data, examine their understanding in light of the evidence and comments by others, and collaborate with peers in searching for the best explanation for the phenomena being investigated.
- *Obtaining, evaluating, and communicating information*—science cannot advance if scientists are unable to communicate their findings clearly and persuasively or learn about the findings of others. A major practice of science is thus to communicate ideas and the results of inquiry in the following ways: orally; in writing; with the use of tables, diagrams, graphs, and equations; and by engaging in extended discussions with peers. Science requires the ability to derive meaning from scientific texts such as papers, the Internet, symposia, or lectures to evaluate the scientific validity of the information thus acquired and to integrate that information into proposed explanations.

Bulloch's introductory comments in *The American Littoral Society Handbook for the Marine Naturalist* (1991a) resonate with the spirit in which the projects in *Beyond the Classroom Aquarium* have been based. These comments illustrate the natural inquiry process in the study of aquatic ecosystem topics and may help us further appreciate the scientific spirit of "going beyond." He writes:

> You need no high-minded justification to go afield to the water, to walk the fringes of the marsh, the tidal flats of the bay, or the edge of a rocky shoreline. Perhaps you simply want the opportunity to relax and watch the waves, see what has washed up in the wrack with the last tide, or wander the back dunes looking at birds or wildflowers. Purposeless observation, far from an idle waste of time, plants the seeds of curiosity and helps develop a quick and questioning eye. It may also lead to pondering on what you see and a craving to know more about the place and its inhabitants.
>
> Once you're there, the surroundings may soon unsettle your idle intentions and fill you with a sense of the delights and mysteries of this place where the land meets the sea. A scuttling crab, small fish darting in the shallows, odd mounds in the exposed mud flat, incrustations on rocks and wrack will arouse your sense of curiosity. And, you will find as many or more intriguing mysteries at sea. Almost every trip will turn up something unexpected.
>
> If you return time after time, mindless looking will not suffice; you will have too many questions needing answers. To satisfy your curiosity, you will fill your knapsack to overflowing with field guides. You will observe, take pictures, and use your guides. As time goes on, your questions will go beyond mere identification. . . .

> You are on the threshold of exploration. If you have not been down this road before, you will need a few words of guidance and access to a few useful tools. What follows includes a mixture of both. (pp. 1-2)

Since the days of Socrates, teachers have striven to carefully guide student observation, discourse, and inquiry so as to lead them to successful formulation and exploration of scientific hypotheses. As most educators today are aware, the processes of observation, discussion, inquiry, and hypothesis formulation are essential learning activities that transcend the specific topic under study in any project. These processes are at the heart of the scientific method and are critical to the success of every student's work on any given project.

"There is almost nothing more exciting in education than being engaged in genuine research – research where the teacher and students are exploring new ground together" (Berger, 2003 p. 75). Educators using the projects and techniques described here will appreciate the power of carefully guiding student observation, discussion, and inquiry based upon operational classroom aquatic ecosystems and specialized modifications of these classroom-housed systems. In project nine, students will be guided to use their powers of observation, inquiry, and discussion (with teacher guidance) to refine their questions into one testable hypothesis.

In some cases, teams of students may each be led to study the same primary topic but to approach the study of it with a hypothesis that is unique to that team. Using this approach, students may be able to formulate a meta-analysis (unifying the unique conclusion of each team) in order to arrive at a more robust understanding of the primary topic of study.

Many public policy principles and state and federal objectives that represent an uncompromising commitment to provide students the technology, information skills, and tools necessary for success in the 21st century, as stated by the International Society for Technology in Education (ISTE, 2010), fit perfectly with activities related to the design, assembly, and maintenance of aquatic ecosystems described in *Beyond the Classroom Aquarium.* "Technology is defined to be the combination of human imagination, inventiveness and the electronic/optical tools to transform ideas into reality" (ISBE, 2002). In *Beyond the Classroom Aquarium*, the potential for technological inclusion is unlimited. The (ISTE, 2010; p.1) public policy principles and state and federal objectives, fully accessible to educators implementing *Beyond the Classroom Aquarium* projects, are based on the following core principles:

1. Technology is an essential element of teaching, learning, and instructional design in effective, engaging 21st-century learning systems.
2. Fluency with technology must be embedded in the learning process and is a prerequisite for attaining necessary 21st-century skills to ensure academic success and workforce preparedness.
3. Pre-service and sustained in-service professional development of teachers and administrators must include instruction on integrating technology into the learning process.
4. The success of student achievement is impacted by data collection, which includes best practices regarding data collection for schools. Analysis of data provides for real-time intervention that is enhanced through technology.
5. Sustained research at the national level is essential to drive innovation in teaching and learning in educational environments.
6. Application of high-quality research-based findings and best practices in educational technology must become part of the education system.
7. Students must have robust Internet and network access no matter where they live and no matter what their socioeconomic circumstances.

In *Beyond the Classroom Aquarium*, student's interactions with technology are multidimensional. As you will read in the description of the math component of this curriculum, I propose that students work in an online computerized and individualized math lab every day. My own recent instructional experience with instructional and course management software allows me to make this recommendation without reservation. In addition, the computerized math lab provides access to many other useful Microsoft Office presentation and data analysis tools, supplementary instructional software such as SketchPad, and CAD-oriented software, to assist students as they create aquatic ecosystem presentations and designs.

Beyond the technologically-oriented and carefully managed individualized math lab, a number of projects in Beyond the Classroom Aquarium provide opportunities for students to gain first-hand experience working with a programmable logic controller (PLC) to develop and implement highly automated control system tasks. As students design, construct, and maintain aquatic ecosystems, development of a fuller awareness of potential PLC applications is encouraged. Activities related to student development of computer-automated control systems are based on a model system that used older Macintosh computers, as described in a paper titled: *A Computer Automated Cold Water Recirculating System for Research in Aquaculture* (Rust, 1996). Information from more recent papers that describe the design and development of low cost PLC workbenches for education purposes (Kheiralla, 2007; Dickinson and Johnson, 2006), and a generous PLC donation from Rockwell Automation, provided the final pieces of information needed to ensure that (after successfully completing activities related to the design and assembly of a microalgae and rotifer culture system and completing suggested introductory PLC activities), students engaging in *Beyond the Classroom Aquarium* projects will be well informed as they are guided through a culminating computer automation activity in which they develop one or more operational computer-automated plankton (copepod) culture and collection system.

The computer-automated control system explained in *Beyond the Classroom Aquarium* includes a PLC and host computer. Students are guided to become familiar with ways a PLC receives inputs from switches and sensors, evaluates these inputs in accordance with programmed instructions, and controls output devices based on the results of this evaluation. The process of designing and assembling a computer-automated plankton culture and collection system in *Beyond the Classroom Aquarium* is intended to be introductory and uncomplicated. However, for some classes, PLC use in aquatic ecosystem design and operation can be scaled up to challenge highly advanced students. Potential applications of computer automation are boundless. While using a PLC unit, students may learn how ladder programs can be written to express different combinations of switches that represent, in turn, various logic functions (AND, OR, and NOT). For some students, teachers will be able to use the PLC to provide guided instruction on topics in instrumentation and measurement. In some settings, work with digital probes and aquatic ecosystem equipment can help students learn principles of supervisory control and data acquisition (SCADA) systems—a type of industrial control system (ICS) that monitors and controls industrial processes that exist in the physical world (Boyes, 2009).

Please excuse the pun—project eight is something you just have to see. In it, teams of students are involved in the assembly of a compound microscope, from scratch.

Students who enjoy working with the PLC may do a project nine research project in which they develop a procedure that may involve: development of applications with an assortment of "field devices" (involving, relays, operator panels, motors, sensors, switches, and solenoid valves) that have been developed and manufactured to measure and control a range of important aquatic parameters such as temperature, light level, pH, and concentrations of many specific ions; using computer-automated technologies to collect and update data digitally; upload data to an Internet-accessible data file; share the data on a local area network (LAN); and access the aquatic ecosystem data remotely.

Beyond daily work with the individualized online math curriculum, technological activities in *Beyond the Classroom Aquarium* include learning basic electrical circuitry and computer programming; the regular use of PowerPoint, Excel, and Word software; digital cameras, digital photography and microscopes; laminar flow fountains, calcium reactors, digital probes, digital analytic balances, hot plates, and stirrers; standard tools of construction trades (saws, drills, pipes, tubes, wires, and GFCI circuits, silicones, paints, epoxies, etc.), CADD software, development of lighting systems, water surge devices; and data collectors and various calculator technologies. Students also have opportunities to explore the development and application of green and renewable sources of energy.

As student teams work on the few final projects, they are challenged to put knowledge and skills developed in the first half of the book to practical use. Students are challenged to incorporate technological applications in their community service projects, school garden projects with ponds and fountains, and development of aquaculture and other related business models. In project thirteen, technological applications challenge students to design, assemble, and operate Remotely Operated Vehicles (ROV's) and Submerged Aquatic Vehicles (SAV's). Opportunities to explore the development of renewable and green energy resources are included as well. Students are guided to look beyond technologies experienced in this program to a vast array of technological applications being developed and described through such organizations as the Marine Technology Society (MTS) and the Institute of Electricians and Electrical Engineers (IEEE).

Many of the projects in *Beyond the Classroom Aquarium* offer uniquely rewarding experiences to students whose teachers want to challenge them with real-world, multidisciplinary design projects (using aquatic ecosystems as a unifying theme). To quote Steve Jobs, "Most people make the mistake of thinking design is what it looks like. People think it's this veneer – that the designers are handed this box and told, 'Make it look good!' That's not what we think design is. It's not just what it looks like and feels like. Design is how it works." In addition to making it possible for students to experience many design and fabrication challenges by incorporating activities from the expanding world of "Do It Yourself" educational resources, technologies, and websites, some projects in *Beyond the Classroom Aquarium* guide students to explore methods for unifying two areas of uncommon study at the secondary level: computer automation and ecosystem fabrication.

To help educators appreciate the value of multidimensional assessment, an adapted list of the scientific assessment criteria listed above is provided below to illustrate how these criteria may be used to assess development of engineering-related knowledge and skill. Bybee's (2011, pp. 2-4) explanation of the rationale for including these same eight categories, to assess student knowledge and skill related to engineering, follows.

- *Asking questions and defining problems*—a basic practice of engineers is to ask questions to clarify the problem, determine criteria for a successful solution, and identify constraints.
- *Developing and using models*—engineering makes use of models and simulations to analyze extant systems to identify flaws that might occur or to test possible solutions to a new problem. Engineers design and use models of various sorts to test proposed systems and to recognize the strengths and limitations of their designs.
- *Planning and carrying out investigations*—engineering investigations are conducted to gain data essential for specifying criteria or parameters and to test proposed designs. Like scientists, engineers must identify relevant variables, decide how they will be measured, and collect data for analysis. Their investigations help them to identify the effectiveness, efficiency, and durability of designs under different conditions.
- *Analyzing and interpreting data*—engineering investigations include analysis of data collected in the tests of designs. This allows comparison of different solutions and determines how well each meets specific design criteria—that is, which design best solves the problem within given constraints. Like scientists, the engineers require a range of tools to identify the major patterns and interpret the results.
- *Using mathematics and computational thinking*—in engineering, mathematical and computational representations of established relationships and principles are an integral part of the design process. For example, structural engineers create mathematical-based analysis of designs to calculate whether they can stand up to expected stresses of use and if they can be completed within acceptable budgets. Moreover, simulations provide an effective test bed for the development of designs as proposed solutions to problems and their improvement, if required.
- *Constructing explanations and designing solutions*—the goal of engineering design is a systematic solution to problems that is based on scientific knowledge and models of the material world. Each proposed solution results from a process of balancing competing criteria of desired functions, technical feasibility, cost, safety, aesthetics, and compliance with legal requirements. Usually there is no one best solution, but rather a range of solutions. The optimal choice depends on how well the proposed solution meets criteria and constraints.
- *Engaging in argument from evidence*—in engineering, reasoning and argument are essential for finding the best solution to a problem. Engineers collaborate with their peers throughout the design process, with a critical stage being the selection of the most promising solution among a field of competing ideas. Engineers use systematic methods to compare alternatives, formulate evidence based on test data, make arguments to defend their conclusions, critically evaluate the ideas of others, and revise their designs in order to identify the best solution.

- *Obtaining, evaluating, and communicating information*—engineering cannot produce new or improved technologies if the advantages of their designs are not communicated clearly and persuasively. Engineers need to be able to express their ideas orally and in writing with the use of tables, graphs, drawings or models and by engaging in extended discussions with peers. Moreover, as with scientists, they need to be able to derive meaning from colleagues' texts, evaluate information, and apply it usefully.

Beyond the Classroom Aquarium contains a variety of highly ambitious design projects. I have seen students use these projects to apply a range of knowledge and skill. Among the design projects addressed are included ideas on how to design, build, and operate at least 14 aquatic ecosystem components. Field trip experiences also provide students with direct contact with a range of tradespersons who may serve as exemplary models as students explore practical applications of electricity, lighting, pumps, plumbing, and construction trades. The team/project matrix illustrates a number of these activities and suggests a possible sequence of projects for a given team of students.

	PROJECT 2 Designs	**PROJECT 3 Designs**	**PROJECTS 4 – 6 Designs**	**PROJECTS 7-8 Designs**	**PROJECT 9 Designs**	**PROJECTS 10 – 14 Designs**
Documentation of Direct Applications of Math and Design Techniques by Team and Project	**System Frame and Lighting**	**System Water Containment, Flow, & Monitoring**	**Final System Set-Up**	**Auxiliary System Framing, Light, and Water Components**	**Research**	**Applications of Literature to Public Policy & Community Service**
Team A	Frame and Lighting for Coral System	Coral System Water Containment, Flow, and Monitoring Components	Coral Propagation and Acclimation of Organisms to the system	Microscopes Calcium Reactor	Surge Devices Wave Makers	Local Ecosystem Data/Records River Watch; Final Assessment – Conference Committee Assignments
Team B	Frame and Lighting for Aquaponics	Aquaponic System Water Containment, Flow, and Monitoring Components	Aquaponic Seed Germination and Acclimation of Fish	Live Food Culture Systems (micro-organisms)	Bell Siphons Aquaponics and Vermiculture	Ponds, School Gardens, Hoop Houses, Grow Beds, Permaculture; Final Assessment – Conference Committee Assignments
Team C	Frame and Lighting for Broodstock & Larval Rearing I	Broodstock and Larval Rearing System Water Containment, Flow, and Monitoring Components	Broodstock & Larval Rearing Acclimation of System Livestock	Juvenile Grow-out Systems & Kreisels	Planktonic Aquatic Larvae; Computer and PLC Automation;	Internships, Jobs Business Development; Reflection on Progress and Future Directions; Final Assessment – Conference Committee Assignments
Team D	Frame and Lighting for Broodstock & Larval Rearing II	Water Purification Reservoir and Filtration Mechanisms	Brochures, Press Releases and Public Information	Lighting and Solenoid PLC connections for Laminar Fountain	Alternative Energies, Arduino Projects	Prairie and Wetland Restoration Coastal Cleanup; Final Assessment – Conference Committee Assignments
Team E	Posters, Signs, Art, Murals, Music, Drama, Fundraisers, Festivals,	Data Files	Frame and Lighting for Juvenile Grow-out	LAN and Data Coordination	Mesocosms; Estuarine Tides Simulation Model; Salinity Variation	Constructed Wetlands; Final Assessment – Conference Committee Assignments

Simulating aquatic ecosystems in the classroom leads students to apply their understanding of math in unique and creative ways. Realizing that the lives of many organisms depend upon their analytical abilities helps foster a sense of urgency in many a students' efforts to study and learn math in order to provide their organisms with an environment that simulates their natural one as closely as possible. Student desire to apply math promptly and correctly adds a compelling educational component of real-life applications of math to this curriculum. The efficiencies and productivities of the ecosystems students design also improve as students apply new knowledge and skill in math while they monitor, maintain and modify their systems.

The aquatic ecosystem projects in *Beyond the Classroom Aquarium* involve students in practical hands-on work. When a decision about a system must be made, all students may be asked to perform an independent analysis of the ecosystem problem and attempt to develop a solution for it. The projects described here invite students to apply math creatively and functionally. Such repeated applications of math, project after project, provide direct empirical evidence that as student proficiencies in math increase students become more capable of designing and assembling high quality aquatic ecosystems. Using math every day either for the purpose of aquatic ecosystem design or problem solving also leads many students to be enthusiastic participants in their mathematical education. Educators may also review books such as *Environmental Math in the Classroom* (Fusaro & Kenschaft, 2003) in order to help them design specific, standard appropriate, math assessment questions that reinforce environmental themes promoted in *Beyond the Classroom Aquarium* learning projects.

Additionally, in many cases, a well-run individualized math curriculum often leads to greater progress in development of academic skill and knowledge. Although individualized math curricula can operate entirely without the use of technology, I recommend that the heart of the individualized math component of this curriculum be based upon a strong online curriculum that is accompanied by user-friendly course management software. It is important to provide training for teachers prior to the new school year so that they may use the online math curriculum software according to their design rather than allowing the software to drive instruction in directions they do not wish to passively follow. Pearson produces an example of online instructional software that some find useful. To begin to explore this software, go to the **pearsonsuccessnet.com** website. With practice, most teachers will comfortably manage student assignments, test scores, and other daily math grades online. Ideally, individualized student accounts will be set up prior to the start of the school year.

To begin, students take an online math placement test during the first week of the school year. After the placement exams have been scored, an individualized math curriculum is prescribed for each student. Each student's curriculum is structured so as to promote an increase of at least one year's worth of academic proficiency by the end of the school year. During this time, each student is provided with an orientation to the math lab and the math lab procedures, including protocols for accessing their individualized online account featuring dynamic personalized instructional math software, working with peers in small study groups, conducting tutoring sessions, and getting individualized academic assistance from the teacher.

After demonstrating the ability to apply their understanding of math to the design, assembly, and operation of aquaponic, coral propagation, and broodstock and larval rearing aquatic ecosystems, the remaining projects in *Beyond the Classroom Aquarium* involve students in many additional activities that require development and application of math skills. As they work through these aquatic ecosystem projects, students develop skills in showing how data tables, graphs, and equations can each be used to represent the same situation. Math knowledge also makes it easier for students to explain their work, justify decisions, and make predictions related to aquatic ecosystems in the classroom and beyond.

Regular and periodic academic assessments of student progress in development of math knowledge and skill are an important component of the individualized math curriculum. When using online computerized math curricula, such academic assessments become routine. Beyond performing a strictly academic assessment of student progress in math, end-of-project assessment activities are designed so as to enable students to document ways they have integrated math skills and performed according to standards while completing the aquatic ecosystem project.

Much of the following description of the math component of the *Beyond the Classroom Aquarium* program is based on a teacher's management of an individualize math curriculum concurrently with the aquatic ecosystem project work. I believe that a well-run and managed individualized math curriculum for each student in the class helps students by giving them power to analyze and solve problems (in more personally relevant ways) while also allowing them to demonstrate their ability to think independently.

I also recommend that an intensive math intervention program, as prescribed by Milgram and Wu (2008), be provided to students not exempted by satisfactory a placement test score. Milgram and Wu strongly suggest that knowledge and success in the following six areas guarantee a solid mathematical foundation that students can rely upon for use in *Beyond the Classroom Aquarium* projects (and beyond):

- Place Value and Basic Number Skills
- Fractions, Decimals
- Ratios, Rates, Percent, & Proportions
- The Core Processes of Math
- Functions and Equations and
- Measurement

It is suggested that students who are placed in the math intervention receive about two hours of mathematics instruction every day following guidelines suggested by Milgram and Wu (2008), including an intense focus on definitions, precision, and abstract reasoning. I believe that the combination of online instructional software and mathematically well-informed teacher will provide an intervention that, far from being a slow class for students with special needs, the pre-algebra curriculum embedded in this program is a more intense and more demanding class. No acceleration occurs when students are not intensely immersed in the doing of mathematics.

In addition to a structured cycle of individualized assessments of academic math knowledge and skill development, an important multidimensional assessment component for use both during and/or after each project includes student presentations. The rubric to help assess student STEM knowledge and skill in a public display of work context identifies ways they have used math to help them design, assemble and/or maintain aquatic ecosystem components. In this STEM assessment rubric, students provide evidence of development in math knowledge and skill by showing how their project work was connected to five process standards. These five process standards are summarized as follows (NCTM, 2011):

A. **Problem Solving**
   - Build new mathematical knowledge through problem solving
   - Solve problems that arise in mathematics and in other contexts

- Apply and adapt a variety of appropriate strategies to solve problems
- Monitor and reflect on the process of mathematical problem solving

B. **Reasoning and Proof**
- Recognize reasoning and proof as fundamental aspects of mathematics
- Make and investigate mathematical conjectures
- Develop and evaluate mathematical arguments and proofs
- Select and use various types of reasoning and methods of proof

C. **Communication**
- Organize and consolidate their mathematical thinking through communication
- Communicate their mathematical thinking coherently and clearly to peers, teachers, and others
- Analyze and evaluate the mathematical thinking and strategies of others;
- Use the language of mathematics to express mathematical ideas precisely.

D. **Connections**
- Recognize and use connections among mathematical ideas
- Understand how mathematical ideas interconnect and build on one another to produce a coherent whole
- Recognize and apply mathematics in contexts outside of mathematics

E. **Representation**
- Create and use representations to organize, record, and communicate mathematical ideas
- Select, apply, and translate among mathematical representations to solve problems
- Use representations to model and interpret physical, social, and mathematical phenomena

The rubric's math assessment information is most useful for purposes of providing further academic instruction and support when current goals of each student's individualized math curriculum are considered. Please note that each student's academic experiences in the individualized math curriculum are directly aligned with common core math standards. The rubric is meant to be a useful tool to assist in the identification of specific standards that students have knowledge of and those that are in need of further development. Math content areas for students working on aquatic ecosystem projects include but are not limited to:

- Measurement (Unit Conversions, Water Chemistry, Design and Construction)
- Algebra (Linear Equations & Systems, Quadratic, Rational, Exponential and Logarithmic Relationships)
- Geometry (Distance, Area, Volume, Graphing)
- Numeracy (Instructional Software)
- Statistics and Probability (Data Collection, Presentation, and Analysis)
- Technology (CAD Design, Probes, Sensors, Computerized Data Collection/Analysis, Microcontrollers and/or programmable logic controllers [PLCs]

The STEM assessment rubric described above is on the following page. It is an assessment tool that is intended for use during the presentation process as students and student teams give a presentation that documents their work on any project between project two and fourteen. Modify it to suit your students' needs when necessary.

Beyond the Classroom Aquarium STEM Self or Peer Assessment Rubric

This self-assessment is valid only to the extent to which you can document and display evidence that supports it. For each category in the left hand column, write a STEM-Oriented example in the box beneath 3 if you can display evidence that your STEM-Oriented example represents work that exceeds expectations. If your evidence shows that your STEM-Oriented statement meets but does not exceed expectations, write the statement in the box beneath 2. Similarly, for STEM-Oriented examples that demonstrate some progress toward expectations.

<table>
<tr><th colspan="2">Name:
Project Number:
Date Project Work Began:
Date of Project Completion:

List Math Topics and/or Objectives Currently Under Study:</th><th>3
Work Exceeds Expectations</th><th>2
Work Meets Expectations</th><th>1
Work Is Progressing Toward Expectations</th></tr>
<tr><td colspan="2">Write examples of questions and defining problems related to your STEM work.</td><td></td><td></td><td></td></tr>
<tr><td colspan="2">Describe how you have developed and used models in your STEM work.</td><td></td><td></td><td></td></tr>
<tr><td colspan="2">Write examples of STEM investigations you planned and carried out.</td><td></td><td></td><td></td></tr>
<tr><td colspan="2">Describe STEM related data that you analyzed or interpreted.</td><td></td><td></td><td></td></tr>
<tr><td rowspan="5">Give examples of how you used mathematics and computational thinking related to the five standards listed to the right.</td><td>Problem Solving (Math Process Standard One)</td><td></td><td></td><td></td></tr>
<tr><td>Reasoning & Proof (Math Process Standard Two)</td><td></td><td></td><td></td></tr>
<tr><td>Communication (Math Process Standard Three)</td><td></td><td></td><td></td></tr>
<tr><td>Connections (Math Process Standard Four)</td><td></td><td></td><td></td></tr>
<tr><td>Representation (Math Process Standard Five)</td><td></td><td></td><td></td></tr>
<tr><td colspan="2">Describe STEM-Oriented explanations and solutions you designed.</td><td></td><td></td><td></td></tr>
<tr><td colspan="2">Explain evidence you used while engaging in a STEM project-based argument.</td><td></td><td></td><td></td></tr>
<tr><td colspan="2">Give examples of STEM information you obtained, evaluated, and/or communicated.</td><td></td><td></td><td></td></tr>
<tr><td colspan="2">TOTAL POINTS</td><td></td><td></td><td></td></tr>
</table>

I first envisioned this aquatic ecosystem project-based program as a fully developed curriculum with a "reach for the stars" scope. A spreadsheet developed to help estimate the cost of materials, equipment, and personnel to realize this curricular aquatic ecosystem vision indicated that it could cost as much as $5000 per student. I saw no way to directly access such funds but believed deeply enough in the vision to begin an effort to expose my students to it in increments that I could afford. Hence, many of my students have been exposed to a seemingly random series of aquariums and system projects over the years (as equipment accumulated, following paydays and the occasional generous donation). Important elements of each project described in *Beyond the Classroom Aquarium* have been shared with students in classes where budgets for educational supplies, equipment, and materials have been nil. In some cases, the scarcity of materials has required individual projects to be conducted in a lab-demonstration format.

As I continued to establish aquatic ecosystems in my classrooms over the years, I eventually had the opportunity to find and work with one local fish store owner who supported my classroom ecosystems in many ways. His shared interest in educating young people about oceans and aquatic life allowed my students opportunities to observe shark hatchlings, larval clownfish, peppermint shrimp, cichlids, and cuttlefish. Because of a caring local shop owner, my students experienced sea urchins and fighting conch. A visit to a not-so-local state fish hatchery allowed students to work with salmon hatchlings in the "classroom refrigerator." I was also fortunate enough to receive a pair of breeding clownfish from a highly esteemed local breeder. Another local businessman donated a 55-gallon marine aquarium system and related supplies to the class. Nurturing relationships with local business people can help open many doors and provide students with many awesome and motivating learning experiences. In addition, students and their families may be able to donate a large inventory of used materials and equipment that they have been holding in storage (garages, attics, basements, sheds, barns, etc.).

Due to variations in educational funding, the range of opportunities available for our students varies greatly between classrooms, schools, district, states, and regions. Despite well-intentioned efforts and adjustments to eliminate gaps and provide optimal opportunities for every student, many teachers, their schools, and districts have budgets that get tighter each year. Conversely, the costs of top-notch experientially based learning activities increase each year.

The benefits of participation in the projects in *Beyond the Classroom Aquarium* make the financial commitment to this program worthwhile, even when it may be deemed financially necessary to establish this curriculum incrementally over a series of years. Where resources are limited, this curriculum is designed to be flexible enough to be modified in such a way that an adapted form of the projects can be successfully presented. Budget-minded educators who implement this curriculum should take note that successful implementation of this curriculum relies critically upon mutual cooperation and support between parents, the local community (especially tradespeople and local businesses), and an enthusiastic teacher. Educators who want to establish a personalized budget for implementation of this curriculum in their own classroom should copy and fill out the form on the following page as they read the descriptions of project implementation provided here and then explore availability and costs of local resources.

Student run fund-raising events and the end of year *Beyond the Classroom Aquarium* Conference provide a public venue to celebrate student work and benefactor donations, as well as highlight needs for additional support. In the future, as school district administrators and educational policy makers support *Beyond the Classroom Aquarium*, comprehensive grants to fund this curriculum (and professional development programs for educators who would like to implement it) may be written. Finally, student participation in a variety of highly reputable, nationwide, "ocean education contests," can generate program resources and funds when work students submit for entry in such a contest is judged worthy.

A School Year Calendar Illustrating Potential Dates for Assessment, Project Duration, and Field-Trips

M	T	W	H	F	S
1	2	3	4	5	
8	9	10	ED	ED	
FT-1	16	17	18	19	
22	23	24	25	26	
29	30				

M	T	W	H	F	S
2	3	4	5	6	
9	10	11	12	13	
16	17	18	19	20	
23	24	25	ED	ED	

M	T	W	H	F	S
		1	2	3	
FT-2	7	8	9	10	
13	14	15	ED	ED	
FT-3	21	22	22	23	
27	28	29	30	ED	

M	T	W	H	F	S
2	3	4	5	6	
9	10	11	12	13	
16	17	18	ED	ED	
23	24	25	26	27	
30	31				

M	T	W	H	F	S
FT-4	4	5	6	7	
10	11	12	13	ED	
FT-5	18	19	20	21	
24	25	26	27	ED	

M	T	W	H	F	S
		1	2	3	
6	7	8	9	10	
13	14	15	ED	ED	
20	21	22	22	23	
27	28	29	30		

M	T	W	H	F	S
FT-6	2	3	4	5	
8	9	10	ED	ED	
15	16	17	18	19	
22	23	24	25	26	
29	30	31			

M	T	W	H	F	S
				1	
4	5	6	7	8	
11	12	13	14	15	
18	19	20	ED	ED	
25	26	27	28	29	

M	T	W	H	F	S
			1	2	
FT-7	6	7	ED	ED	
12	13	14	15	16	
19	20	21	22	23	
26	27	28	ED	ED	

M	T	W	H	F	S
1	2	3	4	5	
8	9	10	11	12	
15	16	17	18	19	
22	23	24	25	26	
29	30				

ED	Assessment
FT	Field Trip

Field Trip	Topic
FT-1	AM: Aquatic Ecosystem Model Sites; PM: Construction Demo
FT-2	Ecosystem specific water-works demo
FT-3	Ecosystem specific technique demo (Optional Division of Labor Field Trip)
FT-4	AM: Plankton tow/algae demo PM: PLC - Computer Automation
FT-5	Planetarium visit/ scope demo
FT-6	Aqua conference a) booths/businesses, b) posters, c) presentations
FT-7	AM: Beach Sweep PM: Aquarium or Botanical Garden

The field trips listed in the calendar on the preceding page are designed to supercharge the aquatic ecosystem design and assembly process, for both teachers and students. They are based on my thinking that coordination of student performance of the many design and assembly activities described in each project in this curriculum - within a classroom setting will proceed much more smoothly with critically timed technical support.

The field trips in *Beyond the Classroom Aquarium* are intended to maximize opportunities for students to experience and apply a range of technical design/assembly skills in a brief period of time. For many educators and students these field trips are critical for the purpose of moving into project activities that involve direct work with system maintenance and protocols for the care and husbandry of the range of species that the class shall have decided to keep in their classroom aquatic ecosystems.

Field trips require advance preparation. To help students receive the proper technical help at the proper time often involves arranging for students to interact directly with a skilled technician or two (in areas of electricity, carpentry, plumbing, programmable logic controller software developers, and others as may be deemed necessary) who will assist the educator in guiding students through each respective design and assembly activity. Parents and local community members are a deep and all too often untapped resource to motivate the education process. Local workshops often have space that they will invite a limited number of students to visit and work in under direct and safe guidance (of course all this information will be part of each student's field trip permission slip).

The advanced planning for such technical (field trip) assistance will be most effective when arrangements for design and assembly materials, equipment, and tools that each respective project may require are also made in advance. Knowledgeable and supportive technical assistants who are provided enough lead time may even like to review the student's design plans in advance recommend proper materials and tools and guide students through development of component parts for the system they will install in the classroom.

In many cases, outside observers (parents, administrators, other classroom educators, and even the public at large) will find the student's work on these classroom systems intriguing enough that the students may be guided to transform the aquatic ecosystem area of the classroom into a living exhibit hall (like a museum or pubic aquarium). Using the classroom as a field trip destination for others to visit can be a fine motivator for students who are being guided to design and assemble the systems that will be on display to others.

Field trip six is designed to guide students through the process of critiquing scientific presentations (to help them reflect on designing their own classroom aquatic ecosystem research project). With advance planning and the school administrator's help it may be possible to conduct a day long, school wide or team-based "in house" conference.

In brief, the seven field trips in *Beyond the Classroom Aquarium* may involve the following respective technicians.

Field trip one – carpenter and electrician
Field trip two – primarily a plumber
Field trip three – public aquarium/hatchery educator/guide
Field trip four – programmable logic controller software technician (in the class' computer lab)
Field trip five – planetarium technician or individual capable of guiding students through optics of microscope design
Field trip six – an "in house" (school wide or team-based) conference for students to attend
Field trip seven – guides to lead student groups to areas where students may perform beach sweep activities (on the shorelines of a lake, river, or ocean or along some riverbed)

Field Trip Feedback Form

Name: Date: Location(s) of Trip:

Describe the purpose of the field trip.

Explain if the field trip served its purpose. Explain how it did or how it did not. How useful was this field trip? When do you think you might use information you obtained?

Do you have contact information from the persons who guided the field trip so they may provide follow up advice and guidance? List that information here.

Write a description of this trip for your project portfolio journal. Please be sure to include information about materials, aquatic ecosystem equipment, and tools needed for this project and important safety practices to follow.

Write additional information you obtained about the trip during the gallery critique of all students' field trip reports.

Note: Provide field trip presenters with design details of the aquatic ecosystem being created in advance of the trip, when possible.

Budget Sheet Guide (Chronologically Itemized Project Cost Assessment [Blank])

Individualized Online Math Software			_______
Computer Lab (PCs and/or Macs)			_______
Hi-Tech Laboratory Equipment (Hardware)			_______
Supplementary Library and Classroom Resources			_______
Classroom System One –	Mechanical Cost		_______
	Livestock Cost		_______
Classroom System Two -	Mechanical Cost		_______
	Livestock Cost		_______
Classroom System Three -	Mechanical Cost		_______
	Livestock Cost		_______
Classroom System Four -	Mechanical Cost		_______
	Livestock Cost		_______
Total Cost to Set-Up Four Classroom Systems			_______
Aquaponics (Field Trips 1 & 2) Costs			
	Carpentry & Electronic Materials	_______	
	Transportation	_______	
	Craftsperson Project Expeditor & Consultant	_______	
	Plumbing Materials	_______	
	Transportation	_______	
	Craftsperson Project Expeditor & Consultant	_______	
Total Aquaponics System Set-Up & Field-Trip Total Cost			_______
Coral Propagation (Field Trips 1 & 2) Costs			
	Carpentry & Electronic Materials	_______	
	Transportation	_______	
	Craftsperson Project Expeditor & Consultant	_______	
	Plumbing Materials	_______	
	Transportation	_______	
	Craftsperson Project Expeditor & Consultant	_______	
Total Coral Propagation System Set-Up & Field Trips Cost			_______
Broodstock Larval Rearing System (Field Trips 1 & 2) Costs			
	Carpentry & Electronic Materials	_______	
	Transportation	_______	
	Craftsperson Project Expeditor & Consultant	_______	
	Plumbing Materials	_______	
	Transportation	_______	
	Craftsperson Project Expeditor & Consultant	_______	
Behind the Scenes Fish Hatchery/Public Aquarium Visit (Field Trip 3)			_______
Total Broodstock Larval-Rearing System Set-Up & Field Trip Costs			_______
Algae Culture System (Including Field Trip 4)			_______
Plankton Culture System			_______
Automated Culture Plankton System			_______
Microscope Project Costs (Including Field Trip 5)			_______
Research Project (Field Trips 6 & 7) Costs			
	Transportation	_______	
	Registration	_______	
Research Project Costs			_______
Juvenile Growout (and miscellaneous larval system) Set-Ups			_______
Community Service Project Costs			_______
Laminar Fountain Costs			_______
Final Project Design Costs			_______
Conference Costs			_______
Total Expense			_______

CHAPTER II: Student Awareness of Universal Aquatic Ecosystem Parameters: Introductory System Design Applications (Projects 1, 2, and 3)

Of the 14 projects in *Beyond the Classroom Aquarium*, the first three projects are presented in Chapter II. In this chapter, students are guided to skillfully apply their knowledge to design and assemble three aquatic ecosystems that are central to the entire curriculum: an aquaponics system, a coral propagation system, and one (or more) broodstock and larval rearing systems. You will also find information about two important field trips that should be scheduled during the time students are working on these projects.

In the weeklong project one introduction, students complete a series of aquatic ecosystem-based drawing activities that familiarize them with aquatic ecosystem settings and vocabulary. Observational skill development is stimulated by assigning students to systematically observe, report upon, and discuss the four fully-operational classroom aquatic ecosystems stocked and managed by the teacher. For practical purposes, the four model classroom systems also serve as stock tanks from which students will conveniently transfer healthy aquatic organisms to one of three student-designed and constructed systems. By serving this purpose, the classroom aquatic ecosystem models may be considered quarantine systems.

Resources to which teachers may refer if they have need of specific assistance in setting-up and stocking the four demonstration systems are provided in project one. I believe these systems are most effective when teachers set them up and stock them prior to the beginning of the school year. Aquatic ecosystems that are operational at the beginning of a new school year generate a lot of important momentum for this class throughout the school community. This momentum is important because despite the rapidity with which students are guided to design and set up three primary aquatic ecosystems, each newly operating system then requires a substantial amount of time to cycle and establish maturity and stability—a "lag" period. The school year time frame in which the projects in *Beyond the Classroom Aquarium* are performed will be most efficiently used when student focus on aquatic ecosystems begins on the first day and is sustained throughout the year. Projects one, two, and three are designed to preserve program momentum by making it possible for the three student-developed primary ecosystems used to drive activities in it to be as mature and biologically active as soon as possible after the beginning of the school year. These first three projects present contrasting but equally important priorities; the value of designing and assembling the systems is essential but so, too, are the bio-active products resulting from mature aquatic ecosystems.

During the concluding activities in project one, students are to be introduced to the three primary aquatic ecosystems they will be developing: a) aquaponics, b) coral propagation, and c) broodstock and larval rearing (possibly two of these – one freshwater, the other salty). In this final process before beginning project two, students are guided to determine specific dimensions to which at least three properly scaled, full-sized operational classroom aquatic ecosystems will be built. In addition to carefully measuring available classroom floor space, students are asked to research, analyze, and report upon numerous organisms in the interest of deciding which will eventually be placed in the aquatic ecosystems they will soon design and assemble. Once the reports are critiqued in class and the organisms to be kept are determined, student groups make research-based decisions related to system framing and lighting component dimensions. Student teams then design their system on paper. The class critiques all designs and votes for the most educationally valuable system designs.

A first field trip must also occur prior to beginning project two. As explained in a previous description, during the AM portion of this field trip each student, in one of three groups, will be given a behind-the-scenes tour of an operational aquatic ecosystem that models one of the three systems they are to build. Beyond this, in the afternoon portion of the field trip, the three student groups will meet with a supportive team of tradespersons (carpenters and electricians in workshop settings), who will demonstrate safety procedures and guide students through introductory aquatic ecosystem design and assembly. As student groups are guided through the development of aquatic ecosystem framing and lighting components during this portion of the field trip, they become familiar with safety guidelines to practice as they continue to work with tools and materials in their classrooms. After the field trip, back in the classroom, students spend time debriefing and critiquing their experiences, contributing tremendously to the quality of the aquatic ecosystems they produce. A form for debriefing and critiquing field trip experiences is provided in the project one assessment section. Copies may be reused following all field trips. Every student in class benefits when students from each aquatic ecosystem design team share fieldtrip ideas/activities their group experienced.

Beyond just installing lighting components in an aquatic ecosystem frame, project two activities are designed to help familiarize students with the nature of light. An important aspect of the lighting systems that they do install is how the students knowingly apply their understanding of light by selecting bulbs and lighting equipment that are suitable for the organisms intended for the aquatic ecosystems they are designing and assembling. By the end of project two, many students shall have discovered that their study of light has led them to work with a range of natural phenomena that relate directly to successful design, assembly, monitoring, and maintenance of aquatic ecosystems. Such concepts that students shall have applied directly to aquatic ecosystem project development include in depth exploration of three primary measures of light (wavelength, energy, and intensity), ways that water influences the light that enters it, specifications of standard and custom lighting systems, available light bulb technologies, spectral requirements of ecosystem lighting, mathematical determination of sufficient light intensity for a given aquatic ecosystem, a line item cost analysis of different customizable lighting systems, and two protocols to follow for designing and assembling aquatic ecosystem lighting components.

Early in project three, a field trip similar to the first must be scheduled. During this field trip, the three student groups will again meet with supportive and expert tradespersons who will provide students in each group with direct guidance in developing appropriate water containment and recirculation components for each respective aquatic ecosystem that students are designing and assembling (an aquaponics group, a coral propagation group, and a broodstock larval rearing system group). It may be of interest to note that by designing and assembling each of the three primary systems in this guide, students gain direct experience with three distinctly unique approaches commonly utilized in processing and managing water flow in many water-oriented industries.

In addition to applying plumbing and mechanical techniques to set up and control water flow in three aquatic ecosystems upon completion of project three, students shall have been guided to perform explorations of a number of “essential to understand” (in aquatic ecosystem contexts) properties of water. Groups of students, in completing and sharing their project related activities will have worked with and solved a range of common aquatic design issues. Such concepts and skills that students shall have applied directly to aquatic ecosystem project development in project three relate to the molecular and chemical nature of water; natural processes of water quality deterioration in recirculating aquatic ecosystems; performance of water quality tests using chemical techniques and digital probes; application of standard methods for sampling water and collecting water quality data and developing water quality data sets and interpreting their data; common techniques for improving aquatic ecosystem water quality conditions, including an appreciation for the beneficial services that bacteria, substrates, and “unsightly” algae perform in well-maintained aquatic ecosystems. In order to determine the percent of unionized ammonia in a sample of water, students are guided to review a 1997 Environmental Protection Agency paper and use calculator curve-fitting technologies to fill in the blank table provided for this purpose.

By the time students finish the first three projects, many are crazy with excitement about introducing living organisms into the systems they have set up (which they get to do in projects four, five, and six).

Before designing and constructing their own aquatic ecosystems within the classroom, students must first be acquainted with important characteristics of natural aquatic ecosystems and properly operating aquatic ecosystem models.

Two great strategies to acquaint students with aquatic ecosystem background knowledge involve a set of art activities (integrated with reading and vocabulary) and a guided observational activity of the teacher's pre-established and fully operational model systems. These activities help establish a common aquatic ecosystem language and a standard set of expectations with which students may begin to assess their progress through the year. The first of these two strategies (a weeklong series of readings and drawings) is described here and presented on the next few pages. The guided observational activity is presented on the pages that follow this art/vocabulary activity.

Selected pages from *The Marine Biology Coloring Book* (Niesen, 2000 – used with permission) provide an enriching weeklong introduction to aquatic ecosystem topics and vocabulary. Text within the page-long essays that accompany each drawing include instructions that guide students through the coloring process while introducing them to many essential aquatic ecosystem concepts and common vocabulary terms. By presenting these coloring activities over a period of five days, it is usually possible to address global topics on the first day, light in aquatic zones on the second day, unique aquatic habitats on the third day, and a range of aquatic plants and animals (from many phyla) on the last two days of the week.

As students follow directions given in the text/story that accompanies each coloring sheet and reflect on their strategic use of colors, many become familiar enough with the range of organisms and their anatomical features in each sheet to successfully identify them and produce an informative summarizing narrative about the subjects of each drawing/coloring work they complete. Some may even become much better artists by applying the coloring directions provided in the text to their drawings. Some artistically advanced students may even be motivated to explore ways to create their own set of coloring sheets with accompanying story lines.

The first drawing in *Beyond the Classroom Aquarium* is designed to help familiarize students with concepts and vocabulary related to 'Tides.' By first presenting a topic of this nature, students are prepared to consider more global-perspectives of many concepts and concerns that are an integral unifying theme throughout the sequence of aquatic ecosystem projects.

The next drawing the 'Photic Zone' may be used to provide a springboard for important project activities related to the role of light in the aquaponic, coral propagation, and broodstock/larval rearing systems. By design, part of the intention is that the 'Photic Zone' drawing sets a tone for further study/exploration of the importance of photosynthesis. The photic zone leads students to reflect on the range of well-lit aquatic environments that exist on Earth and the possibilities they present for future study.

The aquaponics, coral propagation and broodstock/larval rearing systems students create in *Beyond the Classroom Aquarium* represent three (of many) unique aquatic ecosystems. The third drawing, 'Types of Coral Reefs' provides students with more introductory information related to corals. The information in this activity is a valuable asset that will help students successfully participate in project activities related to coral propagation system design, assembly and maintenance.

The fourth illustration, 'Marine Flowering Plants,' guides students to consider interactions between vegetation and water. Although fish will live within the aquaponics system, it must be understood that the plants in this system perform many critical functions. Hopefully, this drawing provides a gentle introduction to ideas students may incorporate into their aquatic ecosystem design plans as they consider the roles that plants (and other less glamorous flora and fauna) perform as important members of most ecosystems on Earth.

On the last day of coloring, students are introduced to the 'Crab Life Cycle' illustration. This illustration provides students with surprising details of these organism's unusual life cycles. After considering information in this illustration, students are guided to begin thinking about techniques for capturing crustacean larvae (and other aquatic larvae over a range of Phyla); and the unique efforts that may be required in order to help these larvae survive to adulthood. And, also, this reflection may guide students to consider design ideas they may require in order to make the system they assemble perform successfully.

1. You will need coloring instruments. Colored pencils or colored fine to medium felt-tip pens are best. Twelve pens or pencils will work, but more is better.
2. The organization of the contents is based on the author's
overall perspective of the subject and may follow the order of presentation of a formal course of instruction. To achieve maximum benefit, you should color the plates in order, at least within a group or section.
3. Once you begin coloring a plate in order of presentation of the titles and reading the matched text, the illustrations will have greater meaning, and relationships of different parts will become clear.
4. As you come to each plate, look over the entire illustration and note the arrangement and order of titles. You may count the number of subscripts to find the number of colors you will need. Then scan the coloring instructions (printed in boldface type) for further guidance. Be sure to color in the order given by the instructions. Most of the time this means starting at the top of the plate with the first title "a," and coloring in alphabetical order. Contemplate a number of color arrangements before starting. In some cases, you may want to color related forms with different shades of the same color; in other cases, contrast is desirable. Where a natural appearance is desirable, the coloring instructions may guide you or you may choose colors based on your own knowledge and observations. One of the most important considerations is to link the structure and its title (printed in large outline or blank letters) with the same color. If the structure to be colored has parts taking several colors, you might color its title as a mosaic of the same colors. It is recommended that you color the title first and then its related structure.
5. In some cases, a plate of illustrations will require more colors that you have in your possession. Forced to use a color twice or thrice on the same plate, you must take care to prevent confusion in identification and review by employing them on separate areas well away from one another. On occasion, you may be asked to use colors on a plate that were used for the same structure on a previous related plate. In this case, color their titles first regardless of where they appear on the plate. Then go back to the top of the title list and begin coloring in the usual sequence. In this way, you will not use a color already specified for another structure.
6. Symbols used throughout the book are explained below. Once you understand and master the mechanics, you will find room for considerable creativity in coloring each plate. Now turn to any plate and note:
a. Areas to be colored are separated from adjacent areas by heavy outlines. Lighter lines represent background, suggest texture, or define form and should be colored over. If the colors you use are light enough, these texture lines may show through, in which case you may wish to draw darker or heavier over these lines to add a three-dimensional effect. Some boundaries between coloring zones may be represented by a dot or two, or dotted lines. These represent a division of names or titles and indicate that an actual structural boundary may not exist or, at best, is not clearly visible.
b. As a general rule, large areas should be colored with light colors and dark colors should be used for small areas. Take care with very dark colors: they obscure detail and texture lines or stippling. In some cases, a structure will be identified by two subscripts (for example, a+d). This indicates you are looking at one structure overlying another. In this case, two light colors are recommended for coloring the two overlapping structures.
c. In the event structures are duplicated on a plate, as in left and right parts, branches, or serial (segmented) parts, only one may be labeled with a subscript. Without boundary restrictions or instructions to the contrary, these like structures should all be given the same color.
d. If you will look at Plate 6 you will see the following symbols. The "&," means to color the titles gray. A title with (a) next to it means that only the titles following it are colored in the art. This usually indicates a category (such as coralline algae) and only the two representations (encrusting and articulated) are to be colored. When you see titles with a^1, a^2, etc., this indicates the parts so labeled are sufficiently related to receive shades of the same color (for example, the two types of coralline algae).
7. In the text, certain words are set in italics. According to convention, the generic name and species of an animal or plant are set in this way (for example, Chaetocerus denticulatus). In addition, the title of a structure to be colored is set in italics where the word first appears.
8. For further guidance on color use, see: Getting the Most out of Color.

As we have seen in the last two plates the ocean's surface is moved by wind and wave and is in constant motion, flowing in slow, ocean-wide currents. There are also more locally experienced currents caused by tidal action. This plate will explore the nature and origin of tides.

Tidal variations are caused by the interplay of the earth, sun, and the moon. As the moon travels around the earth, and as they both, together, complete a larger path around the sun, their combined gravitational attraction along with other factors cause the alternate rising and falling of the surface of the ocean, called tides.

Color the diagram of the earth, the moon, and the sun at the upper left. Then color the three drawings of the earth at the upper right. The top drawing illustrates the moon's gravitational pull on the earth's surface water. The second drawing illustrates the tide caused by centrifugal force, and the third indicates the combined effects of the two forces.

Gravitation (gravity) is an attraction between two masses; it is proportional to the product of their masses and inversely proportional to the square of the distance between them. Thus, although the moon is much smaller than the sun, it is so much closer to the earth that it plays the major role in causing tidal variations.

Visualize an earth completely covered with a layer of water, spinning on its axis. As the moon orbits the earth, that part of the earth directly opposite the moon will be pulled by the moon's gravity, and the water will bulge away from the earth toward the moon.

The spinning motion of the earth-moon pair creates what is known as a centrifugal force. This force causes the water on the side of the earth farthest away from the moon to also bulge out, away from the earth's surface. The gravitational pull of the moon and centrifugal force modify one another, and, at any given time, a bulge of water will occur on the side of the earth facing the moon and the side farthest from the moon.

Because the earth revolves completely on its axis approximately (but not precisely) every 24 hours, every spot on the earth will experience both types of bulge within that 24-hour period. The bulges are called lunar tides; the peaks are the high tides, and the troughs (between the bulges) are the low tides.

In an isolated earth-moon system, each spot on the earth would experience four tides each lunar day (24 hours, 50 minutes); two high tides and two low tides of equal magnitude. But the sun has an important modifying effect.

Color the diagram in the center of the page that demonstrates the position of the earth, sun, and moon during the moon's phases, and illustrates the tidal bulges. Then color the illustration of tidal range and the map that shows tidal variations.

The moon orbits the earth once every 27.5 days (the lunar month). In the course of its orbit, the moon is in a different position relative to the sun every day. At the time of the new moon and full moon, the sun, earth, and moon are in a direct line with one another. The combined gravitational pull causes extra high and extra low tides. These tides are called spring tides and are depicted in the tidal range illustration to the left of the piling.

During the first-quarter and third-quarter phases of the moon, the three masses of earth, sun, and moon are not in a direct line. The sun's gravitational pull minimizes the effects of the moon's gravitational pull, and the tidal variations are of much smaller magnitude. These tides are called neap tides and are depicted in the tidal range illustration to the right of the piling.

There is considerable tidal variation on earth. Geographical position, the shape of the ocean basin, and a host of other local, global, and planetary factors act to modify tidal range and frequency. The tides at the end of a long inlet will be considerably amplified due to the large volume of water pressed into a very narrow space, whereas the tides on a mid-ocean island will hardly be noticed.

Similarly, the frequency of the tides varies from place to place. The coastline of northern Europe experiences two equal high and low tides every day. The coast of California also experiences two high and two low tides daily, but they are not equal. The Gulf of Mexico has only one high and one low tide each day.

TIDES

EARTH $_a$
ORBIT $_{a^1}$
ROTATION $_b$
CENTRIFUGAL FORCE $_{b^1}$ TIDE $_{b^2}$
MOON $_c$
ORBIT $_{c^1}$
GRAVITATIONAL PULL $_d$
TIDE $_{d^1}$

SUN $_e$
SPRING TIDE $_f$
NEAP TIDE $_g$

TIDAL BULGE $_{b^2+d^1}$

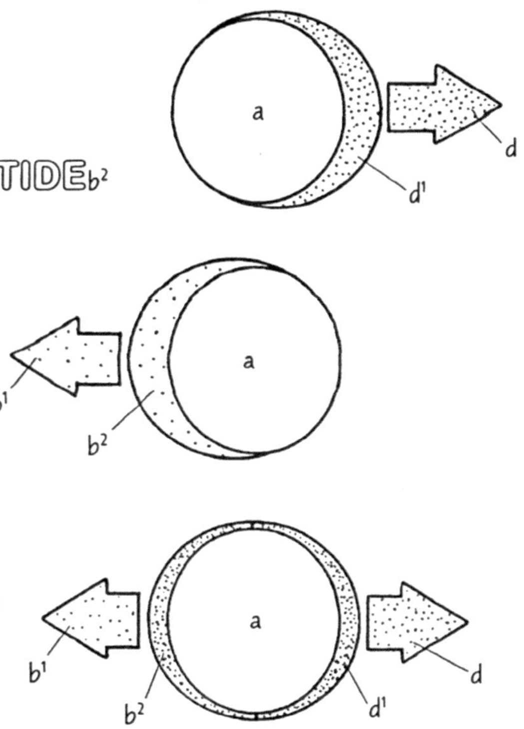

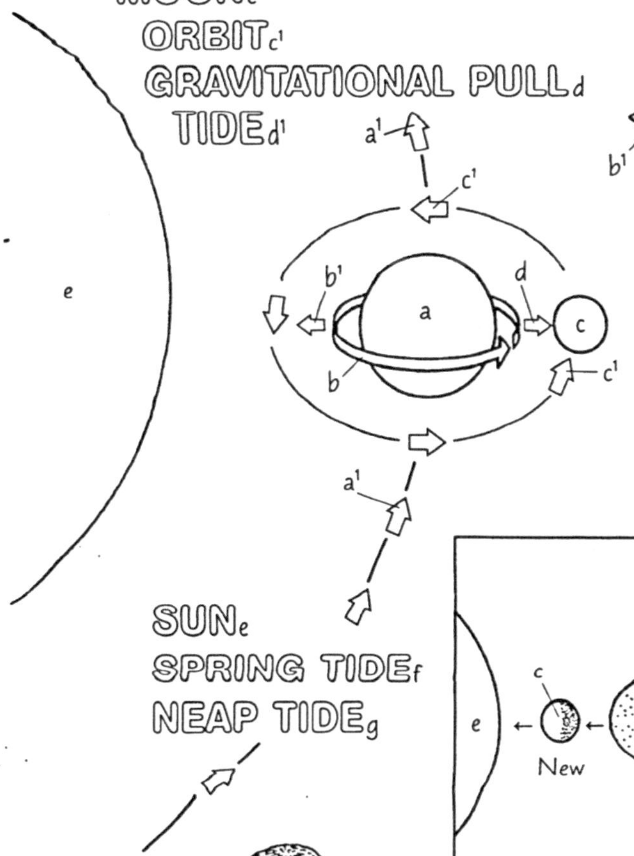

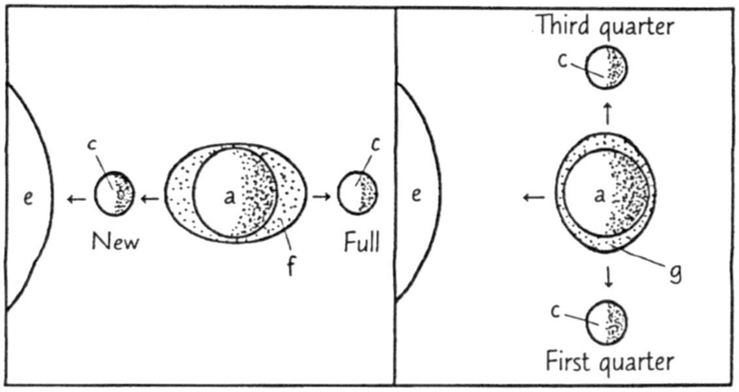

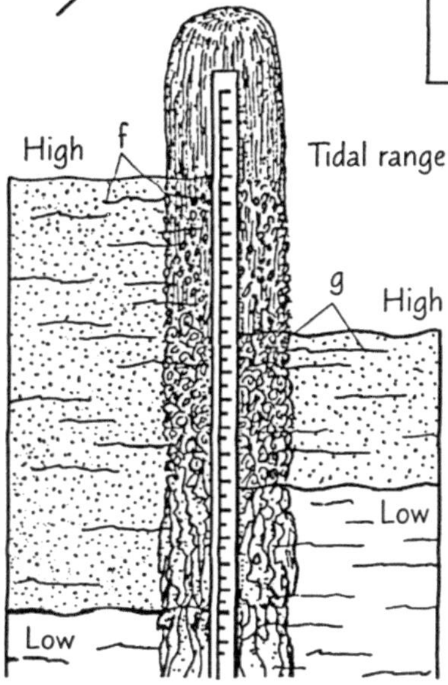

TIDAL VARIATIONS ✿

LARGE $_h$

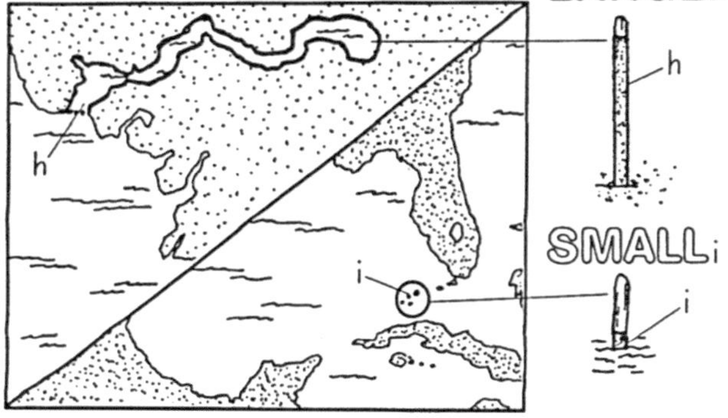

The pelagic zone is the most extensive of the marine environments; it encompasses the volume of water from just above the ocean floor to the surface, and it extends from shore to mid-ocean. The pelagic environment is divided according to its proximity to the continents and depth of water. This plate is concerned with the upper layer of the pelagic zone, where light penetrates through the water, a region called the photic zone.

Light penetration is necessary for the growth of plants and for the presence of animals that must see in order to feed. The water temperature has a great influence on organisms' growth rate and metabolism, and water movement plays an important role in the pelagic zone where organisms either drift with the current or swim within it.

Begin by coloring the three drifting animals belonging to the neuston group as they are discussed in the text.

Neustons are those organisms that float on the surface of the water. Although neustons are moved by both wind and surface currents, the wind is the most important mover of these animals. One of the most common coastal water drifters is called the by-the-wind sailor. The by-the-wind sailor (a cnidarian) is actually a whole colony of individual organisms that live, feed, and float together. As a colony, they set sail to the wind, producing a thin bright blue membrane of stiff material, oriented diagonally across the colony's elliptical body. As wind carries the colony across the surface, one individual in the center of the colony feeds, while others function in defense or reproduction.

Another colonial cnidarian is the Portuguese man-of-war, which utilizes a gas-filled sac as both a float and a sail, while long prehensile tentacles dangle into the water to catch prey.

The violet snail is also a member of the neuston group. The snail secretes bubbles that keep it afloat. This animal feeds on the by-the-wind sailor.

Now, color the enlarged plankton that move by currents.

Plankton includes the drifting organisms that are found at most depths in the water column. Minute single-celled plants are phytoplankton. They are successful only in the uppermost section of the pelagic zone where there is plenty of light for photosynthesis. Phytoplankton are traditionally represented by diatoms and by dinoflagellates (illustrated here, see also Plate 19), larger cells that are captured in standard plankton nets. Biological oceanographers now recognize that much of the photosynthesis that occurs in the photic zone is accomplished by the much smaller phytoplankters that pass through plankton nets. Not illustrated here, these small, flagellated, photosynthetic cells are members of the protozoa: single-celled, eukaryotic organisms that include flagellates, and heterotrophic ciliates and pseudopodia-bearing sarcodinids. These protozoans along with bacteria constitute a "microbial loop," organisms that were previously overlooked because of their small size, but are now thought to play a pivotal role in photic zone food chains.

Zooplankton (animal plankton) such as copepods and euphausiids, feed on the abundant phytoplankton. Two larger, sarcodinid protozoan groups with skeletons are also included in the zooplankton. Radiolarians with siliceous skeletons and foraminiferans with calcareous skeletons extend protoplasmic strands called pseudopodia to trap other plankton. The predatory arrow worm, or chaetognath, is known as the "tiger of the zooplankton" for its voracious appetite. These small animals dart about and devour prey as large as or larger than themselves. Zooplankton play an important role in the food chain and are the main source of nourishment for the young of many fish species, as well as much larger animals that strain the water for food.

Color the nekton, which are self-propelled.

Nekton, are pelagic dwellers that can swim in directions independent of ocean currents. This category includes the earth's largest mammal, the blue whale which can reach 27 meters (90 ft.) in length. The blue whale, together with several other whale species, is a filter feeder, whose principal food source is the zooplankton, especially euphausiids, or "krill." The nekton includes many fast and efficient swimmers, such as the squids, which utilize their own variety of jet propulsion. Albacore prey on the smaller herring. The sharks are a highly successful and predatory group, represented here by the blue shark.

The phytoplankton growing in the sunlit upper layers of the photic zone provide the basis of virtually all life in the entire pelagic zone. The productivity of the phytoplankton thus governs the number of organisms that can survive throughout the vast reaches of the pelagic zone. This explains why, with increasing depth or distance from land and its input of nutrients, fewer creatures are found.

PHOTIC ZONE

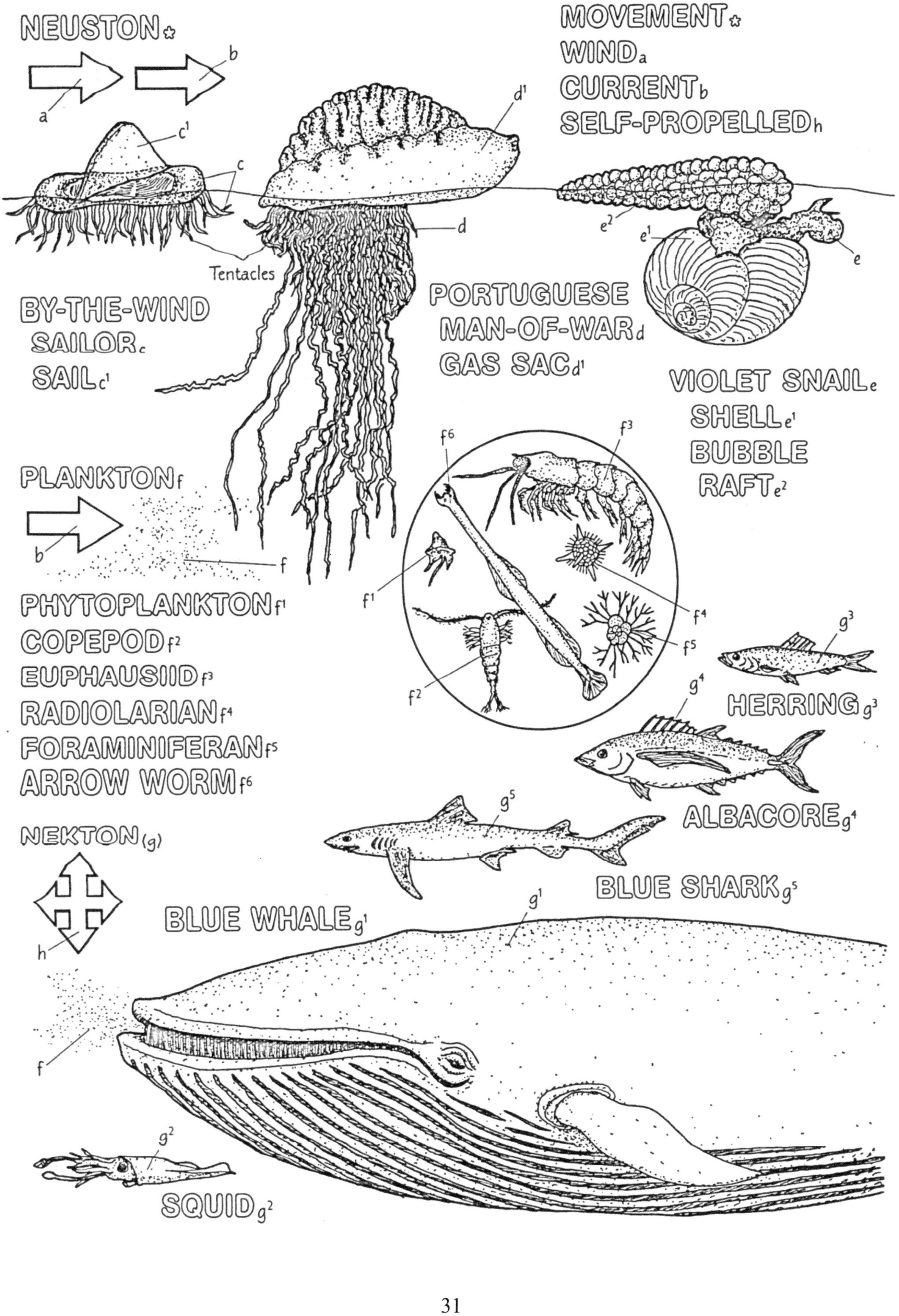
NEUSTON
MOVEMENT
WIND a
CURRENT b
SELF-PROPELLED h
a
b
c1
c
d1
d
e2
e1
e
Tentacles
BY-THE-WIND SAILOR c
SAIL c1
PORTUGUESE MAN-OF-WAR d
GAS SAC d1
VIOLET SNAIL e
SHELL e1
BUBBLE RAFT e2
PLANKTON f
b
f
f6
f3
f1
f4
f5
f2
PHYTOPLANKTON f1
COPEPOD f2
EUPHAUSIID f3
RADIOLARIAN f4
FORAMINIFERAN f5
ARROW WORM f6
g3
HERRING g3
g4
ALBACORE g4
g5
BLUE SHARK g5
NEKTON (g)
h
BLUE WHALE g1
g1
f
g2
SQUID g2

Types of Coral Reef

Coral reefs are spectacular marine habitats. The unique hard bottom substrate is composed of the calcium carbonate skeletons of living and dead corals.

The underlying platform of the reef is formed by the massive skeletons of large coral colonies that have lived on the reef and in death have tumbled to its base. The spaces between these large coral skeletons are filled by a rain of calcareous material including: the skeletal parts of smaller coral species and other coral reef inhabitants, the excavation products of boring animals that attack the corals, and the excreta of parrotfish (Plate 111) and other coral-grazing species. This jumble is lithified into a solid substratum by the precipitation of calcium carbonate dissolved in the sea water. Living coral colonies grow on the surface of this reef platform and maintain the reef's presence against the onslaught of wave action and boring organisms.

Corals are simple animals belonging to the phylum Cnidaria that also includes jellyfishes and sea anemones. The individual coral animal, called a coral polyp, looks very much like a sea anemone and feeds on zooplankton in the water (Plate 23). The coral polyp secretes a calcium carbonate cup into which it can retreat. As polyps grow, new coral cups are laid down over the old ones; and the coral colony grows outward and upward. Only the surface of the coral colony has live coral polyps.

Reef-building corals have very specific environmental requirements. They will grow only in clear, shallow sea water that has an average temperature of at least 20°C (68°F), which limits coral reefs to tropical regions. Modern coral reefs are found between 30° North and 30° South latitudes in the Caribbean Sea, and the Indian and tropical Pacific oceans.

Corals maintain single-celled plants known as zooxanthellae (modified dinoflagellates, Plate 19) in their tissues. The coral-plant relationship is a mutually beneficial (symbiotic) interaction (Plate 91). The plants have a safe place to live, and utilize the coral's metabolic waste products (nitrates and phosphates) as nutrients. The corals receive a substantial portion of their nutritional energy from the photosynthetic products of the zooxanthellae. Without this energy input, corals would be unable to secrete the amount of calcium carbonate necessary to form and maintain the coral reefs. Clear, shallow water is vital for adequate light exposure to the zooxanthellae and, thus, for survival of the reef.

Color the coral reef profile using light colors.

All coral reefs have a similar basic profile. On the seaward side, the coral reef has a sloping reef front that rises up from the depths. The reef front is not a solid wall, but is interrupted by channels that dissipate wave action and allow water and sediment to flow around the reef. Above the reef front is the reef crest that intercepts the onslaught of waves and reaches to the surface or just below it. Behind the reef crest is the protected, gradually sloping reef flat area that ends at the shore on fringing reefs and slopes into the lagoon on barrier and atoll reefs. The reef flat may be narrow or quite extensive with deep channels and a variety of substrates. These three reef areas experience very different exposures to light and wave action and harbor species of coral reef corals specifically adapted to these conditions.

Color the fringing reef, the barrier reef, noting the complex pattern of reef and lagoon areas, and the atoll reef as each is discussed. The three reef types are not drawn on the same scale.

The most common type of coral reef is the fringing reef. This reef projects seaward, directly from the shore. The reef flat area may have coral rubble with small corals as well as sea grass beds where sediments have accumulated.

Offshore of large tropical land masses where a shallow continental shelf occurs, a barrier reef may be formed. The Great Barrier Reef of northeastern Australia and the barrier reef off the east coast of Belize are the two main examples of barrier reefs. These reefs occur well offshore of the land mass and the lagoon between the land and the reef can be many miles wide. Barrier reefs are actually made up of many smaller reef elements, forming a most complex marine habitat. Barrier reefs support a great biodiversity and have served as the evolutionary points of origin for many forms of marine life.

Atoll reefs begin as fringing reefs growing adjacent to steep oceanic islands. The island is actually the top of a submarine volcano that over geologic time has reached the surface due to volcanic activity or a drop in sea level. If the sea level rises, and the change in depth is gradual, the coral reef may keep up with the slow increase in depth by growing upward. As the volcano sinks deeper, the shape of the reef may take on a circular or horseshoe configuration with a lagoon in its center. Extensive and complex patch reefs may be found on the lagoon floor. Reef flat areas of the main reef may become exposed as low-lying islands fringing the central lagoon. Coral sediment accumulates and plants become established from drifting seeds, forming the typical coral atoll island of the tropical Pacific Ocean.

TYPES OF CORAL REEFS

CORAL REEF PROFILE*
REEF FRONT a
REEF CREST b
REEF FLAT c

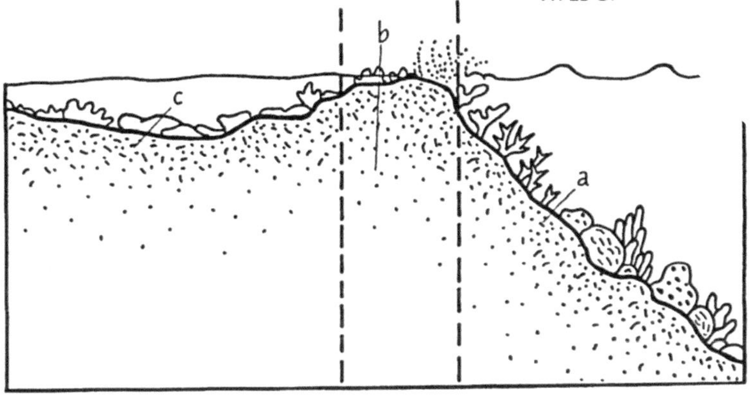

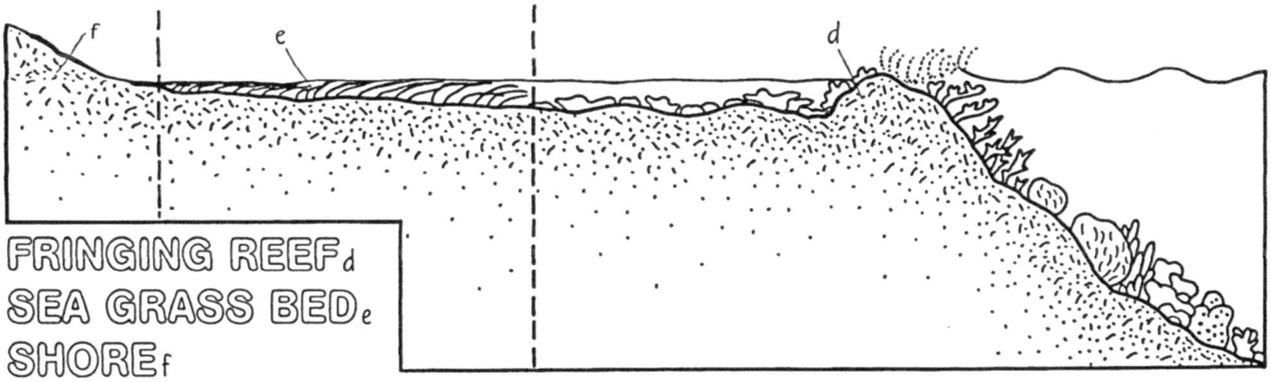

FRINGING REEF d
SEA GRASS BED e
SHORE f

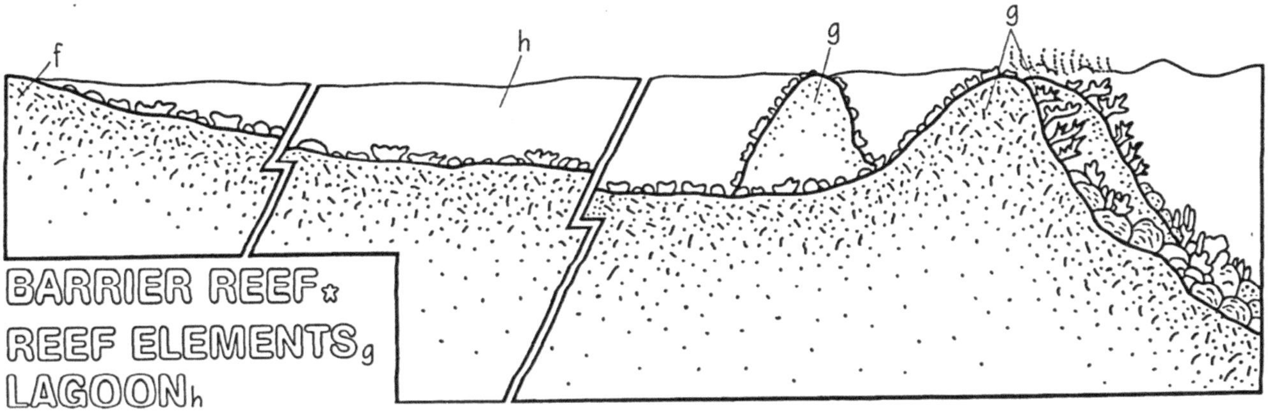

BARRIER REEF*
REEF ELEMENTS g
LAGOON h

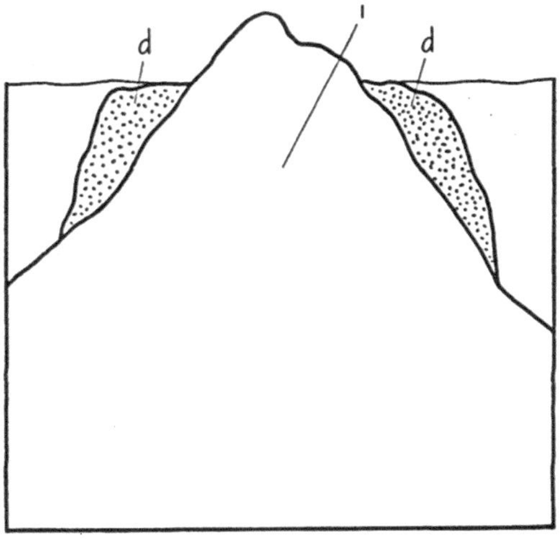

ATOLL REEF*
VOLCANO i
PATCH REEF j
ISLAND k

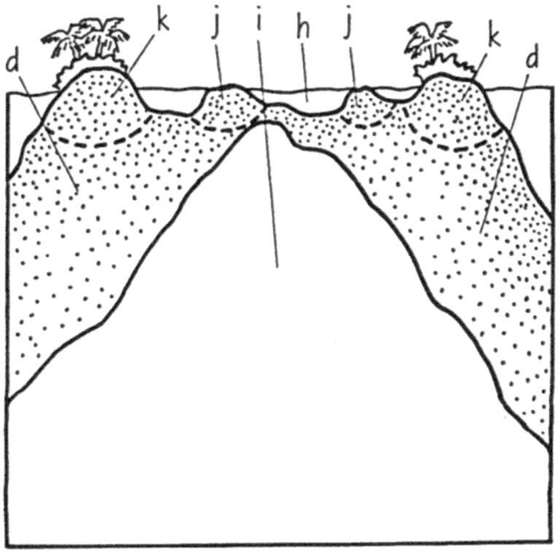

Most flowering plants become metabolically stressed if exposed to sea water, as they are incapable of dealing with the high salt content. This metabolic stress ultimately results in the death of the plant.

A few species of land plants have, however, successfully invaded and adapted to sea-water environments, flourishing in a variety of marine habitats. Unlike algae, land plants can grow only where they can absorb nutrients through their roots. These higher plants also need nearly direct sunlight, so they can only tolerate relatively shallow waters (1-30 meters, 3-100 ft.).

Color the red mangrove, starting with the prop roots at the lower right. Notice the junction of the trunk and roots at the high-tide mark. Color the entire leaf mass green. Color the falling seed and new root protruding from it, and notice the four young leaves emerging from the seed.

One of the largest flowering plants to flourish in the marine environment is the red mangrove. These plants occur in tropical and semitropical regions from Florida to South America and reach tree size (1-3 plus meters, 3-10 plus ft.). The mangrove grows best in the mud bottoms of estuaries, coastal lagoons, and near the mouths of large rivers, where silt is deposited. The mangrove remains stationary in the unstable mud bottom by sinking a mass of large, arching prop roots deep into the mud. The prop roots divide off from the trunk at the high-tide level; the roots proliferate and trap more sediment, raising the substratum level. This, in turn, creates more space where the mangrove can grow, and, in this way, the mangrove colony expands toward the sea.

Mangrove roots are the habitat for a myriad of attached algae and animals such as sea squirts, sponges, and sea anemones. The lattice-work of roots also offers protected nursery areas to young reef fishes and spiny lobsters. The mangrove leaves fall into the water and decompose, providing a detrital food source for many animals residing near the mangrove, and others further out to sea.

Seeds of the red mangrove sprout and begin to grow before they fall from the tree. Along (36 cm, 14 in), slender root emerges from the seed, and several leaves may also grow. If the seed falls at low tide, it may poke, dart-like, into the soft mud near the parent plant, and continue to grow. Seeds that fall during high tide float upright in the water and may be carried far from the parent plant. When the tide recedes, if the seed has been left on the mudflat with the root facing downward, it will be "drilled" into the mud by the jostling of small waves, thus beginning a new colony of mangroves.

Now color the turtle grass and the surf grass. Note that the grasses' stems receive shades of the same color as the mangrove's trunk, and the blades receive shades of the same color as the mangrove's leaves. These structures are homologous (same origin) and have the same function in these higher plants.

The flowering plants known as sea grasses can grow completely submerged. Various sea grass species occur from the intertidal zone to depths of 30 meters (100 ft.). Turtle grass is found in eastern Florida, along the Gulf of Mexico, and in the Caribbean, where it thrives on soft bottoms of mud, shell, or sand. Once established, turtle grass proliferates by extending underground stems (rhizomes) from which leafy blades grow. Turtle grass forms extensive shallow-water meadows, which are frequented by numerous juvenile reef fishes. At night, parrotfish and sea turtles come into these meadows to feed. The stems and roots of turtle grass bind the substratum, which permits the colonization of burrowing animals, such as sipunculid worms and sea cucumbers (Plates 25, 41).

Unlike most sea grasses, which require quiet waters, surf grass grows in the wave-swept rocky intertidal zone of temperate waters. Surf grass seeds possess two stiff, bristled, pointed projections. If the seed lodges on an alga, it can germinate and send out tenacious roots to grab hold of the substratum. The surf grass roots trap sediment, and a patch of long, thin, bright green blades may develop in a tide pool or at the low-tide line. This creates an environment that shelters many species of worms and other small creatures.

MARINE FLOWERING PLANTS

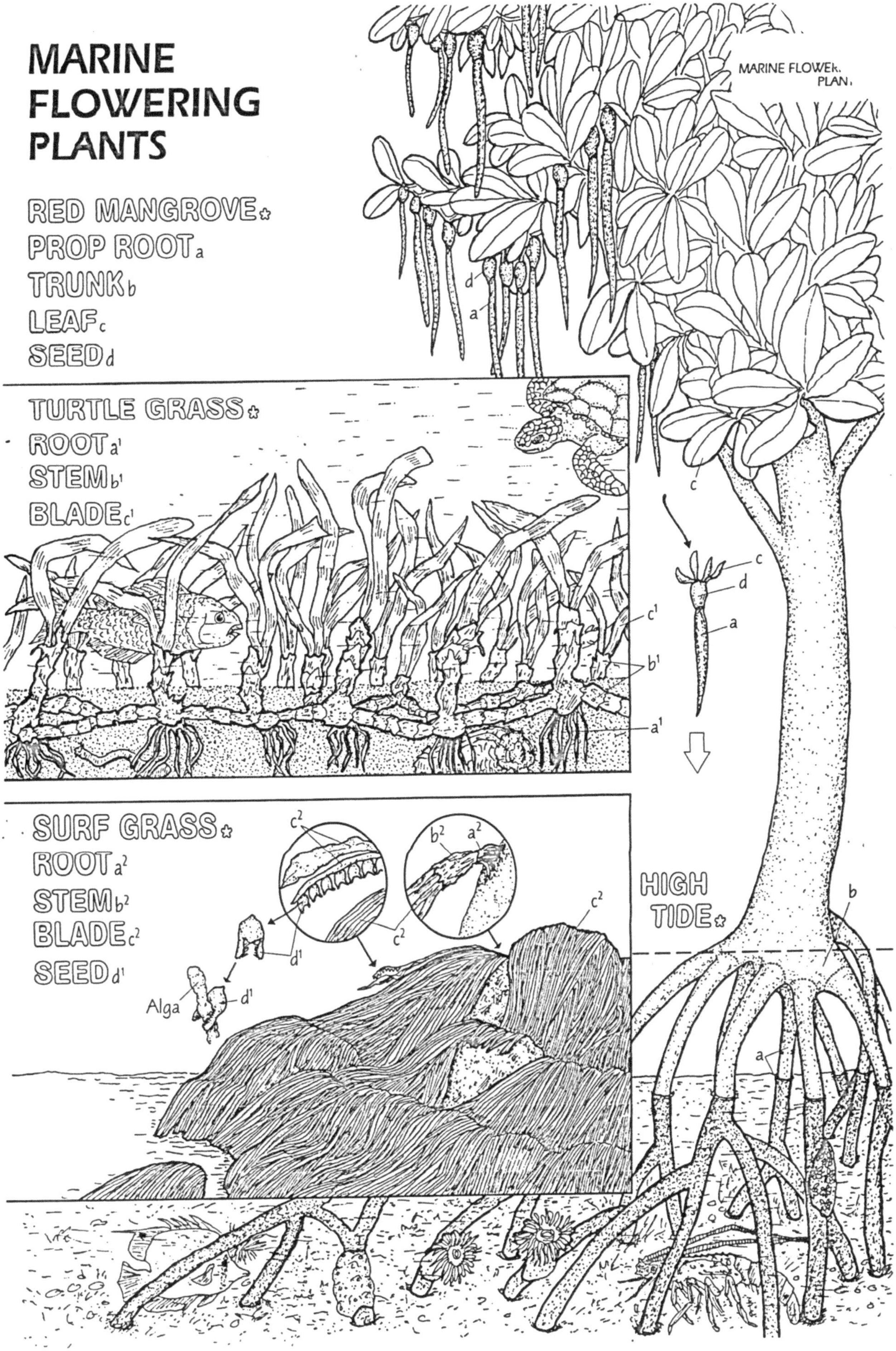

The group of decapods known as the "true crabs" (section Brachyura, Plate 37), are considered advanced crustaceans, and their reproductive biology is quite complex. The life cycle of the red rock crab is discussed here as an example of reproduction in this group.

Begin by coloring the male and female in ventral (underside) view at the top of the page. Locate and color each structure and body region as it is discussed in the text. Note that only a portion of the crab is colored here and that the abdomen is shown bent away from its normal position against the thorax. The normal abdomen position is indicated by using the abdomen color in this location.

Adult male and female crabs vary in the form of their abdomens. Although both sexes have reduced abdomens (when compared with lobsters and shrimps, Plate 36) that are normally kept flexed beneath the thorax, the male's abdomen is much narrower than that of the female. The abdominal appendages (modified swimmerets) are visible in the male as two pairs of shaft-like processes, the smaller pair fitting into grooves on the larger pair. At the bases of the last pair of walking legs of the male are two small openings for the release of sperm from the reproductive tract. Sperm pass from the reproductive openings to the grooves on the larger pair of swimmerets. The smaller swimmerets push the sperm along the groove into the genital openings of the female as described below.

The female genital openings are located on the thorax, and the broad abdomen bears several pairs of feathery swimmerets on which the egg cluster is held during brooding.

Now color the illustrations of the mating behavior and life cycle of the crab. Color each stage as it is discussed in the text. The natural color of the early egg cluster is salmon/coral pinkish, and the late egg cluster is brownish purple. Note that the larval and juvenile stages are greatly exaggerated in size.

During the life cycle of the rock crab, the crabs must molt in order to grow. The crab accomplishes this by resorbing much of the calcium salts from its old skeleton and then swelling with water to burst the old skeleton. The new skeleton is soft, allowing the crab to crawl free from the old encasement. Once free, the crab continues to swell and increase in size, and then gradually hardens the new skeleton by the redeposition of the conserved salts. Typically, the male molts before the mating season. The female, however, must be newly molted for successful mating to occur. It is critical that the male discover the female prior to her actual molting, so that he can be present to fertilize her eggs before her new exoskeleton hardens. The female probably releases a chemical attractant in her urine at the appropriate time. Once together, the mate grasps the female and holds her until she begins her molt. He releases her as she emerges from her old exoskeleton or "molt" (top right diagram of the life cycle). After her new exoskeleton becomes sufficiently firm, she lowers her abdomen and allows the male to insert his modified swimmerets into her genital openings for sperm transfer. Once this has been accomplished, the male holds the female in a protective embrace until her exoskeleton hardens.

The eggs are fertilized by the male's sperm as they are released from the female's genital openings. The sticky fertilized eggs adhere to one another, and the female attaches them to her "hairy" swimmerets. A female rock crab may carry several thousand eggs in her egg cluster. The mass of eggs is light in color when newly formed, but darkens substantially as the embryos develop. After several weeks, the developed crab embryos break free of their egg membranes and emerge as planktonic zoea larvae. The zoea larvae swim, feed on zooplankton, and molt several times until they reach a crablike megalops larval stage. The megalops finally sinks to the bottom, and at its next molt becomes a recognizable newly settled rock crab. Juveniles molt frequently, and the species discussed here may often be seen in the striped phase, with alternating black and red stripes on a light brown background. These crabs mature in three to four years and are then ready to participate in this complex adult reproductive behavior.

CRAB LIFE CYCLE

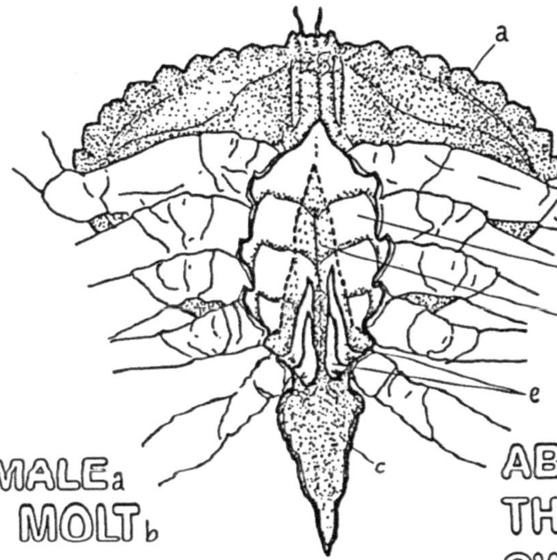

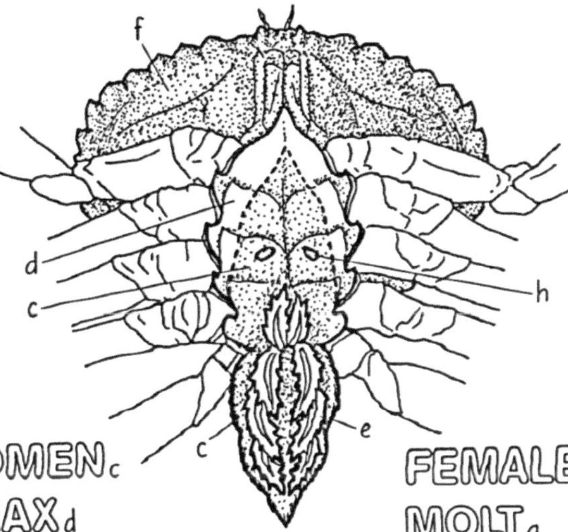

MALE$_a$
MOLT$_b$

ABDOMEN$_c$
THORAX$_d$
SWIMMERETS$_e$

FEMALE$_f$
MOLT$_g$
GENITAL OPENING$_h$

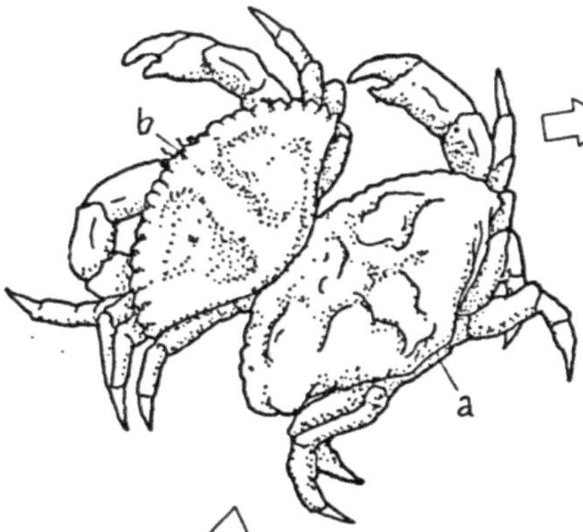

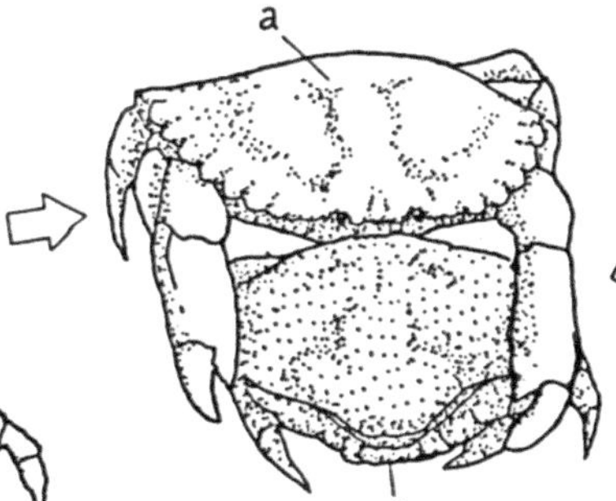

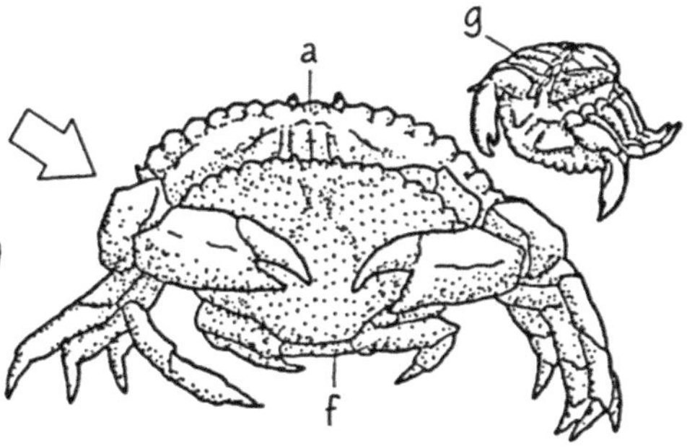

EGG CLUSTER$_{\star}$
EARLY$_i$
LATE$_j$
ZOEA LARVA$_k$
MEGALOPS LARVA$_l$
NEWLY SETTLED CRAB$_m$
JUVENILE$_n$

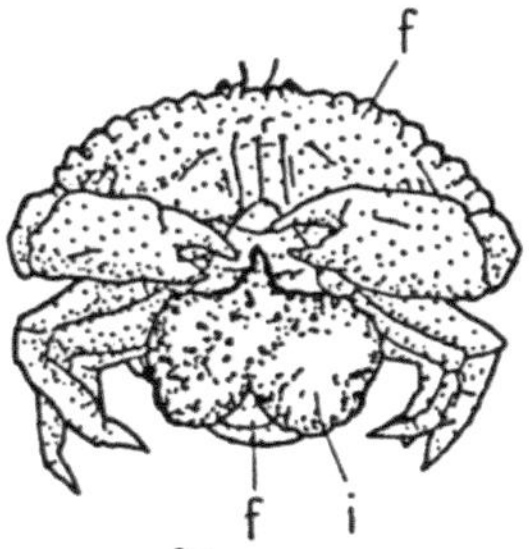

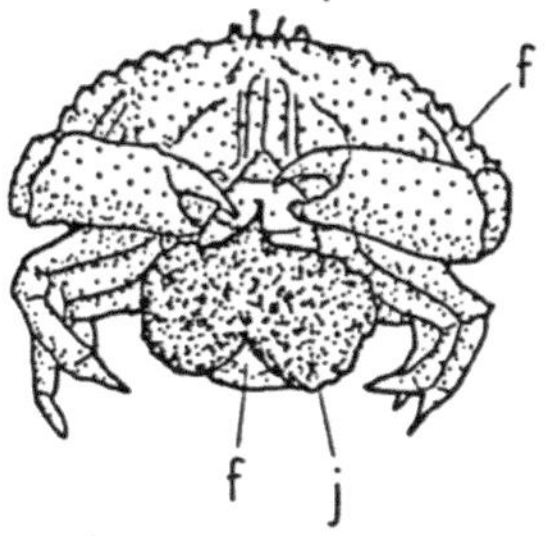

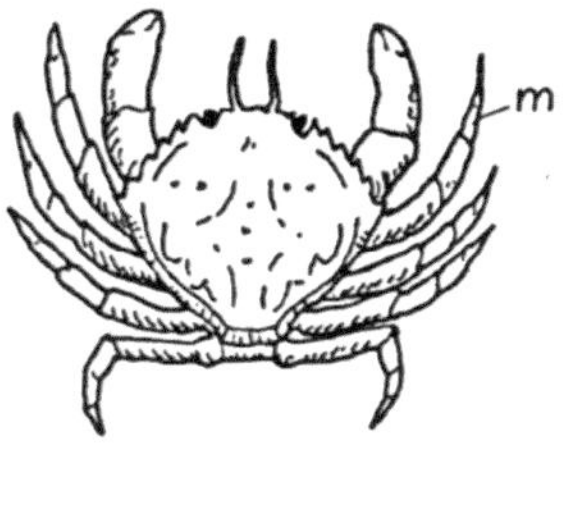

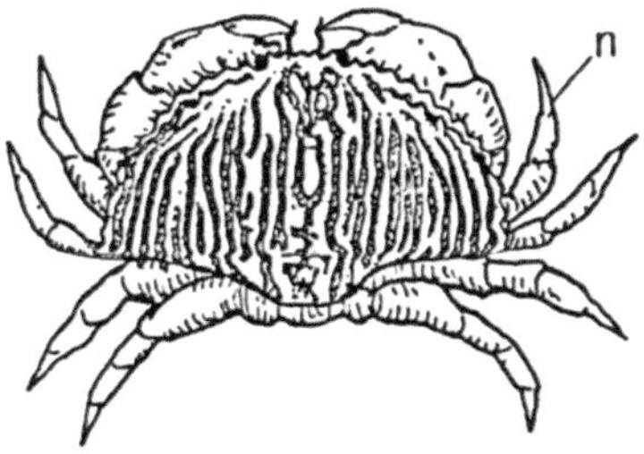

At this point in *Beyond the Classroom Aquarium*, students are guided through an activity to expand their powers of observation. The four operational aquatic ecosystems their teacher has established within the classroom are the focus of this observational activity. These demonstration systems provide students with direct examples that evoke anticipation for the kinds of work they will be responsible for performing as they work through the aquatic ecosystem projects described in *Beyond the Classroom Aquarium*. Student observation, exploration, inquiry, and discussion of these demonstration systems help students expand upon concepts introduced in the first week's artwork activities. Guided class discussions and critiques that emanate from this activity are essential for generating critical motivation and knowledge for students to use in making decisions of how to incorporate light and water components into the three classroom aquatic ecosystems that they will design, assemble, and maintain throughout the next five projects (and beyond).

Introductory observations of the model systems are designed to help students satisfy their initial curiosities about the kinds of animals, systems, equipment, and math applications they will become closely acquainted with in *Beyond the Classroom Aquarium*. The portion of this activity that provides for undisturbed observation of animal behavior (ethology) offers students opportunities to reflect upon the nature of human behavior. For many students, this is a good incentive for studying aquatic ecosystems.

For the teacher's information, an overview of these four operational classroom systems is woven into the description of these systems that is presented here (on pages 38 - 42) before the activity is described. In this overview, the set-up and function of many system components is explained. For educators who need a more detailed set of system set-up instructions than provided here, a list of six resources is provided. The first two contain information useful for freshwater systems. The third and fourth provide information about saltwater systems. The fifth resource raises many interesting ideas about introducing students to aquatic organisms, and the final recommendation is a comprehensive compendium about setting up and maintaining systems—a resource that should provide answers to any question you might imagine. Of the many information-packed websites with which to be familiar, I have always found it useful to review **www.wetwebmedia.com**. Most issues one might encounter are addressed in this website. The books listed on the following short recommended books list is in the appendix so that school resource persons and librarians can procure literature to support project work. I believe the school library should hold a number of such books - if not these specific ones.

- *The Complete Fishkeeper* (Levine, 1991)
- *You and Your Aquarium: A Complete Guide to Collecting and Keeping Aquarium Fishes* (Mills, 1991)
- *The Marine Aquarium Handbook: Beginner to Breeder* (Moe, 1992 & 2011)
- *The Marine Aquarium Reference: Systems and Invertebrates* (Moe, 1992, & 2014 Volumes I & II online)
- *Marine Organisms in Science Teaching* (Hunt, 1980)
- *Dynamic Aquaria: Building Living Ecosystems* (Adey and Loveland, 1991)

At the most basic level of understanding, student inquiry about these demonstration systems should lead them to realize that even the simplest of freshwater systems must consist of more than standing water in a container. These model classroom aquatic ecosystems are intended to reinforce student appreciation for the importance of simulating actions of wind, waves, and flowing currents. When necessary, students should be guided to recognize that as in nature, motion of water in simulated aquatic ecosystems is important for enabling the exchange of gasses that occurs at the water surface because water motion helps disperse these gasses throughout the water column. The dissolved oxygen activity in project three is provided to help students appreciate the fact that oxygen suspended in water is quickly depleted by aquatic life, and that even the simplest systems must incorporate equipment that will facilitate gas exchange between water and air. In order to promote life-sustaining gas exchange in their aquatic ecosystems, student ecosystem designs will often use electronic air pumps and water pumps. Students must be sure to identify these pumps in the aquatic ecosystem demonstration tanks and/or separate equipment demonstration tank.

It is interesting to consider the intuitive disbelief to which many students adhere regarding the idea that many organisms living in water obtain oxygen gas from the water that surrounds them. Aquatic ecosystems can help students appreciate the critical role that respiration performs for aquatic organisms. As we use our lungs to get this oxygen into our bodies, most organisms living in water do not use lungs but use instead other bodily tissues and organs to remove oxygen gas from water.

Another general principle of aquatic ecosystems that students should deduce from observation of their teacher's demonstration systems includes the idea that in order for most aquatic ecosystems to sustain life, water temperatures should not be allowed to fluctuate too greatly. In order to stabilize and regulate temperature of aquatic ecosystems in nature, the sun's interaction with the atmosphere and planetary geology have achieved an effective balance through the most recent few billion years. Students will not have this much time. In student designed aquatic ecosystems, they will use electronic heaters or chillers to control temperatures (or ambient room temperatures). It is important that students make note of the different kinds of temperature control devices that are in the teacher's demonstration systems.

Additionally, while observing each demonstration system, students should recognize efforts to simulate the role the sun plays as a provider of light energy. Aquatic ecosystems that students design and build will usually require the use of some kind of electronic artificial light. Student's attention must also be drawn to recognize the different kinds of lighting systems in the teacher's demonstration systems and the specific purposes for each lighting system.

In addition to some kind of substrate and possible under-gravel filter system, some project ecosystems will also be more efficient and safer if they include some kind of vapor barrier to slow down the process of evaporation.

Although it may be worthwhile and fun to include some exotic organisms in the demonstration systems, it is important that these systems contain many organisms that may be transferred to student developed systems in projects four, five, and six. The value of having these creatures on hand and healthy as soon as your students' systems are ready is inestimable. An observational activity follows the system overview below. This overview provides information about distinct features of each of the four model classroom systems. Students may even describe additional distinctions between these systems during the discussion segment of the activity.

"In House" Classroom Aquatic Ecosystem Model Number One (An Overview)

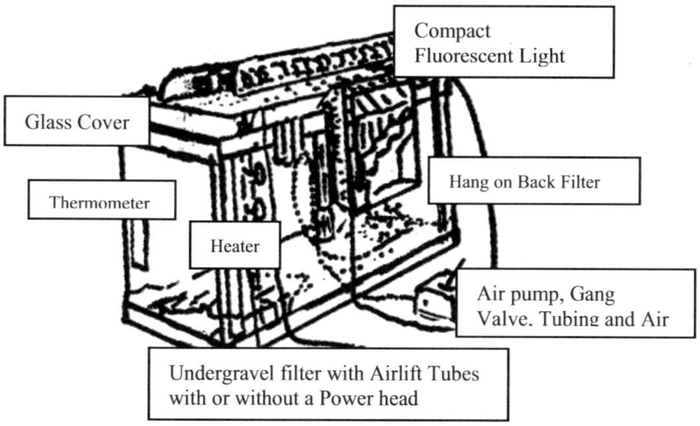

The first system one group of students will explore is a relatively simple freshwater system. Although the first thing most students usually notice about this system is the creatures swimming and crawling around within it, they will, with careful observation, be able to notice many important behind-the-scenes processes that are necessary to keep aquatic organisms healthy and happy.

In discussion about this system in the observational activity, students should note that it includes some kind of pump, heater, light, substrate, and cover/top. For safety, they should notice also how wires connected to the outlet or extension cord are secured above the water level of the system. An image of a simple freshwater microhabitat system is shown above. This first system provides students with a fundamental example to apply in their own projects. With experience they discover additional ways to develop systems that provide environmental stability for successful aquatic survival. A system like this first one is of use for a number of purposes but, unless closely monitored, can become toxic and unbalanced quickly when overtaxed or stressed. When a system becomes stressed, sickness and death quickly result. A system of this type does not closely resemble any natural aquatic ecosystem but is usually balanced enough, with the students' help, to use with simple organisms of interest without stressing them out.

Upon the conclusion of the observation activity, groups of students may be assigned the task of creating a poster that describes this first model aquatic ecosystem. The posters explain the purpose of each piece of equipment and the materials used in the system. The posters also identify the organisms in the system with a brief natural history of them. I recommend that a teacher's simplest freshwater system feature organisms (both plant and animal) that inhabit waters that are soft and acidic. Fish species I recommend include corydora and otocinclus catfish including the pygmy Cory (Corydoras hastatus), blue ram cichlids (Mikrogeophagus ramirezi), cardinal or neon tetras (Paracheirodon axelrodi or P. innesi) and the Discus fish (Symphysodon species). Other hardy and tolerant species such as zebra danios, guppies, and a few small characins and barbs, along with elodea and snails, may also be included. Many of these species can be induced to spawn and will be perfect for transfer to freshwater broodstock and larval rearing systems that students develop in project six. There are a number of ways to condition water so that it will be soft. My favorite method is to place peat moss in storage buckets (used exclusively for water). As students develop their aquaponic system, it will be important to be aware of the fact that roots of aquaponically grown plants function more efficiently when the water is somewhat acidic.

The first two aquatic ecosystems students design and construct will be used to support the lives of numerous freshwater aquatic organisms. As students learn more about natural freshwater aquatic ecosystems (e.g. soft. water, hard water), and the organisms that exist in them, they will be prepared to use their knowledge to design and construct artificial aquatic ecosystems that will be as similar to their natural habitats as possible.

"In House" Classroom Aquatic Ecosystem Model Number Two (An Overview)

The second freshwater demonstration system differs from the first in a number of important ways. Differences between systems that influence their capacities to sustain life should be reflected on the second aquatic ecosystem poster that students create. The poster for this system includes all of its unique design features. Each piece of equipment and the materials used in the system are described on the poster. Inquiry and discussion of this system should lead students to note the utilization of at least three water filtration modalities (physical, chemical, and biological). Explorations of system dynamics, over time, should lead students to conclude that systems like this are usually much more stable than smaller, simpler systems and will not become toxic and unstable as rapidly as smaller systems can. Students should also note, in the interest of safety, the placement of electrical wiring to reduce risk of electrical shock.

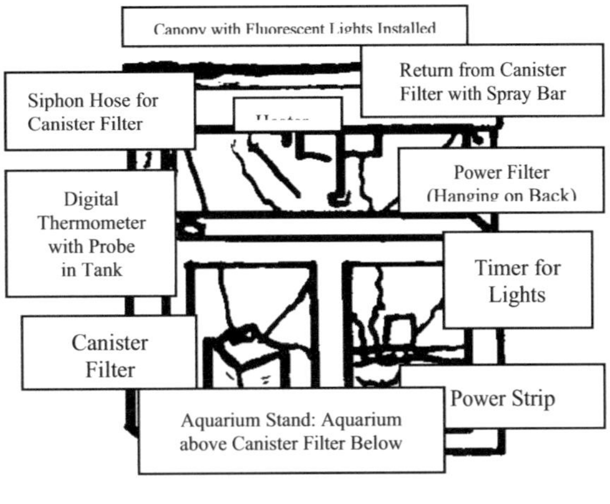

I like to set up this second demonstration system for larger, hardy freshwater fish that thrive in water that may be hard and slightly alkaline. Eventually, these fish (possibly some local game fish) may be transferred into a student developed aquaponics system. Aquaponics systems my students worked with included yellow perch, bluegill, and pumpkin seeds. Larger African cichlids may also be used. Tilapias are a popular cichlid species kept in aquaponics systems. Tilapias grow rapidly enough that many aquaponics producers harvest them for their meat after one year in an aquaponics system.

"In House" Classroom Aquatic Ecosystem Model Number Three (An Overview)

Aside from the different species in the tank, students should observe a few additional materials (live rock and calcium carbonate substrates) and pieces of equipment (protein skimmers and surface skimmers) in the basic saltwater system. Light and heat energy are still required so heaters and artificial light will be part of these systems. Chemical, physical, and biological filtration methods and mechanisms are also essential in saltwater systems. Once again, in the interest of safety, please take a moment to notice the placement of electrical wiring. This system gives students a chance to develop a poster on the basic saltwater aquatic ecosystem. The poster must explain the purpose of each piece of equipment and the materials used in the basic saltwater system. Studying organisms living in saltwater offers the opportunity to learn more about complex interactions between physical, chemical, and biological principles of natural aquatic ecosystems. Depending upon stocking consideration, students may be able to identify living specimens from the following phyla: mollusk, cnidarian, echinoderm, crustacean, polychaetes, and bryozoans—among which are many popular creatures including horseshoe crabs and colorful fish such as clownfish, tangs, gobies, blennies, as well as amphipods, copepods, conch, chiton, live rock, tube worms, caulerpa, sea grasses, mangroves, mithrix crabs, cardinal fish, and cleaner shrimp.

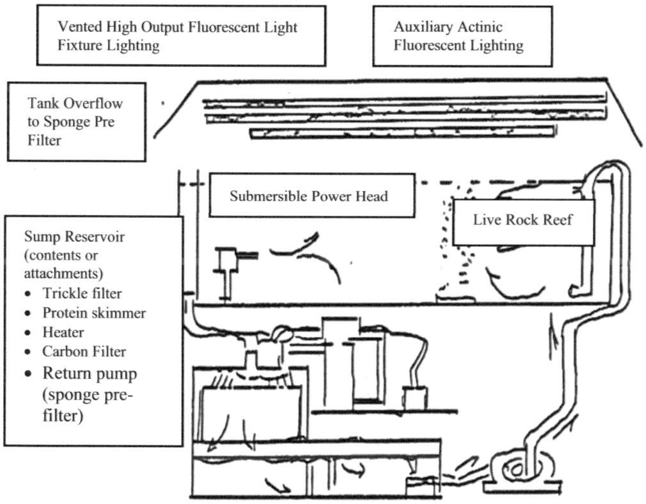

Inquiry into the saltwater systems should lead students to discover that the density of saltwater makes it possible to process the water in ways that are impossible with freshwater. Organic materials, such as wastes in saltwater, rise to the water surface in saltwater systems more uniformly than in freshwater systems, acting as a physical barrier to the entry of light to the organisms below the surface. Also, accumulated organic residue in saltwater allows for algae blooms and other undesirable effects upon the ecosystem. The density of saltwater makes it possible to remove much of the organic waste through a combination of the processes of surface skimming and foam fractionation. In a few weeks, students will learn about the design of surface skimmers, siphon tubes, protein skimmers, and how to construct these pieces of equipment while working with materials like fiberglass and plastic.

Many organisms in this system will be transferred to either a coral propagation system or saltwater broodstock larval rearing system that students complete in projects five and six, respectively. Hopefully, the coral propagation system will contain soft and stony corals, a surgeon fish (young and hardy), and a lawnmower blenny. Organisms to include in the saltwater broodstock larval rearing system include clownfish pairs. My students have taken part in pair development and efforts to induce them to spawn (in time). Many clownfish were raised in my classroom. In addition, peppermint shrimp were also raised in my classroom. I would recommend these species for a third system. Some teachers may find it possible to obtain conch, oysters, and sea urchins with a mind to attempt to study ways to raise them. The possibilities are almost endless.

"In House" Classroom Aquatic Ecosystem Model Number Four (An Overview)

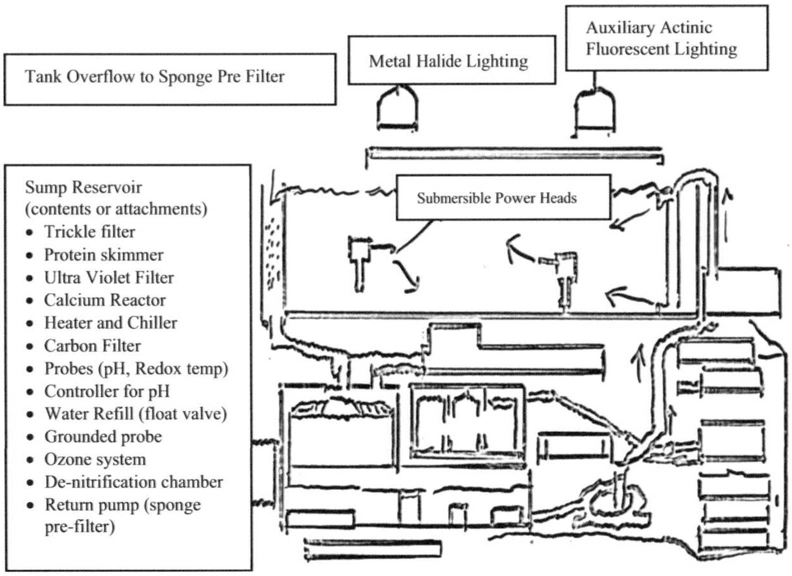

At the conclusion of the observational activity, students will be asked to put information about their classroom's advanced saltwater aquatic ecosystem model on a fourth poster. This poster should explain the purpose of each piece of equipment and the materials used in the system. There is usually more information on this poster than on any of the others. In addition to all equipment previously studied, the advanced saltwater system may incorporate materials and equipment not used in the basic system (e.g. calcium reactor, ultraviolet sterilizer, and externally mounted remote digital camera). Students' awareness of these materials and pieces of equipment make it possible to incorporate them in a system design of their own making. Equipment that is available to maximize the quality of water provided to organisms in artificial saltwater aquatic ecosystems is of particular importance. Such equipment makes it possible to study coral propagation and other dynamic processes in the aquatic classroom laboratory environment. Corals are examples of saltwater organisms that thrive only in artificial saltwater systems that sustain the highest levels of water quality. These systems will also help students learn how various saltwater invertebrates and fish larvae can be sustained and brought to adulthood. In addition to organisms found in the basic saltwater system, organisms in the teacher's advanced system may (after careful consideration) include some but not all of the following: stony corals, linkia starfish, sea anemones, shark eggs, sea fans, parrotfish, Tridacna clams, seahorses, Cassiopeia jellies, and scallops. It may also be interesting to access a few mangrove pods for eventual placement in a small saline aquaponics system.

While equipment used in the simpler systems is also used in this advanced system, be sure to draw students attention to the use of additional equipment and material such as a surface skimmer, sump, timer, digital data collection probes and computer linkages, a remotely operational digital camera frame (from the computer keyboard), an ultra-violet sterilizer, an automatic water replacement mechanism, probes to digitally collect data from the aquatic ecosystem, and aquatic plants. Students should once again notice how barriers are placed between the wires leading to electrical outlets and the system's electronic equipment; these barriers are for safety. For projects eleven and twelve, students may be designing and constructing this kind of system to support their investigation of aquatic ecosystems and related human cultures around the world and to do independent research on an organism that is of interest to them. An important factor in assessing their effectiveness as a developer of aquatic ecosystem designs will be demonstrated by the ethical treatment they provide to all organisms under their care.

Many public aquaria and zoos have departments of education that develop activities and tours to expose their guests to aquatic organisms and systems. In this classroom, the systems and organisms in the teacher's four model systems perform a similar function for your students—providing important introductory experiences in preparation for future work. This activity is also designed to motivate students to become skillful, knowledgeable, and enthusiastic participants in the work they do on project two and beyond.

Divide the class into five groups (A through E). Students should each be given a system observation sheet. Students in groups A through D go to the corresponding teacher's demonstration system. Students in group E go to the chalkboard and put the following table on the chalkboard. Each student should have a pen or pencil. All groups are to remain at their station until asked to rotate.

Once students in groups A – D are comfortably located so as to be able to observe the system to which they have moved, they will be given five minutes to write observations about the system in the spaces provided on the observation sheets. This timed activity is most valuable when students can first find a comfortable position from which observe the system while remaining still and quiet. After five minutes have passed, the teacher will then give the students a few moments to reposition themselves if necessary and then time them for two additional minutes after they have selected one organism in the system that they will observe and write a description about. After students have had time to complete their written description following the timed period of two minutes, the teacher will then ask students in each group to rotate to the next demonstration system. In this transition period, each group of students moves to the next station. The cycle of observations is to be repeated until students in each group have observed each system and had one period at the chalkboard. As students rotate to the chalkboard, they should put their observational data on the board. At the end of the data collection process, all groups should have five or ten minutes to place all of their data on the chalkboard and then completely fill out their observation sheets.

In the following class session, the instructor guides a summary discussion of student observations on the classroom systems. Students use the attached summary form to take notes upon which they will begin to research organisms to keep in aquatic ecosystems they will begin to develop in project two.

Classroom Aquatic Ecosystem Observation Data Format (Chalkboard)

Groups	**Group E**		**Group D**		**Group C**		**Group B**		**Group A**	
observations	**system**	**organism**	**system**	**organism**	**system**	**organism**	**system**	**organism**	**system**	**organism**
Station A										
Station B										
Station C										
Station D										

Name:	Observations		Summary	
Observations	System	Organism	System	Organism
Station A				
Station B				
Station C				
Station D				

Guidelines for Deciding Which Organisms to Keep in Student Designed Aquatic Ecosystems: A Library Assignment

The next important decisions upon which students must come to agreement involve selection of the aquatic species to place in their classroom systems. Each species selected will impact their work throughout the school year. Consequently, it is important that students are capable of making effective use of the many available sources of helpful information that they can access as they make decisions about their aquatic ecosystems.

In the following "What species of marine fish can we culture in our classroom?" activity (Moran, 2001), students are given a list of 13 species of fish that might be good to culture in a hatchery. Each student is to research the fish information including each species' natural history (where it normally lives, its spawning season, and what are its requirements for temperature, salinity, light, and food.) In their research, they are also to determine if this fish has been raised previously in aquaculture. Students are to then use this information to fill in the *Fish for Aquaculture* form (below). Finally, students share their findings and are guided through a critique about which fish are (or are not) good classroom aquaculture candidates.

Fish for Aquaculture

Name of fish species (include scientific name)

Native home range

Spawning season

	Requirements for survival
Temperature	
Light requirements	
Salinity	
Egg incubation time	
Feed requirements	
Life stages and life history	
Other important information	

Fish Descriptions

Common Name	Scientific Name

Tautog — **Tautoga, onitis**

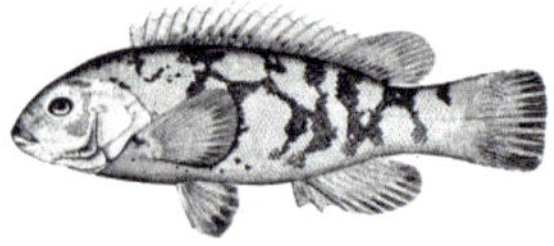

Wikipedia Commons

Popular food fish in the Northeast; its range is from Nova Scotia to South Carolina. The fish are structurally oriented and can be found in rocky areas and around pilings, piers, wrecks, etc. Feeds on shellfish; males are dark olive to grey with little blotching; white chin. Females and young are blotched with darker grey and black on a paler olive to brownish background.

Black Sea Bass — **Centropristis, striata**

Wikipedia Commons

Ranges from North Carolina to NE Florida; Gulf of Mexico and the Florida Keys; These fish prefer hard bottom habitats. Their head and dark areas of the body are bluish in color.

Summer Flounder — **Paralichthys, dentatus**

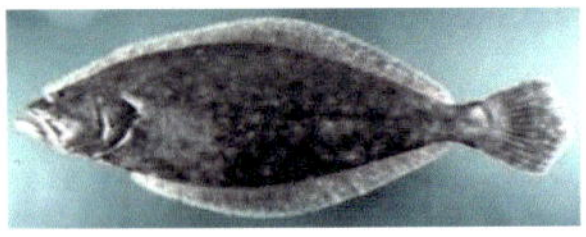

Wikipedia Commons

Important food and sport fish. Ranges from Maine to northern Florida; Left-eyed flounder

Cod — **Gadus, morhua**

Wikipedia Commons

One of the world's most popular commercial fish; they range from S. Greenland to Cape Hatteras and the East Atlantic; brownish body color with a white belly; well-developed barbel protruding from the chin.

Haddock — **Melanogrammus, aeglefinus**

Wikimedia Commons

Important commercial fish now greatly depleted by overfishing. Ranges from N. Newfoundland to Cape Hatteras and East Atlantic; dark grey body with purple reflections; they have a black blotch on their side and a black lateral line; small barbel protruding from the chin.

Grouper — **Hyporthodus, nigritus**

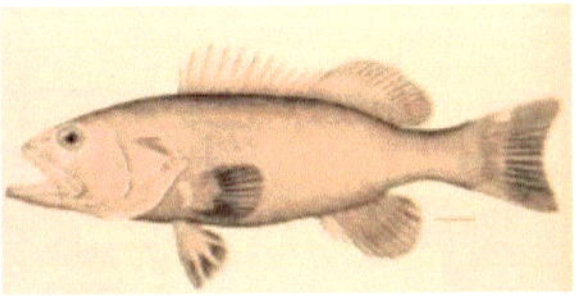

commons.wikimedia.org

There are several species of grouper that are popular commercial fish. Grouper are found in warm-temperature and tropical coastal waters. Most species occur on rough bottoms, around pilings, rocks, coral reefs, and shipwrecks. Sex reversal occurs in all species. When the fish is small, it is female and then transforms into a male later in life.

Snapper **Lutjanus, campechanus**

Important food and game fish; this species is wide ranging, mainly of tropical shore waters. Reef species are often tan and yellow in color whereas species that inhabit deep, rocky slopes are usually red.

Wikipedia Commons

Red Drum **Sciaenops, ocellatus**

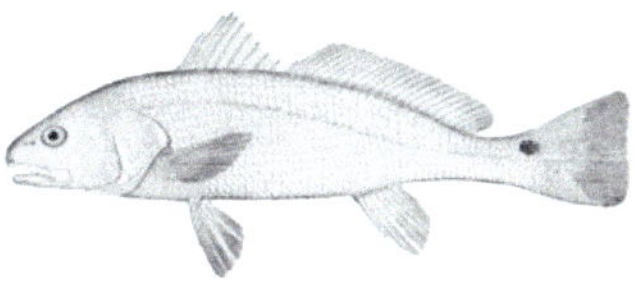

Important game fish ranges from Massachusetts to N. Mexico Including S. Florida; bronze colored fish, darker above; n chin barbels

Wikimedia Commons

Great Barracuda **Sphyraena, barracuda**

Gray with a greenish cast above, whitish below; ranges up to 6 feet and 100 pounds; important game fish and are widely eaten although they are known to cause the disease ciguatera; has been known to be dangerous to swimmers and divers although attacks are rare.

commons.wikimedia.org

White Shark **Carcharodon, carcharias**

Heavy bodied, slate blue or lead grey, becoming dirty white below; blunt snout, large head; teeth unmistakably large and triangular with serrated edges; can reach 26 feet; worldwide distribution in temperate and tropical waters; aggressive dangerous and may attack man.

Wikipedia Commons

Bluefin tuna **Thunnus, thynnus**

Pectoral fin short, tip does not reach point below beginning of 2nd dorsal fin; 2nd dorsal and anal fins dusky with some yellow; finlets bright yellow with dark edges; head relatively long and pointed; eyes relatively small; can reach 14 feet and 1500 pounds; found worldwide in tropical and temperate waters; important game and food fish with great economic importance.

Wikimedia Commons

Swordfish **Xiphias gladius**

Upper jaw sword-like prolonged and flattened; pelvic fins absent; no teeth in first jaw; a famous game and food fish; found round the world in tropical and temperate oceanic waters; may reach 15 feet and 1300 pounds.

Wikimedia Commons

Southern Stingray **Dasyatis *americana***

Disk body shape almost perfect rhombus with pointed corners; mid dorsal row of low spines; a few short spines in short rows near shoulder; whip-like tail with barbed spine; up to six feet across disk; tropical waters.

Wikimedia Commons

This final activity in project one unifies and finalizes student decisions about the species that they will keep and (in some cases) attempt to culture in the aquaponic, coral propagation, and broodstock larval rearing systems. Divide students into four groups: A through D. Each student in group A is given an aquaponics system form. Each student in group B is given a coral propagation system form. Each student in group C is given a broodstock and larval rearing form and THEN asked to circle saltwater. Each student in group D is given a broodstock and larval rearing form and THEN asked to circle freshwater. Students in group A will use their form to help select species to keep in the aquaponics system. Students in group B will use their form to help them select which species to keep in the coral propagation system. Students in group C will use their form to help them select the species to keep in the saltwater broodstock and larval rearing system. Students in group D will use their form to help them select which species to keep in a freshwater broodstock and larval rearing system.

- First, students disperse and use the Internet and other sources of information to help them fill out their form (including the supplementary Species Specific Lifecycle Requirements Form).
- Students reassemble in groups and critique one another's recommendations. In order to fairly consider each group member's recommendations, students use the rules and guidelines below (from Berger, 2004 p. 94) to critique each other's recommendations. Following this critique, students in each group arrive at a consensus about which aquatic species to recommend to the class for placement in the classroom aquatic ecosystems.

Critique Rules and Protocols

In this critique process, each member of the group takes turns offering suggestions explaining his/her ideas and goals and, when necessary, what particular aspects of the work are still unclear.

- Group members critique the suggestions, not the person.
- Group members try to begin critique comments with positive responses to the suggestions and then move on to ideas about the need for using constructive feedback to revise the suggestions.
- Group members try to use "I" statements when possible:
    - "I'm confused by this," rather than "This makes no sense."
- Group members try to use a question format when possible:
    - "I'm curious why you chose to begin with this…?" or
    - "Have you considered including…?"
- Group members emphasize proper use of topic-related vocabulary

- Each group reports their consensus about their species recommendations and asks the class to critique them.
- Following these critiques, all students vote for the species to be placed in the classroom aquatic ecosystems.

Aquaponic Species Selection and Lighting System Proposal Form

In the ideal light intensity category – state the kind and number of light bulbs that each plant in this system will require.

	Species One	Species Two	Species Three	Species Four	Species Five	Species Six
Aquaponic Species (Plant/Fish Name and Description)						
Ideal Water pH						
Ideal Light Intensity						
Ideal Temperature						
Seed Germination Protocol for Plant Species						
Time to Harvest						
Potential Pests						
Other Factors						

Coral Species Selection and Lighting System Proposal Form

In the ideal light intensity category – state the kind and number of light bulbs that each plant in this system will require.

	Coral. One	**Coral Two**	**Coral Three**	**Coral Four**	**Coral Five**	**Coral Six**
Coral Species						
Ideal Water pH						
Ideal Light Intensity						
Ideal Temperature						
Ideal Calcium						
Growth Rate						
Compatibility						
Other Factors						

Broodstock Species Selection and Lighting System Proposal Form

In the ideal light intensity category – state the kind and number of light bulbs that each plant in this system will require.

Freshwater/Saltwater Circle	**Species One**	**Species Two**	**Species Three**	**Species Four**	**Species Five**	**Species Six**
Broodstock Species						
Ideal Water pH						
Ideal Light Intensity						
Ideal Temperature						
Age to Maturity						
Other Important Water Parameters						
Dietary Concerns						
Other Factors						

Species Specific Lifecycle Requirements Form					
		Use Numbers 1 – 4 to number each species on your list in column two	**Unique Features of Life Support Systems at Different Life Cycle Stages**		
Reproductive Adaptations	**Species List**	**Accessibility to Fertilized Eggs**	**Adult**	**Larvae**	**Juvenile**
Live Bearer **Or** **Mouth (Pouch) Brooder**		1. Broodstock System 2. Teacher's System 3. Nature Collection 4. Commercial Lab			
Egg Broadcaster		1. Broodstock System 2. Teacher's System 3. Nature Collection 4. Commercial Lab			
Egg Cases		1. Broodstock System 2. Teacher's System 3. Nature Collection 4. Commercial Lab			
Adhesive Egg Scatterer		1. Broodstock System 2. Teacher's System 3. Nature Collection 4. Commercial Lab			
Nest Builder		1. Broodstock System 2. Teacher's System 3. Nature Collection 4. Commercial Lab			

It is intended that as students formulate and critique their observations, they are being guided through a process leading to the establishment of a shared classroom vision related to the natural structure and sequence of projects and activities in which they will engage from this point forward.

Having looked closely at the four model systems—the equipment and organisms in them—and after determining which species to keep in the classroom, class activities are now designed to guide students through the processes of designing and assembling the three classroom aquatic ecosystems. Logistics for handling this expeditiously in a classroom setting must be considered. Students should carefully review the approach described in chapter II with their instructor. It is essential to capitalize upon the excitement that group decision-making activities engender. Once all understand and agree on the value of next designing aquatic ecosystem frames and lighting systems, (guided by the teacher and additionally the often generous and essential support of local tradespeople [carpenters, electricians, plumbers, engineers, digital technologists, etc.] and public or private enterprises that do business that is related to aquatic ecosystems), it is now time for students to participate in field trip number one.

As described previously, the trip has an AM and PM component. Upon returning to class, students are to conduct a general whole-class critique of the trip, using the field trip feedback form in the project one assessment activities section of *Beyond the Classroom Aquarium* (see page 21). In the AM component of the field trip, students are to visit/tour one of three local operational aquatic ecosystems (an aquaponics, coral propagation, or a broodstock and larval rearing system). The PM portion of the field trip may or may not involve additional travel. The purpose of the PM portion is to introduce students to local tradespersons who may serve as resource persons throughout the remainder of the school year. At this two hour session, carpentry and electrical safety protocols are demonstrated as students are guided through the process of working with tools, materials, and aquatic ecosystem equipment. With guidance from the teacher (and assistants), these tradespersons may be able to lead students through an overview of the entire mechanical process of developing a specific classroom aquatic ecosystem. As a result of some foresight and guided hands-on work with tools and materials, each student field trip group may return to class with all the precut lumber (and wiring diagrams) they could need in order to assemble their respective classroom aquatic ecosystem frame and lighting system.

Beyond the Classroom Aquarium Project One Summary

Students have now seen their teacher's systems, studied basic components of natural aquatic ecosystems on Earth, and been introduced to aquatic organisms and equipment that is used to keep aquatic ecosystems up and running. Students have also been involved in introductory discussions related to aquaponics, coral propagation, and broodstock/larval rearing systems. They have been guided to consider how they might be called upon to develop and apply scientific, technological, engineering, and mathematical (STEM) skills. In addition, they have also determined which organisms they will be placing in the systems they design and assemble. In chapter II, students will learn about framing, lighting, and water components of the three aquatic ecosystems. They will also begin to establish protocols for introducing aquatic organisms to their respective system.

Write a list of **science-based** activities you have participated in during this project and briefly describe evidence that verifies your interaction with and knowledge of scientific concepts for each item on your project one list.

Write a list of **technologically-oriented** activities you have participated in during this project and briefly describe evidence that verifies your interaction with and knowledge of technologically-oriented concepts for each item on your project one list.

Write a list of **engineering-oriented** activities you have participated in during this project and briefly describe evidence that verifies your interaction with and knowledge of engineering-oriented concepts for each item on your project one list.

Write a list of **mathematical** activities you have participated in during this project and briefly describe evidence that verifies your interaction with and knowledge of mathematical concepts for each item on your project one list.

Write a brief list of **Non-STEM-related** activities you have participated in during this project and briefly describe evidence that verifies your interaction with and knowledge of Non-STEM related concepts for each item on your project one list.

Rate your level of involvement in STEM-related activities in this project (1=lowest, 10=highest).

1—2—3—4—5—6—7—8—9—10

Beyond the Classroom Aquarium Daily Journal Guidelines

Entries students place in their daily journals should be:

1. Dated accurately and logged chronologically.
2. Kept in a notebook and written neatly in ink or typed.
3. Written or typed in complete sentences using correct capitalization, spelling, and punctuation.

In their journals, students should keep a daily record of:

1. Who they worked with each day by including the names of school staff, fellow students, community members, and volunteers.
2. What they did each day, such as lecture topic, activities or project design, assembly, or maintenance.
3. What they learned each day by asking questions, following instructions, and/or observing activities.
4. How *Beyond the Classroom Aquarium* experiences relate to past, present, or future life experiences.

For bonus points, students may complete these journal options, including illustrations, diagrams, or creative expression relating to the journal entry. For project one, journal writing bonus topics may include answers to the following questions or descriptions of a project activity or assignment. Reflection on these topics may also help students prepare for exams, quizzes, and other assessments.

- Describe similarities and differences between the four model aquatic ecosystems.
- Using information gathered from discussions about the classroom's four model systems, work with a partner to, create a poster that informs readers about one of the four model systems and the organisms the system contains. If possible, create a poster for each system.
- Describe one organism from each system and how the organisms are similar and how the organisms differ.
- In what ways do you think these systems involve science, technology, engineering, and math?
- What would you like to know about the organisms and systems that you observed?

- What do you want to better understand in order to contribute more to your team's productivity as it independently designs, assembles, and then manages an aquatic ecosystem?
- Draw a picture and create a story about the organism or system that is most interesting to you.
- Create a journal entry that tells what information you learned during project one. Also, write down what you would like to know more about and if you know someone else who might help you learn more if you asked them.
- Select a book or magazine to read that is related to a topic in project one. Write a report on this book and include your report in your *Beyond the Classroom Aquarium* portfolio.

Journal entries may be evaluated on a point system, with four points being the maximum number a student can earn. A half-point will be awarded for each of the first eight journal requirements above. Bonus points are awarded when journal options are completed. Journal points will be totaled and divided by the number of journal dates.

A = 4.0-3.5 average; B = 3.4-2.6 average; C = 2.5-1.6 average; D = 1.5-1.0 average; F = 0.9-0.0 average.

Additional Project One Assessment Activities and Questions

How many days are in a lunar month? Explain what a lunar month is.

Explain the causes of tides.

What is the photic zone in a marine environment?

Describe three kinds of plankton that exist in the photic zone and explain how they survive.

What phylum do corals belong to and, in English, what does the name of this phylum mean?

Explain the symbiotic relationship between zooxanthellae and corals.

Describe three different kinds of coral reefs.

Explain the life cycle of the red mangrove.

Describe turtle grass habitats and how turtle grass grows.

Describe the aquatic ecosystem equipment that all four classroom model systems have in common.

Select a fish listed in any of the four classroom model system descriptions. Develop a presentation about this fish including a description of ways that it has been successfully bred in captivity. Be sure to describe ways the species you selected is being used for research purposes. Prepare this report for class presentation and critique.

Review activities and information that interested you in this project and determine if this information may fit on your "Research-ability" Potential Topics Form on page 59. If so, add it to the list.

Beyond the Classroom Aquarium STEM Self or Peer Assessment Rubric				
This self-assessment is valid only to the extent to which you can document and display evidence that supports it. For each category in the left hand column, write a STEM-Oriented example in the box beneath 3 if you can display evidence that your STEM-Oriented example represents work that exceeds expectations. If your evidence shows that your STEM-Oriented statement meets but does not exceed expectations, write the statement in the box beneath 2. Similarly, for STEM-Oriented examples that demonstrate some progress toward expectations.				
Name: **Project Number:** **Date Project Work Began:** **Date of Project Completion:** **List Math Topics and/or Objectives Currently Under Study:**		**3** **Work Exceeds Expectations**	**2** **Work Meets Expectations**	**1** **Work Is Progressing Toward Expectations**
Write examples of questions and defining problems related to your STEM work.				
Describe how you have developed and used models in your STEM work.				
Write examples of STEM investigations you planned and carried out.				
Describe STEM related data that you analyzed or interpreted.				
Give examples of how you used mathematics and computational thinking related to the five standards listed to the right.	Problem Solving (Math Process Standard One)			
	Reasoning & Proof (Math Process Standard Two)			
	Communication (Math Process Standard Three)			
	Connections (Math Process Standard Four)			
	Representation (Math Process Standard Five)			
Describe STEM-Oriented explanations and solutions you designed.				
Explain evidence you used while engaging in a STEM project-based argument.				
Give examples of STEM information you obtained, evaluated, and/or communicated.				
TOTAL POINTS				

Fill out the following summary form to demonstrate your preparation to responsibly keep organisms in classroom aquatic ecosystems.

Introductory Aquatic Ecosystem Inhabitant Form				
	Project Four	**Project Five**	**Project Six**	
	Aquaponics	Coral Propagation	Freshwater Broodstock Larval Rearing	Saltwater Broodstock Larval Rearing
Important things I learned while studying information about organisms to consider for introduction to the system:				
Important protocols for healthful transfer of organisms and "code red" quarantine protocols				
Organisms in the system – A final decision				

Observation Protocols and Quarantine – Identifying Healthy Fish (supplementary reading)

Are Your Fish Really Suffering From Disease? - A guide to non-disease causes of illness *

You have a fish, or several fish, that appear sick. They are scratching against decorations, lethargic, unusually nervous, or gasping at the surface of the water. Clearly, something is wrong! But before reaching for your preferred medication, stop for a moment and evaluate the situation. If there are no outward signs of disease (white spots, cottony growths, red streaks, etc.) your fish may not be suffering from an infectious disease at all.

More often than realized, seemingly sick fish are actually suffering from some negative environmental conditions or introduced toxins. While many of the substances in our daily lives do not affect us adversely, it must be remembered that fish live within their medium. Any foreign substances that enter the tank can and will be breathed in, swallowed, or even penetrate the skin, causing poisoning.

Many of the symptoms we so often associate with disease are in reality general symptoms that can have more than one cause. For instance, there are indeed a few diseases that can cause flashing (scratching) in fish, but there are also many toxic substances that can cause the same behavior. Flashing can be equated with coughing in people. You may cough when you have a cold, but you may also cough if you walk into a smoky room. It's simply a sign that your lungs have been irritated. The same applies to the gills of the fish.

Please take a quick look at the other possibilities presented on this page before you treat your fish for an uncertain disease. A brief list of symptoms, an explanation, and a recommended course of action is offered for each circumstance.

Before beginning please note:

- Fish that are showing clear signs of a specific disease should be treated immediately with the appropriate medication. This list is only intended to help track down potential problems when the cause of certain symptoms is not clearly diagnosable. Be aware that in the early stages of many diseases, outward signs may not yet be visible.
- Fish exposed to toxic substances or conditions may, during or after exposure, develop disease due to stress and lowered resistances. Treat if necessary.
- It is assumed that the reader has a basic knowledge of the aquarium hobby. If you are completely new to the hobby, please familiarize yourself with standard aquarium practices.
- Underlined passages indicate fairly diagnostic symptoms of specific health issues, or other important information.

Water Quality Issues	Breathing Issues	Introduced Substances	Fumes	Other Issues
Low pH (acidosis)	Hypoxia	Metal	Smoke	Injuries
Hi pH (alkalosis)	Gas Bubble Disease (nitrogen narcosis)	Cleaning products	Paint fumes	Poor nutrition
Ammonia	Carbon Monoxide Poisoning	Sprays	Glues & epoxies	Behavioral Stress
Ammonium	Excess Carbon Dioxide	Air Fresheners		Stray Voltage
Nitrite	Hydrogen Sulfide	Pesticides		Bacterial Bloom
Nitrate		Soaps & Detergents		Temperature
Chlorine		Dirty hands, used buckets, containers,		
Chloramine		Contaminated equipment		
		Medication poisoning		

* The information on this page is abstracted with permission from the following Aquarium Guys website: **http://www3.sympatico.ca/drosera1/fish/illness.htm**

Getting Ready for Project Nine – "Research-ability"
Potential Topics Form (Foreshadowing)

As you encounter interesting information, resources, activities, and topics in this and upcoming projects, place ideas in this table that you could possibly explore more fully during your research project in project nine. Having this list will be helpful to you when you are preparing to design your aquatic ecosystem research project.

"Research-ability" Potential Topics Form			
Possible Topics (General)	**Sources of Related Information**	**Specific Researchable Issue**	**Potential Value of Research Findings**

After the first field trip and many carefully guided classroom discussions about design and assembly of classroom aquaponics, coral propagation, and broodstock larval rearing aquatic ecosystems, most students agree that in this second project, aquatic ecosystem lighting is the topic they should study next. Teachers will find the information in project two useful for leading students to complete their work related to design and assembly of full-scale lighting system components for these three aquatic ecosystems. The aquatic ecosystem design and assembly activities of this project proceed more smoothly when the lead teacher was able to provide student-developed classroom design specifications to local tradespersons who were recruited to serve in supportive advisory roles during the field trip experience in project one. In this case, much of the framing work could be completed already.

In project two, important aspects of aquatic ecosystem lighting and related electrical concepts are studied and applied. By addressing topics of light and working with electricity at this early point in the year, students receive many opportunities to reinforce safety procedures that are to be practiced throughout the rest of the year. The electrifying and enlightening topic of light captures student interest and imagination by addressing at least three real-world applications that they will work on in project two: 1) design; and 2) assembly of an aquaponics, coral propagation, and (possibly two) broodstock and larval rearing lighting systems; and 3) an introduction to the function and use of many other electronic pieces of aquatic ecosystem equipment.

By working on lighting system components for three aquatic ecosystems, students develop a better understanding of the nature of light and ways that it can be measured and used. One of the first tasks in project two requires student teams to develop a lighting plan proposal that will directly influence the success of each aquatic ecosystem project. In this proposal, they are challenged to address the question: what kind of light bulbs and how many of them are necessary for each aquatic ecosystem? A System Frame and Lighting Report form is provided in each of the three system design and assembly sections of project two (pages 80, 83 and 87). Students may use the information they enter in these forms to contribute to class discussions, analyze their options, and make good decisions about species to be placed in each system. Through careful instructional guidance and additional lighting information sources (e.g. Moe, 1992a, 1992b; Weast, 1987; Lundegaard, 1985; Osborne, 1983; Reef Aquarium Farming News, 2000; Taylor, 2000; Chase, 2006; Carner, 1997), students have ready access to much of the information they will need to successfully design and develop lighting systems for each of their three aquatic ecosystems. As students work on lighting systems in this project, they will use information related to ballasts, transformers, bulbs, LEDs, starters, safe and efficient lighting fixtures, volts, watts, amperes, and alternating versus direct current.

Light, Vision and Color – Project Two Opening Activity

Students enjoy the NOAA ocean explorer lesson, “All that Glitters” (Chase, 2006). The activity requires that students construct goggles, obtain a sheet of black construction paper (craft foam or felt) to represent the darkness of the deep sea, and cut out four two-centimeter square pieces of craft foam or felt in each of the following colors: red, orange, yellow, green, blue, and black/dark brown.

Construction of the goggles requires the use of **blue color filter gels** for the goggle lenses. The gel plastic is called **Roscolux #80 primary blue**, **Lee #079 just blue**, or **Gam #850**. The gel plastic comes in sheets that are 20 x 24 inches. Six sheets should produce 31 goggles that have blue plastic lenses that are four layers thick.

Each set of goggles is composed of three pieces of flexible folder paper. The front of all goggles measure seven inches. Students use the goggles template on the next page to determine the size of the left and right side of their goggles to trace, cutout, and attach to the front piece. Students modify the size of their goggles to fit their own heads by attaching left and right extensions onto the 7-inch front piece of their goggles. Students first use a tape measure to determine the circumference of their head. Next, students trace each side piece so that it is long enough to extend behind the student's head, at which point a slot on the right side piece of flexible folder paper can be carefully clasped into a groove on the overlapping left side piece of flexible folder paper.

If the circumference of a student's head (and hair) is about 21 inches, then students cut a 9-inch length to staple carefully to the left front of the goggles and a 7-inch length for the right side to staple carefully to the front piece (including a 1-inch overlap on each side).

Circumference	21 inches	22 inches	23 inches	24 inches	25 inches	26 inches	27inches
Left (slot)	9 inches	9.5 inches	10 inches	10.5 inches	11 inches	11.5 inches	12 inches
Right (tab)	7 inches	7.5 inches	8 inches	8.5 inches	9 inches	9.5 inches	10 inches

Once the three goggle pieces have been connected, students then cut the eyeholes and enough blue gel plastic to make goggle lenses four layers thick. If necessary, students may attach strips of stretchable elastic material to the goggles in order to hold them in place. Students may proceed with the activity once they have their goggles, four squares of foam or felt ($2cm^2$) of each color (dark brown, red, orange, yellow, green and blue), and a sheet of 12 x 12 inches black paper or craft foam.

Each student is to receive a "Color in the Sea" data sheet (provided below) to tally and record their observations.

Class discussion of results and possible explanations is conducted after students finish the activity.

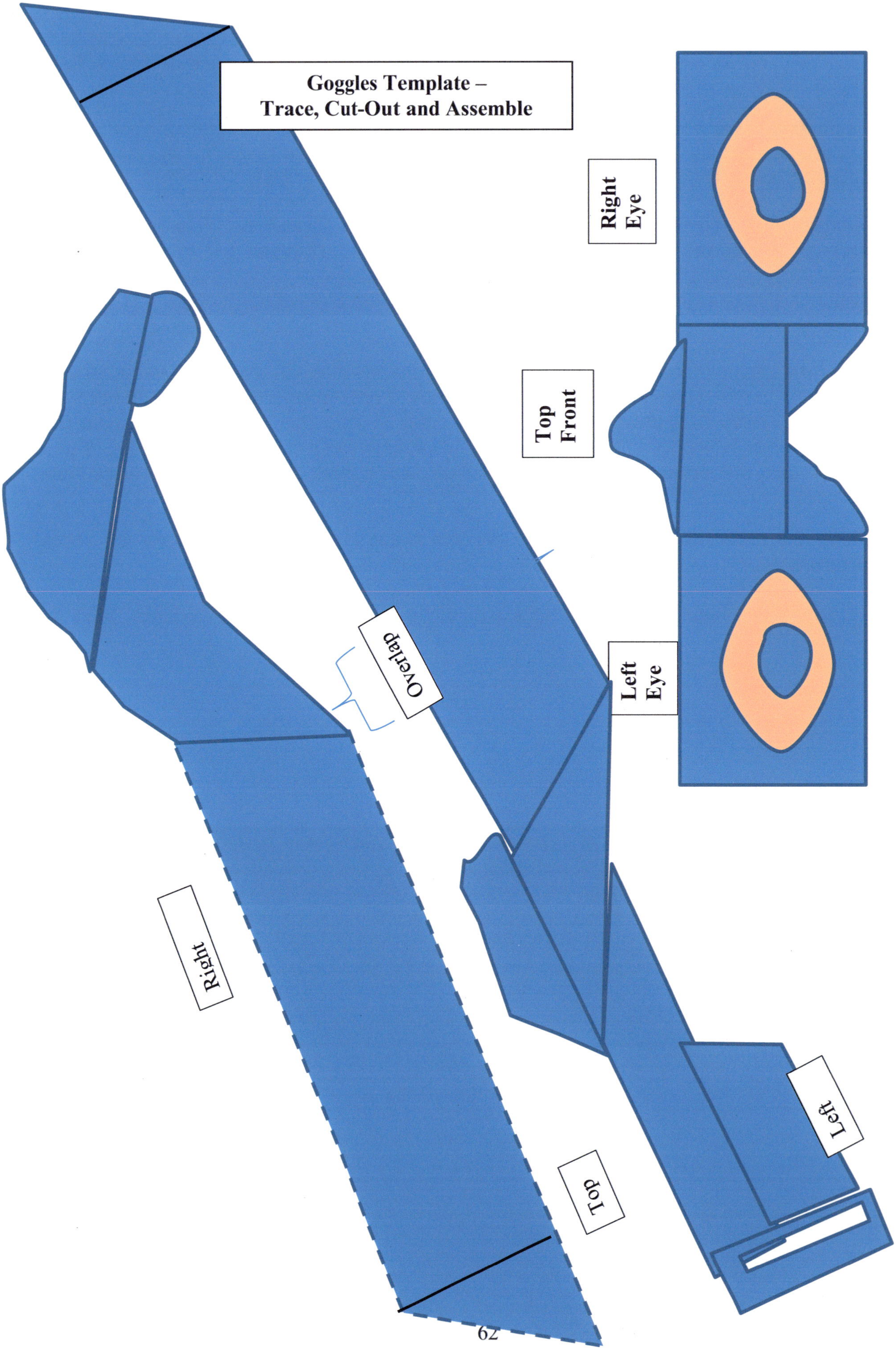
Goggles Template –
Trace, Cut-Out and Assemble
Right Eye
Top Front
Left Eye
Overlap
Right
Top
Left

You are hunting for bits of food in the "twilight zone." As you look away, your partner will randomly arrange your colored foam or felt squares on the black foam. Put on your blue goggles to simulate light conditions in the twilight zone using two or three layers of blue gel over each eyehole (depending upon the ambient light in the room). Quickly pick up the first 10 pieces of food you see. Record your results. Exchange places with your partner and repeat the activity. Record your results. Analyze overall findings.

Partner	Red	Orange	Yellow	Green	Blue	Violet/Black
1						
2						
3						
4						
Total						

The Illuminating Engineering Society of North America (IESNA) defines light as "radiant energy that is capable of exciting the retina and producing a visual sensation" (Taylor, 2000). In this project, students will learn to apply their understanding of electricity, wiring, light, and light bulbs in order to design aquatic ecosystems that meet the light needs of specific aquatic organisms, including, but not limited to corals and others that use zooxanthellae, aquatic algae and sea-grass, and common garden plants and trees (among these – mangrove trees). Student documentation of work on this project includes projections of electric costs as part of a design budget. Documentation must also include reports on system light and wavelength, light color, light temperature, and development of light systems to use for projects holding specific organisms.

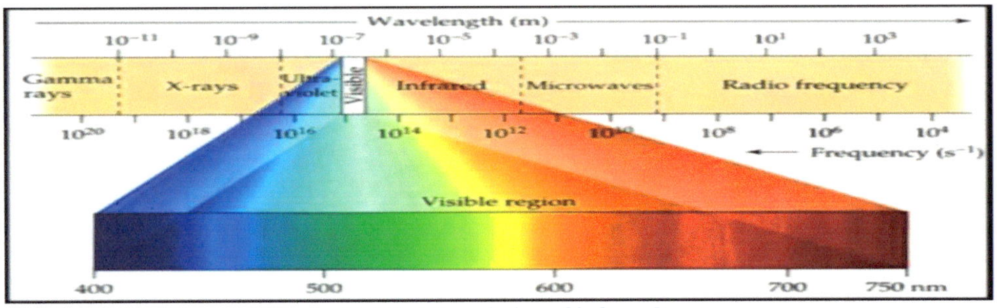

Used with Kind Permission of infraredtv.com

According to Moe (1992b), "along with water and oxygen, light is of critical importance to the marine environment, and we should know something about it." Moe's exposition of light for the layperson aquarist is phenomenal. I would not attempt to improve upon his presentation, so most of the light related information in Beyond the Classroom Aquarium is provided comes to you from Moe's work (1992b, pp. 59 - 71).

> Actually light is just as important as water and oxygen, although marine organisms can live a little longer without light than without water and oxygen.... Light is radiant energy that travels through space and is selectively absorbed by various substances. The fundamental unit of light is the photon, a massless particle of spin 1. When an atom absorbs energy (heat, light, electricity, nuclear reactions), the electrons move away from the nucleus of the atom and the atom gains energy. The atom is unstable in this energized or "excited" state, however, and the electrons quickly move back to their stable state of their original inner orbital shells. Excess energy is then released as a photon that shoots away at the speed of light. Under constant stimulation, the atom bounces back and forth between energy states and pumps out photons continuously. The human eye can detect as little light as a single photon, but usable light levels are only produced when billions and billions of atoms release photons.
>
> Nuclear reactions, fission and fusion, release tremendous amounts of energy, much of it in the form of light. The nuclear fusion of hydrogen in the sun produces the abundant radiation energy that we perceive here on earth as sunlight. Scientific experimentation has shown that light waves travel at 186,000 miles per second in a vacuum and drops to 140,000 miles per second when travelling through the dense medium of water. As well as speed, the composition and direction of light changes to some degree when it passes through any medium, including glass, plastic, and water, depending on characteristics of the medium. For example, ordinary glass absorbs ultraviolet waves (UV), which is why the glass tube of a UV sterilizer must be made of a special quartz rather than ordinary glass.
>
> ...In the natural environment, the sun is by far the greatest source of radiation. We don't get very much of it, however. The Earth only receives about one half of one billionth of the total output of the sun. Most of the suns' radiant energy that enters the marine environment is in the very narrow part of the electromagnetic spectrum that consists of visible light, although the infrared and ultraviolet wavelengths that bracket the visible spectrum are also present. Visible light is that part of the electromagnetic spectrum that ranges from deep red, at a wavelength of about 700nm, to violet, at a wavelength of about 400 nm. This visible radiant energy has two important functions in the marine and terrestrial environment. One, it provides information about the environment by allowing animals

to form images (sight), and two it provides the energy that allows plants to produce food (photosynthesis), the very basis of life. If we are only interested in providing enough light to allow fish and sighted invertebrates to find their way around an aquarium, then there is little to consider. An ordinary fluorescent light of whatever color and brightness is most pleasing is all that is needed. However, if we are concerned about the growth of algae, the well-being of animals that contain algal cells in their tissues (primarily corals and anemones), and the subtle effects of intensity and quality of light on animals in general, then we must carefully consider the nature of the light we give to our captive marine environment.

[Properties of light are measured in] wavelength, energy, and intensity. Wavelength is simply a measurement of the distance from crest to crest or trough to trough of a particular wave. The metric meter (100cm or 39.37 inches) is the fundamental unit of measurement of electromagnetic wavelengths. There is an astounding variation in the length of waves of the various types of radiant energy that make up the electromagnetic spectrum. Waves that are in the range of one meter include microwaves, television, and radar. The longest waves include long radio waves at 30,000 meters (almost 19 miles) in length to waves that are about a million kilometers (620 miles) long. The shortest are gamma rays at wavelengths as short as one tenth to one hundredth of a nanometer (nm), which is a billionth of a meter.

Visible light is only a small portion of the electromagnetic spectrum. Wavelengths of visible light are usually measured in nanometers (nm), billionths of a meter, or in angstroms, (Å), a unit that was developed by Anders Jonas Angstrom and is equal to one tenth of a millimicron. There is only one decimal point difference in these units, 7200 Å is equal to 7320 nm. Just to give you an idea of how short a nanometer is, the thickness of a page in a book is approximately 125,000 nanometers.

The colors of light have different wavelengths and different frequencies (energy levels). This makes it possible to separate the colors that make up white light and create rainbows with glass prisms, diffraction gratings, and the natural prisms of raindrops. The wavelength of red light ranges from 700 to 650 nm; orange, 650 to 590; yellow, 590 to 575; green, 575 to 490; blue, 490 to 425; and violet, 425 to 400.

Infrared waves are longer than 700 nm, and ultraviolet waves are shorter than 400 nm. …the boundaries of the wavelengths of colors are not precise (discrete) and wavelength color boundaries from different references may vary. The different wavelengths of light are important to marine scientists for two reasons. One, water absorbs the longer wavelength, at the lower energy end of the spectrum rather rapidly. Red light is 90% absorbed within the first 15 feet of water. Blue and violet light, however, penetrate hundreds of feet in clear water. The wavelength of maximum light intensity in the underwater environment is 475 to 500 nm, green with a rapid shift to blue as depth increases (Dawes, 1981). And two, the various types of chlorophyll present in different algae use certain wavelengths of light more efficiently than others, so we can enhance the growth and survival of algae, corals, and anemones by controlling the wavelengths of light that illuminate the aquarium.

The energy of an electromagnetic wave is the property of its particulate nature. The energy of a light wave varies with wavelength, and long, red waves have less energy than short, violet waves. Ultraviolet light has enough energy to alter the genetic chemistry of organisms, create mutations, and even kill viruses, bacteria, and other microscopic life. Ultraviolet light (UV–C) kills bacteria at 220 to 300 nm, with the greatest bactericidal effect at about 260 nm (2600 Å). The tanning effect on human skin occurs at 280 to 300 nm (UV–B).

Visible light has relatively low energy. Orange light, for example, with a wavelength of 600 nm, has an energy content of only one third of a billionth of an erg (an erg is a centimeter-gram-second unity of energy, a very small unit of work energy). The energy of light is important because different substances give off different types of light at different energy states, which are measured by the temperature of the substance. Everything above the temperature of absolute zero 0°K or -273°C (add 273 to any centigrade temperature to change it to the absolute temperature Kelvin scale) emits energy called "black body radiation." A warm sidewalk for example, is radiating low frequency infrared energy as heat that can be felt but not seen, except with special infrared sensing devices. When a temperature of 950°K is reached, the frequency of radiant energy released is high enough that dull red visible light is released.

As the temperature of a radiating substance increases, the frequency of the light that is emitted also increases. Low temperatures produce low frequency light in the red end of the spectrum and high temperatures produce high frequency light in the blue end of the spectrum. The temperature of the surface of the sun is 6000°K and its peak light emission is in the yellow orange range at 509 nm.

Normal daylight, filtered through the atmosphere, has a color temperature of 5500°K, and it is at this color temperature that most colors appear normal and that the color rendering of a light is best. Light bulbs of various types often give their output in color temperature, from 2800°K for ordinary tungsten light bulbs to 10500°K for certain types of fluorescent bulbs. According to Lundegaard (1885), fluorescent light sources with color temperatures of 5000°K to 7000°K provide an excellent color appearance to marine aquariums.

Another aid to selecting bulbs that will give good colors to an aquarium is the "Color Rendering Index" (**CRI**). This is the standard for comparison of the colors of objects lit by artificial light. The **CRI** of sunlight is 100, and the nearer an artificial light comes to replicating the colors revealed in sunlight, the higher it's **CRI**. Most fluorescent bulbs that give a natural appearance to marine aquariums have a **CRI** of 90 or higher. A high **CRI** index however, does not mean that light will supply the wavelengths that a marine aquarium requires. It only means that the human eye will see the colors in the aquarium close to the way they appear in natural sunlight.

The intensity, the amount of light or "brightness," is also very important. Light intensity is measured as the amount of light per unit area and is usually expressed in units called lumens and lux. To understand what these units measure, however, we have to go back to the standard measure of light intensity or brightness, the candlepower. The original measure of one candlepower is the total light produced by a single candle of specific form made from the wax of the sperm whale. Candlepower or candles, is the original standard measure of light intensity, the basic unit for measurement for the output, the brightness of a light source. The light output of a lamp is usually given in candlepower.

Science is now more precise in the measure of light intensity. The candela (cd), a very precise measure of light radiation, was adopted in 1979 as the standard scientific unit of luminous intensity. A candela is a measure of luminous intensity, in the direction of normal, of a black body surface 1/600,000 of a square meter in area, at the temperature of solidification of platinum (2042°K) under a pressure of 101,325 **N/M²**. Although it is a far more precise measure of luminous intensity, one Candela is basically equal to one of the old Candle units. Candlepower is still the expression of this unit in common use.

Candlepower, however, does not measure the amount of light actually reaching a particular area, thus a different type of measurement is also necessary. One foot-candle (fc) is a unit of measure of the illumination over an area of one square foot at a distance of one linear foot from the candle flame. Light spread outward in all directions, thus the further we are from a light source, the less light we see and the dimmer the source appears. Expressed mathematically, the intensity of light varies by the square of the distance from the source. Since area also has to be taken into consideration, the lumen, a unit of luminous flux, is the amount of light that falls on one square foot of area with all points located one foot from the flame of a standard candle. Flux, incidentally, means that the energy is in continuous flow. One lumen per square foot is the light of one foot-candle. The inside surface of a sphere with a radius of one foot has a surface area of 12.56 square feet, so one candlepower is equal to 12.56 foot-candles or 12.56 lumens. A lumen, then, is a measure of particular amount of light incident upon a surface of a specific size. Since it is a specific measure of a particular amount of light (one foot-candle), light intensity can also be expresses in lumens, and the output of most fluorescent and incandescent bulbs is listed in lumens.

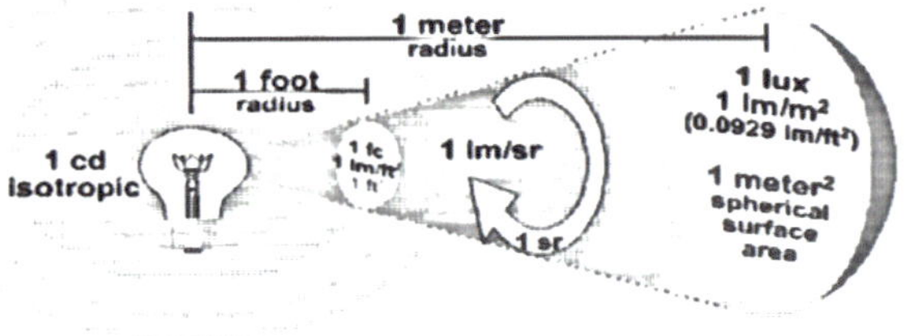

Ryer, A.; Light Measurement Handbook, 1997

Lux, the International Standard unit, is the same type of measure in the metric system and is also termed a meter candle. One lux is defined as one lumen falling perpendicularly on a surface of one square meter. One lux is equivalent to 0.0929 foot-candles (lumens) and one lumen is equal to 10.76 lux. Sunlight falling on a particular surface is measured as energy per unit area. The basic measure is gram calories per square centimeter (gm.cal/cm^2), which is equal to the Langley, a common measure of electromagnetic radiation incident upon a surface. One gm.cal./cm^2 per hour is equal to about 770 foot candles, which is about 8288 lux.

In the natural environment, there is great variation in the amount of sunlight that strikes the water's surface in different seasons, from day to day and even hour to hour as weather changes occur. Time of day is very important because the smaller the angle between the sun's rays and the water's surface, the more light is reflected off the surface and the amount of light entering the water is reduced. Thus light intensity is much lower under water than at the surface in early morning and late afternoon. The angle of the sun's rays and the number of daylight hours vary with season and with distance from the equator. The winter sun is low in the sky and less light penetrates the surface at any time of the day. Maximum light intensity in the aquatic environment occurs neat the surface on a calm summer day when the sun's rays are perpendicular to the surface and only a few white clouds dot the sky. On such a day, over tropical seas, the intensity of light just above the surface may exceed 130,000 lux.

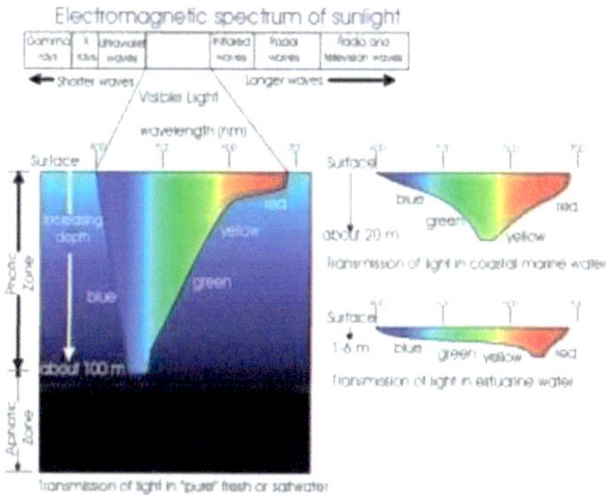

Buchheim (1999c)

Light intensity under the surface is decreased to a greater of lesser extent by factors other than angle of the sun's rays and cloud cover. Due to reflection and absorption only about 100 gm.cal/cm^2/hour (77,000 lux), penetrates the sea surface when the sun is at zenith. Surface turbulence increases reflection of light off the water's surface. Only about 4% of the light is reflected on calm days and up to 25 percent is reflected when the surface is disturbed by moderate winds. Once light enters the water, it is absorbed and scattered by water molecules and by particles and organic dyes in the water. Light penetration in turbid coastal waters is severely restricted due to these conditions, while light transmission through clear, offshore waters is much greater. The water soluble yellow pigments (gelbstoff or gilvin) common in marine coastal waters are most responsible for the extinction of the blue end of the spectrum in coastal waters.

Under the most favorable of conditions (clear, offshore waters), only about 20% (up to 15,000 lux) of total incident energy penetrates 10 to 15 meters below the surface. Light intensities at depths of 5 to 10 meters in temperate coastal waters at noon are only about 7500 lux, dropping to lower levels at other times of the day. Thus, since average daily illumination is more important than peak noontime levels, many algae adapted to low light levels typical of coastal areas (including many microscopic red and blue-green algae) can live in aquariums with only 5000 lux per hour for 12 to 14 hours per day. Coral reef algae require more light. According to Homes (1957), maximum production form photosynthesis occurs at about 10 meters in tropical seas, and sunlight saturation in these algae occurs at about 35,000 lux; which means that may tropical algae, especially green macroalgae, and corals are adapted to very intense lighting. A marine aquarium usually gets too little light for growth and survival of the desired algae and corals. Red and some brown algae often grow to the exclusion of the desired green algae. Even if the light includes the proper wavelength and color temperature for growth of macroalgae, and the zooxanthellae of corals and anemones, the intensity may not be great enough even if the proper water quality is present. Individual opinions and purposes will vary but the following specifications should provide adequate lighting for most marine invertebrate aquariums. Light intensity is the most important factor in lighting an aquarium, providing wavelength peaks and color temperature are reasonably proper. An intensity of about 12,000 to 16,000 average lux at the water's surface should maintain most organisms that utilize light for photosynthesis. Some reef tanks may require a higher surface level of 18,000 to 20,000 lux. This light intensity will decrease through passage through the water, only about 10,000 lux may be present at 12 inches, and perhaps 5,000 lux may remain at the bottom at 18 to 22 inches deep. Given proper light intensity, light with a color temperature of 5000 to 7000°K and a wavelength pattern with peaks in 475 and 650 nm wavelength ranges will provide the right quality of lighting.

Photoperiod - Basically, photoperiod refers to the number of consecutive hours of light that occur within one day, a 24-hour period. This is also known as the diel periodicity. Except for cave dwellers, animals and plants do not live in nature under constant dark of constant light. They do not survive well in captivity if reasonably natural cycles of light and darkness are not provided. In nature, intensity and periodicity of lighting is not consistent. Sunlight intensity and day length vary with season and weather conditions. Night lighting varies not only with season and weather, but also with phases of the moon. Nighttime is not absolutely dark on the reef. Bright moonlit nights provide considerable light for sighted creatures in shallow waters, and even starlight and the light generated by some marine organisms (bioluminescence) keep the nights from being totally dark. A dim light somewhere in the room is a good thing for a marine aquarium, but it is not absolutely necessary.

The number of daylight hours varies with latitude and season. This variation is most pronounced above the arctic and Antarctic circles where the sun does not set in mid-summer and does not rise in mid-winter. Night and day in the tropics on the other hand are almost equal the year round. If animals and plants from temperate waters are kept 8 to 10 hours of light per day is good, and if tropical organisms are kept, 12 to 14 hours is good; although they will also do fine with fewer light hours. There are two things that are important to consider under the general topic of photoperiod. One, it is not always noontime on the reef, and two, the sun does not go on and off with the flick of a switch. Light levels on the reefs slowly increase during morning hours and slowly decrease during afternoon hours. The crepuscular periods, dawn and dusk, are slow transitions between light and dark in the natural environment.

Dawn and dusk can be simulated in a marine aquarium with an incandescent bulb or two set up on a timer that will turn these bulbs on 30 minutes before, and off a few minutes after, the main tanks lights come on in the morning. The same timer turns these crepuscular period lights on just before the main lights go off at night and then turns them off 30 minutes later. Such a lighting setup provides a natural environment and reduces the shock between light and dark, but most typical marine and reef aquarium systems seem to get along well enough without a dawn/dusk period.

And Moe continues on page 147,

There is one important factor to consider when planning the tank display under any type of lighting. Light on the reef is not uniform. The light intensity any organism receives is greatly dependent upon its depth and placement on the reef. A coral or algae on the top of the reef in only a few feet of water receives far more intense light than one in deeper water or in the shade of other reef structures for all or most of the day. Therefore, whatever type of reef lighting is placed over a tank, it is important to provide areas of varied light intensity over an aquarium. Overhangs and shaded areas are important for some invertebrates.

Some organisms such as anemones and algae can move to, or grow toward, areas where lighting suits them but others such as corals must be placed in the right areas. Learning where an organism is usually found on the reef – shallow with high light intensity or in the shade or deep water with low light intensity – is a good start for finding out where it should be placed in an aquarium. Each tank is different however, and information on the place of natural occurrence of the organism may not be available, so some experimentation on the part of the aquarist may be necessary for proper placement of various light requiring organisms.

[By using a standard aquarium light hood, an] aquarist's control of above tank lighting is limited to the selection of the type of fluorescent bulb to place in the fixture. The fixture usually holds only one bulb over a small tank and two bulbs over a large tank. The basic specifications of the bulbs in these standard fixtures are usually either 18 inch (15 watt ~ 800 mean lumens); 24 inch (20 watt ~ 1200 mean lumens); or 36 inch (30 watt ~ 2100 mean lumens). These units also usually come with bulbs selected by the manufacturer already in place. These are bulbs manufactured for use with aquaria and usually have a spectrum designed to stimulate the growth of freshwater plants and enhance the colors of freshwater fish.

[Moe (1992) reports that] bulbs commonly found in these fixtures are commonly made by a manufacturer such as Osram, Sylvania, Westinghouse, or General Electric expressly for the aquarium company and are labeled with the hood manufacturer's trade name. The "Gro-Lux **WS®**" wide-spectrum bulb manufactured by Sylvania is a good example of this type of bulb. There is a broad spectrum peak in the orange/red range of 600 to 675 nm with a much sharper peak in the violet 440 nm range. Although this bulb and others like it are fine for well-planted freshwater aquariums, they tend to give an unnatural look to a marine aquarium.

There are a number of bulbs made expressly for marine aquariums that have a spectrum closer to natural sunlight with an additional stronger peak in the violet/blue range. Some of the trade names for these bulbs include: Sun-glo ®, Aquarilux ®, Sealux ®, Naturalux ®, Actinic White ™, Actinic Day ™, Triton ®, Ultralume ®, Paralume ™, Radiant White ™, and Vibran-Sea ® along with a number of others. These bulbs have strong spectral peaks in the violet/blue and red wavelength ranges and generally give a good, natural light to a marine aquarium. Some of these bulbs even combine blue actinic light with a high temperature white light (6000°K) to provide both blue and daylight from the same bulb. Full spectrum bulbs that emit light in the full range and color of natural sunlight (not to be confused with wide spectrum bulbs that emit an extended red light) are available.

[Martin Moe recommends] using the bulbs that come with the light hood at least during initial setup and "run in" of a marine aquarium. Then, if you are displeased with the general appearance of the tank, look for a bulb type that will provide the color rendering that is desired. Conventional aquarium setups are limited in type of invertebrate life they can support. One or two fluorescent bulbs, even with a "high tech" filter will not allow survival of corals and most marine macroalgae. Off the shelf types of light hoods contain the ballast in the fixture with the bulb. The ballasts are heavy metallic electrical equipment and generate considerable heat. They also have to be protected from saltwater and salt spray. This means that the bulb and fixture should be separated from the water surface by a glass plate and that the aquarium cover hood should be strong enough to support the weight of the light fixture.

The light and ballast must also be well ventilated and the plastic heat resistant – to avoid warping. The glass plate that separates a light from the aquarium surface quickly accumulates salt build-up and even algal growth. The encrusted salt and algae growth decrease and change the quality of light that enters the aquarium. If you notice that the light in the aquarium is not as bright as it used to be and has a particular greenish hue, then the glass plate is probably coated with a green algal growth and needs to be cleaned.

A marine aquarist with an interest in invertebrates may feel a bit discouraged or even intimidated by the recent development in reef tank technology; especially the explosion of high technology applications in lighting and filtration. If you do not want to make the investment of time, space and money that such systems require, don't let it bug you. While it is true that these developments make it possible for the home aquarist to keep certain invertebrates more successfully than ever before, they have not driven the traditional marine aquarium into obsolescence. Not all invertebrates require the reef tank environment, and a traditional marine aquarium can support many invertebrates that do not require high intensity lighting and super water quality. Also, inexpensive modifications to traditional systems in lighting and additional filtration often greatly increase the number and kind of invertebrates that can be maintained. Time and effort spent in acquiring and observing invertebrates suited to the traditional marine aquarium will be well spent.

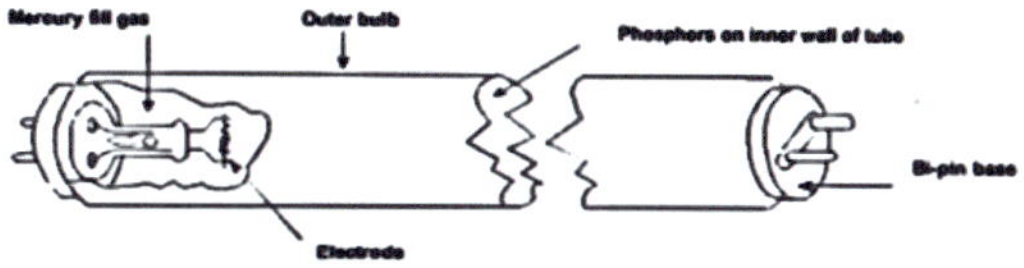

Leslie & Conway (1993)

Custom lighting hoods - Assuming an adequate filtration system, even if frequent water changes are necessary, one of the most important custom additions to a marine tank to help support more invertebrate and algal life is improved lighting. The easiest and most inexpensive way to accomplish this is with fluorescent bulbs. Incandescent bulbs may find some uses on a marine aquarium, i.e., a spot of intense light in a small area for an anemone in an otherwise poorly lit tank, but incandescent bulbs have a low color temperature of 2800°K and generates far too much heat for the amount of light produced (low efficiency) to be generally useful for aquariums.

Fluorescent bulbs are glass tubes coated on the inside with various types of phosphor powder. This phosphor is made to glow with emitted light by passing an electric current through a mercury vapor inside the tube (do not breathe the fumes if a bulb breaks!). Electrodes at each end of the tube generate the electric current for the bulb. A choke coil, condenser, and starting switch make up the ballast that accompanies each lamp, and this transforms the household electric current into energy the bulb can use. The bulb and ballast can be designed to emit particular intensities and spectrums of light, which gives this type of lighting great variability at reasonable cost.

A general estimate of the effective life of a fluorescent bulb is 20,000 hours, which figures out to about three and a half years if it is on 15 hours a day. The initial lumen output of a standard fluorescent bulb is 10 to 15 percent higher than its output after about 4000 hours of use. Therefore, one should use 85 percent of the initial lumen rating to get the mean lumen rating, and this will provide a more accurate estimate of the light the aquarium will receive over the life of the bulbs. A 40 watt, cool white fluorescent bulb is rated at light output of 3000 to 3200 lumens. The light output of a fluorescent bulb is related to wattage, but different bulbs of the same wattage may produce a different amount of lumens. For example, a standard cool white is rated at 3150 lumens and a standard 40 watt Vita Lite® is rated at 2400 lumens. Generally, however, the higher the wattage the more lumens produced.

It is possible to get high output (HO) and very high output (VHO) bulbs that double and triple the initial lumen output of a standard bulb, but these brighter bulbs require more watts, higher amperage, and a, special, more expensive ballast than standard bulbs. Four high output bulbs can take the place of eight standard bulbs, so if you want the greatest light intensity possible from fluorescent bulbs, it is worth the greater expense for the HO or VHO lighting setup. It is a good idea to replace all fluorescent bulbs after six months to one year of service to make certain that the intensity and spectrum are always as intended for the aquarium. This is particularly important for reef systems with hard and soft corals. . . .

The purpose of the custom fluorescent light hood is to provide the most optimal light quality possible for the marine aquarium at a reasonable cost. Since the quality and intensity of light in the aquarium is the primary consideration, concessions have to be made that favor the life in the aquarium over ease of maintenance and construction. Different materials (wood or plastic) and construction methods can be used to build a light hood, but the following design elements should be included.

No metal should be included in the construction. Certainly no metal of any type should be exposed to salt spray or condensation. Also, all plastics must be nontoxic. Polyethylene is about the only flexible plastic that can be safely used in and over a marine aquarium, but most rigid plastics are safe. Paints on any wood surfaces must also be nontoxic.

The bulbs should be only a few inches from the water surface and there should be no glass or plastic between the bulbs and the water (except for narrow braces). Although some transparent plastics are said not to change the character of the light that is transmitted through them, the need to use the full spectrum and as much light intensity as possible, takes precedence over a condensation and spray shield between the bulbs and the water. If the tank has an open air lift that releases air bubbles in a relatively restricted area of the surface, a small plastic spray shield can be placed over this area, but it should be cleaned frequently as slat spray and algal growth will reduce light transmission over this area. The bare bulbs should also be cleaned frequently of any salt spray crust.

The bulbs need not be contained in a rigid fixture by may be loosely supported by plastic or glass bars above the tank or by supports attached under the hood. The electrical connections on the end of the bulbs must be waterproof plastic caps that seal tightly on the ends of the bulbs. The open plastic clips that hold the ends of fluorescent bulbs in typical fixtures will soon accumulate salt and create an electrical shock and fire hazard and must not be used when the bulb is exposed over the tank water. If professionally made, waterproof end caps cannot be obtained it is possible to make a waterproof connection for the ends of fluorescent light bulbs. Get a 1.25-inch PVC pipe cap and about a 2-inch length of 1.25-inch thin-wall PVC pipe. Thin wall PVC pipe of this diameter fits closely over a standard fluorescent bulb. Glue the short section of PVC pipe into the PVC cap, with PVC glue, and then drill a small hole through the end of the cap and pass the plastic coated electrical cord through the hole. Attach the cord to the pins on the end of the light bulb with strong clips or solder. Put about an inch of silicone sealer into the bottom of the cap/pipe assembly and push it slowly over the end of the bulb while gently pulling the electric wire from the hole. Once the bulb is seated in the end of the cap, make certain the hole in the end of the cap is well sealed. Clean up the excess silicone sealer and put the assembly aside for a day or two until the sealer has set. Note that this method is not much less expensive the commercially available water-tight caps, considerably more trouble, and the caps can't be removed and used on another bulb.

The ballasts for the bulbs must be remotely located, far away from any exposure to salt spray and saltwater and where the heat from the ballast will not affect the tank. The wires leading from the bulb to the ballast should be carefully installed so that the water cannot condense on the wire and then run down to the ballast.

The top of the aquarium need not be totally enclosed; however, the bulbs must have a white or silver curved plastic reflector above them along their entire length. White, plastic rain gutter material is good for this purpose. Plexiglass mirror, 1/8th inch thick – mirrored plexiglass, can be cut and attached to the hood above the light with silicone sealer to make an efficient reflector. Without a reflector, most of the light from the bulb does not enter the aquarium. A total enclosure constructed with a nontoxic, white or mirrored surface over the aquarium is possible because fluorescent bulbs without the ballast don not create much heat, and beneficial because it will reduce evaporation and spray and reflect additional light down into the water. Be sure to provide a few vent holes to allow whatever heat the bulbs do produce to escape.

The length of the bulbs should be about the same as the length of the aquarium. Obviously, the bulbs should not be longer than the length of the aquarium. If the bulbs that must be used are shorter than the length of the tank, the freedom of placement allowed by the lack of rigid fixture allows bulb placement to be staggered so that each end of the tank receives light from ends of half the bulbs.

Spectrum - There are two considerations, equally important, when the spectral quality of light in the marine aquarium is decided. The first is the spectral quality that is best for life in the aquarium and the second is the spectral quality that is best for human observation. The visible spectrum for normal human sight ranges from violet at 400 nm to red at 700 nm, but the eye sees "best and brightest" on the green yellow range of 500 to 600 nm. Colors appear most normal when light is rich with wavelengths in this range. Thus, the cool white fluorescent bulb, designed to provide efficient light for human use, produces most of its radiant energy in this wavelength range. Photosynthesis in plants, however, including algae and coral symbionts, makes best use of other wavelengths. The absorption of light by chlorophyll has strong peaks in the ranges of 400 to 450 nm and

600 to 650 nm. Another important point is that blue/green light in the 400 to 550 nm range is dominant at depths where many corals occur in nature. The "best" lighting for a marine aquarium would be sunlight since it contains a relatively equal amount of all wavelengths and certainly has the intensity to provide as much light as would be necessary for all purposes.

Natural sunlight is the standard. The **CRI** of sunlight is 100. The higher the **CRI** of a bulb, the more all colors appear natural to the human eye. A **CRI** of 90 or better is best for reef tanks. To get natural sunlight, some aquarists have installed plastic skylights in the roof of their homes over the aquaria. These tanks are usually 3,000 gallons or more to warrant such an exceptional installation. Supplemental lighting is also used for night and cloudy day illumination. The inside location of the tank and separation of the direct sunlight by the plastic skylight provides enough temperature control to keep the tank cool enough in the simmer and a heater keeps the tank water warm enough in the winter. (Another way to use sunlight for a marine aquarium is to mount a sunlight collector on the rood that automatically tracks the sun, collects the light and then transmits the light with plastic tubes down to a fixture over the tank where the ends of the tubes release the sunlight over the tank. Cloudy days and night present a problem with this arrangement, and since this technology is still under development, it is probably better to use artificial light.)

The spectral mix that works for all general purposes – a strong blue peak, 425 nm for corals and photosynthesis; adequate green/yellow, 575 nm for viewing; and an orange/red peak, 650 nm for photosynthesis – is best obtained from a variety of bulbs in combination....

Many marine labs use only cool white bulbs for their microalgae cultures, full-spectrum bulbs, or bulbs aimed at marine aquariums, provide some of all of the above including the orange/red and blue spectrum. A cool white or full-spectrum bulb gives more of the range needed for viewing and an actinic 03 and or a blue bulb provides the blue range for coral growth.

On a four fluorescent bulb set up, adequate for most marine tanks, but not for a serious coral reef tank, Moe's recommendation is to use two full spectrum bulbs, one cool white bulb, and one actinic 03, though even this setup may not produce enough light for some organisms. Some of the new bulbs on the market combine actinic (blue) light, a bright white light component and an orange-red peak from within the same bulb. For most marine aquarium setups between 50 and 100 gallons, four of these 40W multipurpose bulbs will produce adequate lighting. Actinic bulbs were developed for various industrial and medical purposes and produce a high output in a narrow blue wavelength range. 03 bulbs peak at about 425 nanometers and range between about 375 and 475 nm. Despite the blue glow of the light, there is no germicidal UV component (about 256 nm), and although it does contain some high UV-A radiation, it can be viewed with safety, although one should not stare at the lit bulb for a long time. (Note that UV-C light 200 to 280 nm is most damaging to living organisms, UV–B 280 to 315 nm is also dangerous but UV–A 375 to 400 nm is not particularly harmful to humans. Glass and water will filter out UV–B and UV–C, but a special UV filter is required to filter out UV–A.) The term actinic refers to the property of radiation that produces chemical effects and it is used in treatment of some human ailments. These bulbs seem to be a good supplement to full spectrum lighting in the marine aquarium, but should not be used as the only light source, unless the aim is to reproduce conditions on the deep reef. Of the actinic bulbs available, only the 03 actinic bulbs or a blue bulb are suitable for aquaria. Blue fluorescent bulbs that peak at 450 nm are now also available and in combination with high intensity daylight fluorescent bulbs that emit light with a color temperature over 5000°K and a CRI of about 90, at the proper intensity for the size of the tank, provide god lighting for keeping corals in most reef tanks.

Students may wish to explore different lighting setups with various combinations of light sources. The spectrum provided by a number of bulbs is available for general reference so that students can use this information to make well-informed decision about lighting their projects effectively.

Intensity - The basic question is: how many bulbs of what type should be placed over a particular system, to produce the amount of light necessary for the system being designed? First of all, students must decide on the spectral quality (type of bulbs) and the total surface lux of the system. If uncertainty and concern for expense plague this decision, go with about 10,000 to 12,000 lux of full spectrum lighting and make changes based on future experience. The main problem with fluorescent lighting over a reef tank is that the intensity is often too low for many types of corals. For this reason, the extra high intensity of metal halide lighting is often preferred for large reef systems.

If we assume that a standard four foot, 40 watt fluorescent bulb of any particular type puts out roughly 2500 lumens (a reasonable estimate for full spectrum bulbs) and multiply the lumens by 10.76 to convert to lux, we get 26,900 lux from one standard bulb! Wow, does this mean that one bulb will do a whole system? Unfortunately, it does not work that way. That 2500 lumen is the total output of the entire bulb, outward in every direction, right at the surface of the bulb. Our concern is in the amount that reaches a specific organism, not the amount that leaves the bulb.

There are many factors that diminish the light between the bulb and an anemone. These include the distance between the bulb and the anemone, the type of reflector and its distance from the bulb, the total surface area of the system, the characteristics of the water such as color and amount of surface disturbance, the depth of the water, shape of the system, and the environment. The only way you can be absolutely sure of the amount of light at any point above or in a particular system is to measure it at that point with a light meter or lux meter.

There is a way to develop a working estimate of the number of bulbs required over a system of a specific size. Even though the patterns of light distribution and light absorption over various systems may vary greatly, the formula presented by Spotte (1979) can be used to develop an estimate of the total wattage required to produce a specific amount of lux over a system of a specific size.

$$\text{Watts} = \frac{(\text{lux}) \times (\text{area of system in square meters})}{(\text{bulb efficiency}) \times (\text{utilization factor})}$$

Or,

$$W = \frac{(L) \times (A)}{(E) \times (U)}$$

- L is the total lux desired over the surface of a system
- A is the total surface area of the system in square meters
- E is the efficiency of the bulbs in lumens per watt
- U is the utilization factor, a constant (0.5) that represents the light from the bulb that actually strikes the surface of the water

As an example, let's say we want 15,000 lux over the surface of a 75 to 80 gallon system that measures 48 inches in length, 20 inches in width and 20 inches in depth. In this case, depth does not matter since we need only surface area of the system. The surface area has to be in square meters, however, and we have inches. We can convert to meters in the linear measure by multiplying 48 and 20 by 0.025 to convert inches to meters (1.5 and 0.5 meters), and then multiply these figures to get the surface area in square meters; or we can get the square inches first (960 square inches) and then divide by the number of square inches in a square meter, 1, 5500. Either way, the surface area of the system comes out at 0.6 m^2. The efficiency of the bulb can be found in the data table on bulbs or the lumen output divided by the wattage of the bulb will also give this value. If all cool white bulbs were desired, the efficiency would be 79; if all daylight bulbs were desired, the efficiency would be 65. A 1 to 1 ratio of cool whites and daylight would give an efficiency value of 72. All Vita-Lites ® would give an efficiency value of 60. For the example, we will use a 3 to 1 ratio of Vita-Lites® to cool whites (an efficiency value of 65). The utilization factor is the constant 0.5.

Substituting proper values into the formula gives:

$$W = \frac{(15{,}000) \times (0.06)}{(65) \times (0.5)}$$

$$W = \frac{9{,}000}{32.5}$$

$$W = 277$$

So, we have a requirement of 277 watts, which is, seven 4-foot long, 40 watt bulbs. And at a 3 to 1 ratio of Vita Lite to cool white bulbs, this is 5.25 Vita Lite to 1.75 cool white bulbs. In this case, students must justify the choice to go to 8 bulbs (6 and 2), staying at 7 bulbs (5 and 2 or 6 and 1), or dropping to combinations of 6 or 5 bulbs if they feel that perhaps they do not need quite as much light as 7 bulbs will produce. If only 10,000 lux is required (185 watts), we can get by with 4.6 bulbs, and for this system, 4 bulbs would probably be enough if good reflectors are used above the bulbs. The decision usually comes down to using either 4 or 6 fluorescent bulbs above a tank, and with a reef tank system, 6 bulbs (240 watts) is usually the better choice, especially for deep tanks. This is adequate for macroalgae growth, but note that some aquarists use up to 12 fluorescent bulbs (4 actinic and 8 daylight) over reef tank setups (almost 500 watts) to get the intense, blue supplemented, full spectrum light that they feel is best for many corals. Students may find this formula and helpful for making important decisions about system lighting. Note that the formula cannot be applied to metal halide lights.

Another option is to install the higher wattage, High Output (HO) or Very High Output (VHO) bulbs, and produce about the same intensity of light with about one half or one third of the bulbs; although getting the proper spectral quality may require a mix of standard and high output bulbs. A small marine reef aquarium (20 to 50 gallons) is probably best lit by a custom hood with two or more of the new types of bulbs designed for marine aquariums.

Metal Halide Lighting

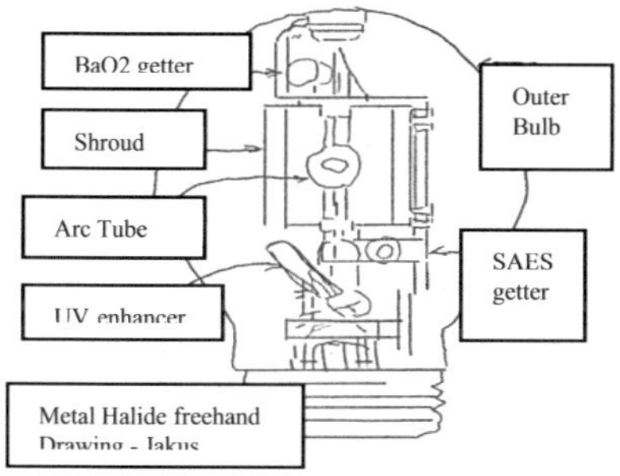

The difference between the "super" setup and the fluorescent light hood is the use of metal halide (Multi-Vapor) lamps and/or the less expensive mercury vapor bulbs. These lamps produce a great deal of light with an excellent spectral range, but they also produce a lot of heat. They must be positioned some distance above the reef tank, and this requires that the tank top be open to the lights. The tank can have a glass or plastic cover to reduce evaporation and contain salt spray, but this will reduce the spectrum and alter the effect of the high intensity lighting. Such a cover will also increase maintenance since both surfaces will have to be cleaned regularly. A partial cover or shield that extends a few inches inward around the top of the system gives protection from water splashes and can support various additional lights, reflectors, shelves, filters, etc. The higher the light intensity lamps are above the system, the less light enters the system. The lamps must be high enough so that they will not be affected by water from the tank. Depending on tank size and water movement, this may be 18 inches to 36 inches. Closer is dangerous and further away is ineffective. Metal halide lighting is very intense and the inclusion of a dawn/dusk cycle reduces the shock of immediate strong light and immediate dark. An actinic 3 fluorescent bulb that comes on 30 minutes before the metal halide lights and extends to 30 minutes after the main lights go out provides a more natural, gradual light cycle and adds a strong blue light element to the daily lighting of the system. The addition of the blue Actinic 03 is not very noticeable to the human eye while the metal halide lights are on, but many corals seem to benefit from this light.

A high intensity lighting system can be composed of only mercury vapor bulbs, only metal halide lamps, a mix of the two, or any of these in combination with certain fluorescent bulbs. Since the cost of high intensity lamps is high, a combination with fluorescent bulbs may be the answer for some students. Mercury vapor bulbs are widely used by European aquarists. The color rendering is god and the spectrum is broad with peaks in the right places, a bit high in the yellow range however. They are less expensive, but produce less light than the metal halide lamps. The color rendering of metal halide light is excellent and the lumen output is very high. Sodium vapor lamps should be avoided since the spectrum they produce is not suitable for a marine aquarium.

One interesting effect obtained by metal halide light over a reef tank is the play of shadows across the reef structures caused by the light passing through the ripples on the surface. This rippling effect of light bands on the reef is just as one observes while snorkeling over a shallow coral reef on a calm sunlight day. Metal halide lights also produce ultraviolet light, and this can cause problems if the lamp does not have and ultraviolet screen or filter. Corals may become burned if UV light is too intense in the system. And a highly unfiltered **UV** component is also harmful to the human eye. If an aquarist decides to install metal halide lighting, it is best to go with a system that has been designed and manufactured expressly for marine aquarium installation. (pp. 147-164)

The following lighting idea is from Carner (1997). His work addresses development of lighting for coral reef systems. Since any other aquatic ecosystem requires much less light intensity, lighting systems for other primary aquatic ecosystems can thus be modified from this plan. Carner calls his design a quick and simple method of adding fluorescent lighting to an existing or new tank. Other lighting designs exist and students will research and report upon them for class consideration. In addition, students will also research appropriate applications for more recent lighting technologies like T-5 and LED lighting systems.

- Determine the length and width of the system to be illuminated.
- Determine the intensity and spectrum of light to be provided above the system.
- Determine the optimal combination of bulbs to provide the desired intensity and spectra over the area to be illuminated.
- Design and construct a frame to hold the properly positioned number of rain-gutter light cabinets.
    - A. Purchase vinyl rain gutter at Home Depot or similar outlet.
        1. Basic brown exterior with white interior.
        2. Purchase end caps available for the gutter material.
    - B. Cut the material with a hack saw to the length of your frame, MINUS the additional inch(s) that the 2 end caps occupy.
        1. Install the end caps onto the gutter. They snap on and have a seal.
        2. Attach the light strips to the frame.
- Purchase a standard, 40 watt ballast (for 48" tanks), 30 watt ballast (for 30 and 36" tanks), etc.
    - C. Wiring diagrams are printed directly on the casing.
        1. Purchase pre-wired rubber end caps for fluorescent tubes. Coralife and others manufacture these.
        2. Purchase some plastic tie-wraps and 3/4" black electric tape; wire nuts as well. Buy the orange colored wire nuts as they are sized for the gauge wiring you're about to work with.
        3. You will also need a "lead-out." This is nothing more than an electrical cable, 3 pronged, that you wire up to the ballast.

"The bulbs that you purchase are very important. If wanting to keep Mushrooms and other soft corals, or even a stony or two, 50/50 Fluorescents are best. Matched with an Actinic blue tube, the set provides all spectrums of light that these varieties of animals need for survival. The rest is up to you! I recommend 2 strips on a 40 to 55 gallon and 1 strip on smaller tanks. 6 bulbs on a 55 will give great results, but the gutter material is too wide to allow a third strip to be installed! So use two 50/50 and 2 actinic, one each to a strip" (Carner, 1997).

Carner (1997), reports that this light setup is "ample for hardy stony corals and that soft corals explode with this set-up." He reminds us that aside from Triton Bulbs that "provide great color rendition and automatically SHUT OFF when their useful life is exhausted, we should be aware that all fluorescent tubes lose their true spectral output about 6 to 8 months after initial powering up." He states that many other good bulbs are available but does not have personal experience with them. The following diagram and instructions come directly from Carner (1997).

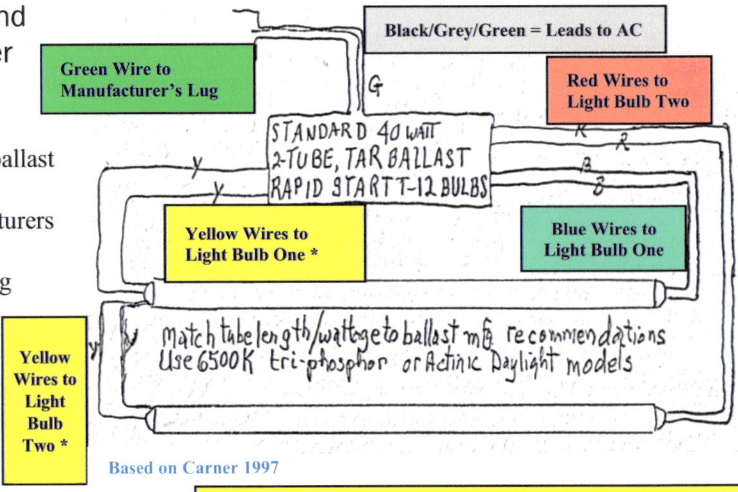

Based on Carner 1997

* Note: Yellow leads form a "T" connection (4 Wires into 2 Leads)

1. Cut the colored lead wires wrapped around the ballast to a length of about 6".
2. Spread the leads apart as per the manufacturers diagram on the casing.
3. Wire nut the end cap leads to the corresponding colored leads from the ballast. Most pre-wired end caps have 9' leads attached and the 2 smaller wires inside the casing are color coded as well.
4. Match the colors, substituting your end cap's colors for those on the ballast.
5. Make certain the harness is correct and that all wires are attached properly.
6. Torque the wire nuts finger tight.

7. Once all the wiring is completed, the AC lead-out installed and grounded to the ballast case, carefully separate the leads and attach the end caps on the bulbs.
8. Make sure that the color-coded leads all match your harness and connections just as shown on the ballast!
9. Plug it in, stand back and marvel at the light.

Many students find this to be a relatively simple endeavor. Cut or carve a notch in the end, near where the gutter attaches to the cap. This is where your wire leads will go after attaching the two tubes into the gutter.

Some people water-proof the electrical contacts with liquid electric tape and place their wired-up bulbs directly on aquarium frames without first attaching them solidly to the section of gutter material being used as a frame. In school-based aquatic laboratory settings, this is not in the best interest of student safety. To make the lighting strip safer, attach bulb clips inside the gutter and snap the bulbs into them in. Finally, attach the light strips to your system frames and plug them in. Carner reports he contact-cemented aluminum foil to the inside of these gutters but did not detect any difference in light intensity between foil-coated light strips and molded white strips.

Mount ballasts on the back side of your systems if they are wood. If you use metal stands or acrylic cabinets, mount ballasts on a piece of plywood and place them on the floor. Carner (1997) reminds us that "ballasts get hot and you may want to place them away from inquisitive fingers!"

General Protocols for Lighting System Design and Construction of Three Primary Aquatic Ecosystems

From this point forward, considerations involved in designing lighting components for each of the three primary system frames (aquaponic, coral propagation, and broodstock and larval rearing) include many details for which students must account in their design proposals. In this project, student teams must design a frame for each system. Although each system's frame is configured differently, they serve many common purposes. First, the frames help establish the footprint of the system. Second, the frames provide strength and support for all system components. Third, the frames must be designed to easily attach and access lighting components for each system. Fourth, the frames must be designed to promote easy and stable water drainage and flow. Fifth, the frames must provide substantial space for water filtration equipment and materials to be safely implemented and accessed. Frames for each system must be designed to safely incorporate electronic equipment in aquatic and possibly wet settings.

Each system frame is designed to require minimal dependence upon water pumps and take up a minimal footprint. The frames are also designed to make maximum use of gravity to promote water transport, permit maximum access to serviceable components, and permit maximum levels of illumination and growing space for system inhabitants. Please note that dimensions for each of these systems are expected to vary between teachers, schools, classes, and student groups. The aquatic ecosystem dimensions provided in *Beyond the Classroom Aquarium* are used to exemplify the process of system development. For each project, student teams must scale their systems' dimensions to meet the spatial standards established by each class at the end of project one.

Designing and assembling the aquaponic system frame and lighting components described here, enables students to operate a fully functional aquaponics aquatic ecosystem by the completion of project four.

Photo bv Robert Jakus

The first step in the process of developing an aquaponics system frame to which lighting will be attached requires students to review aquaponic system frame and lighting information found in this text and obtained during field trip one as students were guided through the process of using carpentry tools and materials to turn their plans into an actual classroom aquaponics ecosystem frame. Following this review, students are given the assignment of developing a report that demonstrates their understanding of the knowledge, skills, materials, and equipment that development of this system's frame requires. Reports about aquaponics system materials should contain dimensional details about lumber used to assemble tables and cuts in lumber that are necessary. Reports should include information they reviewed about aquaponic lighting systems and a team developed proposal that gives detailed dimensions, design layouts and system modifications that students would follow to complete their frame design. Completed aquaponic system frame and lighting designs that students propose for development must demonstrate that their design is capable of providing enough light so that organisms in each section of the system will be provided with all of the light they require. In their report, students are to include contact information for tradespersons with whom they are familiar and whom they might invite to donate guidance, support, or possibly further demonstration of tools, equipment, and materials students must safely work with as they complete the development the aquaponics system frame and lighting components.

Once presentation of reports is complete, the class is guided through a critiquing process to help assess the merits and possible drawbacks of each design proposal. Following the critique, students vote for the completed aquaponic system frame and lighting design they think will best meet the needs of the class.

Teachers may develop their own Aquaponics System Frame and Lighting Report or use the one provided below in order to guide student efforts to identify information they could include in their reports.

Aquaponics System Frame and Lighting Report

	Materials Needed	Design and Assembly Steps AND Skilled Tradesperson	Function Performed by this Piece	Safety Concerns
Two Tables 30" by 72"				
A-Frame				
Shelves				
Two Tables 48" by 96"				
Lighting				

From this point forward in project two, student activities transition to those that require direct hands-on work with equipment and materials to assemble (and interconnect) four aquaponic frame and lighting sections – implementing plans critiqued and approved by the class. As this work is completed, students shall have developed an aquaponics system frame capable of providing organisms in this system (both plant and animal) with optimal conditions for growth and development.

Per the model system described below these four aquaponic system sections include:

1. two 36-inch high tables to elevate and support the A-Frame (each 30 x 72 inches)
2. a cube in which to fit an A-Frame and attached 5 foot long, 4-inch diameter growing pipes
3. two 4 x 8-foot tables (placed side by side) upon which an 18-inch deep 48-inch radius pool may be placed and from which water may easily drain to the reservoir and to which it may return as A-frame grow tubes drain
4. adequate lighting components optimally and safely attached to the A-frame

Each section of the aquaponics system must lie at an elevation that is properly positioned (vertically and with respect to water levels) in relation to other components of the system. By constructing the frame in sections, students are more able to ensure that their aquaponics system will make efficient use of different elevations to promote the flow of water through it while also optimizing illumination above the growing tubes. In this model system, the first section of the aquaponic frame establishes a standard 36-inch elevation to which other components of the system may be positioned. The position of all other sections will be relative to the first. Student teams may scale their aquaponic system according to dimensions and limits of their own classrooms

The dimensions of the aquaponics system provided here are used with the understanding that actual aquatic ecosystem dimensions will vary between teachers, schools, classes, and student groups. Consequently, it is expected that for each classroom aquaponics system, student teams will re-scale and redesign the given dimensions to fit the footprint that the teacher of that class has approved. The standard dimensions suggested here as an aquaponics example for students to follow, are based on my interpretations of work by Nicol (1990).

In the example system design presented here, the footprint for sections one and two of the system occupies an area determined by placement and spacing between the two supporting tables. For this example, the first section is comprised of two sturdy tables; supported by 4 x 4 weight bearing beams – to a height of thirty six inches with a ½" thick, rough wooden plank surface. The surface of each table has dimensions of 30" by 72." Carpentry protocols for safe and efficient construction of generic rectangular tables as described for sections one and three are readily available. Students should be guided to locate such information and use it to guide them in the design and assembly of tables to be used in their classroom aquaponics system.

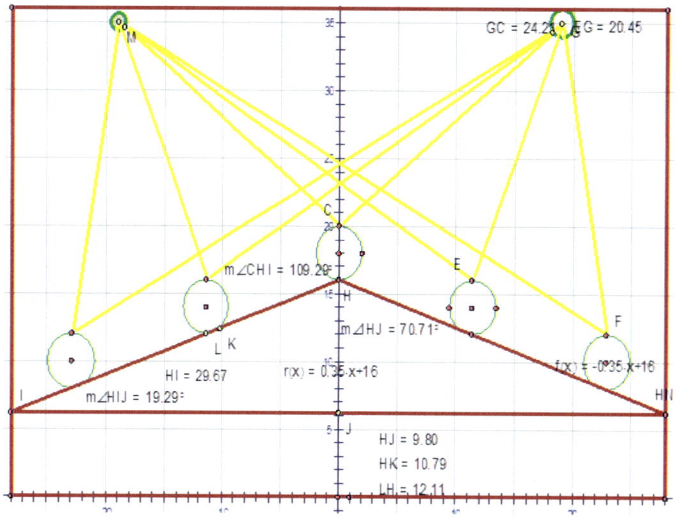

Placing the two tables side by side lengthwise, they are intended to provide a space upon which to place the aquaponic A-frame (section two). Section one tables may be positioned so that students may position a 50 gallon or larger sump/reservoir in an open space between the tables. The top of the sump may extend above the surface of the tables so that drainpipes may be connected to the upper rim area of the sump. The spacing between the two tables makes it possible to position the sump underneath the tables and (as necessary) set upon cinder blocks and lumber to achieve a satisfactory height. Leaving an open space between the tables and above the sump simplifies access to materials and submersible pump that will be placed in the sump.

Once the tables are complete, students may be guided to follow the directions below to assemble the (section two) A-frame they will place upon their (section one) aquaponic tables. To create a 60" by 41" top-frame for the A-frame students may

- Attach 41 inch lengths of 2 x 4 flush with each end of a 56 inch length of 2 x 4 in such a way that they extend perpendicular to the 56-inch board.
- Next, place another 56 inch length of 2 x 4 in the space between the ends of the two 41 inch pieces of 2 x 4 (already attached to the first 56 inch length) and connect these.

Students should now have a rectangle that measures about 60 inches by 41 inches. This frame will be used as the top of a cube in which the A-frame will be encased. For this example system, grow lights will be securely attached to this portion of the cube after the A-frame has been encased within it.

As represented in the 2-D cross-section image of the A-frame shown above, the 2 x 4's that determine the slope of the A-frame must be cut to about 0.35 (approximately 20 degrees). While sufficient enough to allow water to flow vertically from pipe to pipe and back to the system (via gravity), a 0.35 slope also permits more efficient illumination above the entire growing area than a steeper slope would. The A-frame example provided here uses a slope of 0.35. For this example, the following pieces of lumber are used: six 56 inch lengths of 2 x 4 are used for the sides of the A-frame. The wider sides of these boards are used for the base of the frame. Additional lumber needed for the sides of the A-frame include four 38 inch lengths of 2 x 4 (for the vertical frame to which the lighting frame will be attached), six 4 inch blocks of 2 x 4 and two 8 inch blocks of 2 x 4.

- Stand a 4 inch block of 2 x 4 at each end of a 56 inch length of 2 x 4. Connect these blocks to the 56" length of board and then position a third 4 inch block in the center of the 56 inch length board and attach it also.
- Connect another 56 inch length of 2 x 4 to these blocks – also connect the vertical 38 inch legs of the rectangle at each end.
- Connect an 8 inch block of 2 x 4 (extending vertically) from the center of the upper 56 inch 2 x 4.
- Prepare two 30 inch lengths of 2 x4 for the frame by mitering a19.3° angle from one flat end, and a 19.3° from the cross section of the board. On the top side, the 2 x 4 will measure 29.7 inches, on the other side the 2 x 4 will measure about 25.5 inches.
- Connect these two 30 inch mitered boards to the 56 inch length of board and rest the ends that meet upon the 8 inch 2 x 4 block. According to Pythagoras, this should fit well enough.

Students will repeat the above process to create the second side of the A-frame. After each side of the A-frame has been assembled students may be guided to follow the directions below to complete the cube in which the A-frame will be encased and to which lights will be attached.

- Attach front and back sides of the A-frame cube (at the top and bottom corners – vertical 38 inch lengths) of the A-frames sides with 60 inch lengths of 2 x 4.
- Add additional wooden braces across the front, back and corners for support as necessary.

Before attaching lights to the top of the A-frame, students may use the cross section above to determine where the PVC pipes should be placed. After temporarily installing the PVC pipes on the 0.35 (approximately 20°) sloped A-frame, students may perform calculations to describe how lights may be positioned above the frame to maximize illumination on all pipes (while minimizing bulb use and distance from bulbs to growing plants). Students will install the pipes permanently in project three.

The 2-D overhead view of the five PVC pipes on the A-frame (depicted below - by circles whose radii are connected) shows the direction water flows from the central pipe – into each of the two PVC pipes that are directly beneath the central pipe on the A-frame. The water then flows out of these pipes and into the lowest pipes on the A-frame. As water exits the lowest pipes, it is directed back to the part of the system to which students will eventually add fish. The drainage of water from one aquaponic growing tube to the next and drainage of water back to the fish-containing pool is promoted by inserting a one inch elevation booster beneath the end of each pipe at which water enters it so that the pipes output is one inch below its input. In addition, to promote the return of water from the growing tubes to the fish containment section of the system (by gravity), the elevation of these tubes (and thus the entire A-frame) must be such that they are elevated above the surface of the water in the adjacent fish containment section of the system.

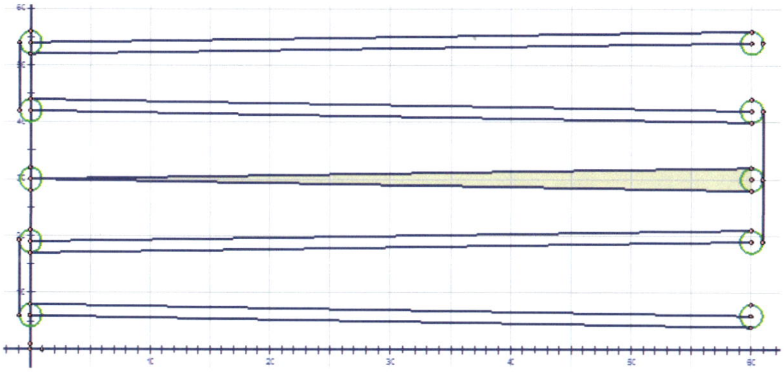

Now that the aquaponic system frame has been assembled, students may be guided to review class decisions about the fluorescent lighting component they decided to adopt for their system and again transition from academic analysis activities to hands-on implementation of design plans.

- For this example system, the lighting fixtures will be comprised of four foot long T-5 bulbs (two of these). Student analysis of distances from the light bulb locations to the growing tubes should lead them to suspend the bulbs from the top of the A-frame - 9 inches off-center (and parallel to the tubes.)
- After students properly install lights on the A-frame, their work on aquaponic system lighting is complete. They will complete the aquatic sections of the aquaponic frame in project three.

Lights for this system will be attached to boards in the 41 inch wide frame that were installed across the top of A-frame cube. In addition, after installing appropriate lights in the A-frame, students then place the A-frame upon the previously constructed tables. In project three the five 5ft PVC pipes will be permanently attached to the A-frame, and nutrient-filled water, from the fish habitat, will flow through these pipes (from the sump) before being returned to the fish habitat. Plant seedlings will grow upward from the pipes toward the lights overhead.

Photo by Robert Jakus

Completing the Design and assembling the coral propagation frame and lighting component described here will enable students to operate a fully functional coral propagation aquatic ecosystem by the completion of project five.

Photo by Robert Jakus

The first step in the process of developing a coral propagation system frame to which lighting will be attached requires students to review coral propagation system frame and lighting information found in this text and obtained during field trip one as students were guided through the process of using carpentry tools and materials to turn their plans into an actual classroom coral propagation ecosystem frame. Following this review, students are given the assignment of developing a report that demonstrates their understanding of the knowledge, skills, materials, and equipment that development of this system's frame requires. Reports about coral propagation system materials should contain dimensional details about lumber used to assemble tables and cuts in lumber that are necessary. Reports should include information they reviewed about coral propagation lighting systems. Completed coral propagation system frame and lighting designs that students propose for development must demonstrate that their design is capable of providing enough light so that organisms in each section of the system will be provided with all of the light they require. In their report, students are to include contact information for tradespersons with whom they are familiar and whom they might invite to donate guidance, support, or possibly further demonstration of tools, equipment, and materials students must safely work with as they complete the development the coral propagation system frame and lighting components. For many students, motivation to assemble high quality aquatic ecosystems increases when skilled tradespersons and parents contribute their time.

Once presentation of reports is complete, the class is guided through a critiquing process to help assess the merits and possible drawbacks of each design proposal. Following the critique, students will compile and assess all information they have acquired and conduct class discussions or perhaps an election to arrive at a consensus regarding the adoption of each section of the coral propagation system frame design and lighting system that the class will design and assemble.

Teachers may develop their own Coral Propagation System Frame and Lighting Report or use the one provided below to guide student efforts to identify information they could include in their reports.

Coral Propagation System Frame and Lighting Report

	Materials Needed	Design and Assembly Steps AND Skilled Tradesperson	Function Performed by this Piece	Safety Concerns
Vertical Legs				
Cross Braces on Legs				
Shelves				
Middle Shelf Lighting				
Top Shelf Lighting				

From this point forward in project two, student activities transition to those that require direct hands-on work with equipment and materials to assemble coral propagation system frames –

implementing plans critiqued and approved by the class. As this work is completed, students shall have developed a coral propagation system frame capable of supporting attachable system components (lights, tanks, pumps, heaters/chillers, and pipes) that are necessary to provide organisms in this system with optimal conditions for growth and development.

In order for the coral propagation system frame to perform the functions described above, students must design and assemble it in three separate stages each of which will ultimately be connected to the others. These sections include the frame's vertical legs and cross braces, three shelves to attach to the frame, and lighting components installed in the frame above the two coral propagation tanks.

Space above the upper two tanks must allow for access to corals and safe access to lights while also allowing maximum illumination to encourage the coral pieces to expand and grow in the system. I like to guide students to consider some form of pendant lighting for the top shelf tank and a combination fluorescent lighting system on the middle shelf. Lights for the center tank may be attached to wooden frame material in the upper shelf above the middle tank. Students should also have devised a way to suspend a light frame from chains in the ceiling to illuminate the tank on the top shelf. Lighting a coral propagation system introduces a new set of concerns about providing organisms in the system with sufficient illumination.

The dimensions of the coral propagation system described here are used with the understanding that actual aquatic ecosystem dimensions will vary between teachers, schools, classes, and student groups. Consequently, it is expected that for each classroom's coral propagation system, student teams will re-scale and redesign the given dimensions to fit the footprint that the teacher of that class has approved. The system consists of three interconnected and intensely lit water components.

In the sample coral propagation system described here, the footprint of the system is about 14" by 24." With all component pieces of the system in hand, the unit can be assembled and disassembled by one person. The systems side legs are made so that the shelves fit into slots routed into cross braces as shown in the image below. The systems shelves lock into the side legs without the use of screws. Students should be guided to make all interlocking cross braces and shelve frames the same size to simplify setup of the coral propagation system frame.

Photo by Robert Jakus

The first step in the process of developing this model coral propagation system is to create three shelves upon which to place the three water containing coral-propagation aquaria. Students may be guided to make these shelves by framing 2 x 2 pieces of wood (in a 12 inch by 24 inch configuration). A ¼ inch thick piece of plywood is placed upon the 2 x 2 frame. Students may use waterproof glue and wood screws are used to secure the plywood to the frames. Two 12-inch pieces and two 22-inch pieces of 2 x 2 are used for each frame. Thus, six pieces of each length 2 x 2 must be used for the purpose of making the shelving frame. The bottom outer piece of each 12 in 2 x 2 that forms the width of the shelf must have a half inch square groove routed out along the board's entire 12 inch length. This groove must be indented a half inch or more from the outer edge of the board

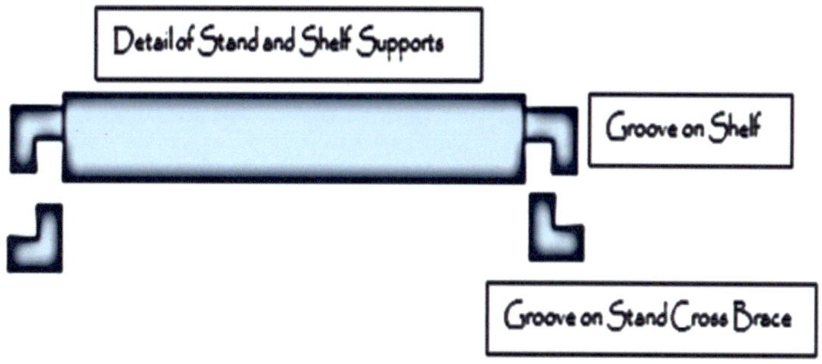

The second section of the coral propagation system frame that student teams will construct are the left and right sides of the wooden stand to which cross braces and shelves will be attached. For this small model system, the weight of the water and other system contents will be slight enough to permit students to use a pair of one by threes as front and back legs on each side of the stand (left and right). Two by two's may also be used as cross braces connecting the pair of legs for each side of the system. Students should be guided to use a router on these cross braces between the front and back legs on each side. Each of six cross braces must also have a half-inch square routed out of it so as to provide a space into which the shelf may fit. As shown in the image above, the routed out piece in each shelf will be placed upon the routed section of a cross brace.

Wood screws should be used to attach each cross brace to a stand leg. The half inch routed out square should be designed to fit comfortably with the corresponding half-inch groove in the shelf. Before attaching the cross braces to the right and left legs, the height of each shelf needs to be determined. The bottom shelf will hold the tank with largest volume of water. The bottom tank must be deeper than the tanks on the two upper shelves in order to serve as a water processing and filtration sump (not much actual coral propagation with bright illumination is needed for the bottom tank but some space above it will make accessing filtration and other components more convenient). For this example, the six cross braces on the boards used for the legs are attached to the legs at lengths from the bottom at 3 inches, 33 inches, and 63 inches. The length of each piece of wood used for a leg (four in total) was 76 inches

Students may now be guided to review information and class decisions regarding lighting requirements that will be necessary to successfully propagate corals. Students with some technical knowledge and skill may be encouraged to design and assemble their own lighting systems. For the model system described here, students may acquire a variety of metal halide and actinic lighting combinations to illuminate the coral in the tank on top shelf. Students may be able to safely suspend some of the lights above this system by using chains and or hooks that are securely attached to the ceiling above the system. For the coral in the tank on the middle shelf, students may acquire different (possibly more spatially compact) combinations of fluorescent bulbs and high output (10,000 K) actinic bulbs or LEDs. The design for the hood for the actinic bulbs, to be placed above the tank on the top shelf is described below.

In project three, students will design and assemble plywood tanks to be fitted upon the three shelves in the coral propagation system frame. This includes two coral propagation tanks of wood painted with water-proof epoxy (including bulkhead holes), and a wooden sump tank painted with epoxy, water flow connections, and filtration equipment.

Photo by Robert Jakus

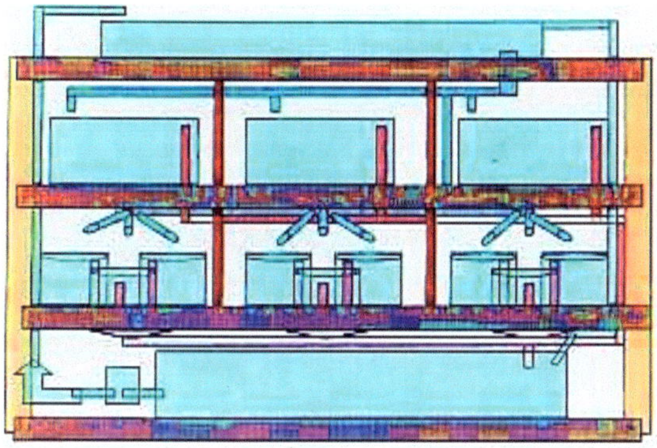

By Robert Jakus

Completing the design and assembly of the broodstock and larval rearing system frame and lighting components described here will enable students to operate at least one fully functional broodstock and larval rearing aquatic ecosystem by the completion of project six. I find it valuable to have students develop two such systems (one for freshwater and the other for saltwater). Unlike many organisms in the aquaponic and coral propagation systems that require high intensities of lighting, many organisms in the broodstock system may not rely so critically on high levels of illumination.

Adult fish commonly eat their eggs and babies. Many techniques for separating adults from their offspring exist. Higher rates of egg viability and larval survival result when flowing water from parent tanks is used in circular hatchery tanks. Use of broodstock larval rearing system water ensures that stresses to unhatched eggs, from removal and relocation, are minimized. The broodstock and larval rearing system described here is designed to make use of this advantage. It is also capable of holding numerous broods of eggs and recently hatched larvae in such a way as to optimize survival. Larger systems promote water stability and also increase the odds that students will have the opportunity to participate in actual hatchery activities because the system has space to house a number of pairs of brood stock species.

As in the process of developing the two previous system frames and lighting components, students are similarly asked to review broodstock and larval rearing system frame and lighting information found in this text and obtained during field trip one as students were guided through the process of using carpentry tools and materials to turn their plans into an actual classroom broodstock and larval rearing ecosystem frame. Following this review, students are given the assignment of developing a report that demonstrates their understanding of the knowledge, skills, materials, and equipment that development of this system's frame requires. Reports about broodstock and larval rearing system materials should contain dimensional details about lumber used to assemble tables and cuts in lumber that are necessary. Reports should include information they reviewed about broodstock and larval rearing lighting systems. Completed broodstock and larval rearing system frame and lighting designs that students propose for development must demonstrate that their design is capable of providing enough light so that organisms in each section of the system will be provided with all of the light they require. In their report, students are to include contact information for tradespersons with whom they are familiar and whom they might invite to donate guidance, support, or possibly further demonstration of tools, equipment, and materials students must safely work with as they complete the development the broodstock and larval rearing system frame and lighting components. For many students, motivation to assemble quality aquatic ecosystems increases when tradespersons and parents contribute their time.

Once presentation of reports is complete, the class is guided through a critiquing process to help assess the merits and possible drawbacks of each design proposal. Following the critique, students will compile and assess all information they have acquired and conduct class discussions and/or perhaps an election to arrive at a consensus regarding the broodstock and larval rearing system frame design and lighting system that the class will design and assemble.

Teachers may develop their own Broodstock and Larval Rearing System Frame and Lighting Report or use the one provided below in order to guide student efforts to identify information they could include in their reports.

Broodstock and Larval Rearing System Frame and Lighting Report

	Materials Needed	Design and Assembly Steps AND Skilled Tradesperson	Function Performed by this Piece	Safety Concerns
Vertical Legs				
Cross Braces on Legs				
Shelves				
Brood Tank Lighting				
Hatchery Lighting				

From this point forward in project two, student activities transition to those that require direct hands-on work with equipment and materials to assemble one or more broodstock and larval rearing systems – implementing plans critiqued and approved by the class. As this work is completed, students shall have developed a broodstock and larval rearing system frame capable of providing organisms in this system (both plant and animal) with optimal conditions for growth and development.

The dimensions of the broodstock and larval rearing system described here are used with the understanding that actual aquatic ecosystem dimensions will vary between teachers, schools, classes, and student groups. Consequently, it is expected that for each classroom's broodstock and larval rearing system, student teams will re-scale and redesign the given dimensions to fit the footprint that the teacher of that class has approved. Generally, directions for design and assembly of broodstock and larval rearing system frames are modifications of systems used commonly in local aquarium shops.

In order for the broodstock and larval rearing system frame to perform the functions described earlier, students must design and assemble one four-shelved broodstock and larval rearing frame. Different organism habitats on the four shelves will ultimately be connected to one another. In this system, students will use standard sized aquariums on the broodstock shelf and the size of the system can be scaled so as to comfortably fit the sizes of the tanks selected for use. In the sample broodstock and larval rearing system described here, the footprint of the system is 18 x 36 inches. The height of this prototype BSLR model is seven feet.

I encourage students to design and build systems that can accommodate 30-gallon aquariums on the broodstock shelf. Since each 30-gallon aquarium is 36 inches long, the full-scale design for this system is 13 feet in length, eight feet in height, and two feet in width. Tanks with smaller lengths will allow design and assembly of smaller systems. Smaller aquaria limit the choice of organism that may be kept in the system. Many organisms will not exist comfortably in a 10-gallon aquarium but feel quite comfortable within tanks of larger volume. The goal for each class is to develop a system that is sized to allow their broodstock system inhabitants to feel comfortable and behave as naturally as possible.

Regardless of the dimensions, the BSLR system described in *Beyond the Classroom Aquarium* consists of numerous interconnected water components, six of which require moderate levels of regular daylight. Students can frame four 18 x 36 inch shelves and place a 18 by 36 inch piece of plywood to surface each frame. The height of each shelf is determined primarily by accessibility to the system components. The bottom shelf (in some cases the floor) beneath the system will hold a sump reservoir for filtration of system water. The top shelf (at an elevation of 70 inches) hold an eight-inch deep tank (30-inch long, 18-inch wide ~ 10 gallons) for use as the main distributor of fresh filtered water to the individual units that house broodstock and eggs/larvae. The hatchery shelf should be positioned 30 inches above the floor. This shelf will hold circular tanks in which eggs and larvae will hatch and grow. Students will design plumbing to each of these egg/larvae tanks that will allow for a wide range of flow rates and aeration. Students will place the broodstock shelf above the hatchery shelf. This shelf will be placed at a height of 50 inches above the floor. This shelf holds tanks that house mature broodstock pairs. The broodstock shelf holds a separate and spacious domicile (a regular glass aquarium) for each of three broodstock pairs. Students may choose to acquire and use standard aquarium hoods to light these broodstock tanks. Lighting for the hatchery containers will be temporary lighting units that can be accessed, repositioned, and removed easily. All in all, once the frame for this system is complete, the lighting does not require additional fitting and can be installed after students complete the plumbing work in project three.

Photo by Robert Jakus

The information in the following chart may be assigned for students to fill in after they have adopted each system design. This chart can provide a quick overview of dimension and scale in each system. Information about each ecosystems footprint and essential lighting materials and equipment needed to adequately illuminate each system can be listed here.

Some students may prefer to assemble their systems without devoting much effort to learning about important properties of light that must be considered to efficiently design each system to operate successfully. These students may learn at least as much by assembling their ecosystems (without in-depth background knowledge) and then "backward-engineering" their operational ecosystems to improve system performance over time. Some educators may encourage system design to evolve from such "student discovery." As information to improve system performance is needed, these educators may guide their students to seek information as needed; a process through which students often also learn many significant but less directly relevant concepts.

Summarization of Student Decisions about Three Classroom Aquatic Ecosystems Form				
	Project Four	**Project Five**	**Project Six**	
	Aquaponics	Coral Propagation	Freshwater Broodstock Larval Rearing	Saltwater Broodstock Larval Rearing
Primary Sources of System Information				
Important System References				
Organisms in the System				
System Design and Full-Scale Dimensions; Framing Details				
System Lighting				
System Water Management Components				
Livestock Transfer				
System Monitoring and Control				
System Log Books Documentation				
Organism Log Book and Brood Pair and Nest Data				

Write a list of **science-based** activities you have participated in during this project and briefly describe evidence that verifies your interaction with and knowledge of scientific concepts for each item on your project two list.

Write a list of **technologically-oriented** activities you have participated in during this project and briefly describe evidence that verifies your interaction with and knowledge of technologically-oriented concepts for each item on your project two list.

Write a list of **engineering-oriented** activities you have participated in during this project and briefly describe evidence that verifies your interaction with and knowledge of engineering-oriented concepts for each item on your project two list.

Write a list of **mathematical** activities you have participated in during this project and briefly describe evidence that verifies your interaction with and knowledge of mathematical concepts for each item on your project two list.

Write a brief list of **Non-STEM-related** activities you have participated in during this project and briefly describe evidence that verifies your interaction with and knowledge of Non-STEM related concepts for each item on your project two list.

Rate your level of involvement in STEM-related activities in this project (1=lowest, 10=highest).

1—2—3—4—5—6—7—8—9—10

Beyond the Classroom Aquarium **Daily Journal Guidelines**

Entries students place in their daily journals should be:

1. Dated accurately and logged chronologically.
2. Kept in a notebook and written neatly in ink or typed.
3. Written or typed in complete sentences using correct capitalization, spelling, and punctuation.

In their journals, students should keep a daily record of:

1. Who they worked with each day by including the names of school staff, fellow students, community members, and volunteers.
2. What they did each day, such as lecture topic, activities or project design, assembly, or maintenance.
3. What they learned each day by asking questions, following instructions, and/or observing activities.
4. How *Beyond the Classroom Aquarium* experiences relate to past, present, or future life experiences.

For bonus points, students may complete these journal options, including illustrations, diagrams, or creative expression relating to the journal entry. For project one, journal writing bonus topics may include answers to the following questions or descriptions of a project activity or assignment. Reflection on these topics may also help students prepare for exams, quizzes, and other assessments.

- What do you want to understand better in order to contribute more to your team's productivity as it independently designs, assembles and then manages an aquatic ecosystem?
- What did you learn during project two? Write down what you would like to know more about and if you know someone else who might help you learn more if you asked them.
- Select a book or magazine to read that is related to a topic in chapter II. Write a report on this book and include your report in your *Beyond the Classroom Aquarium* portfolio.
- Create posters and signs so that people who visit your aquatic ecosystems will be able to read about specific and unique system components and the organisms in them.

- Create a new picture and related story about the organism or system that is most interesting to you.
- Select a book or magazine to read that is related to a topic in project two. Write a report on this book and include your report in your *Beyond the Classroom Aquarium* portfolio.
- Three rules required of every student during a "critique," are: 1. be kind, 2. be specific, and 3. be helpful (Berger, 2003). Describe how your fellow student's kindness during a critique session helped you feel safe from any hurtful comments, including sarcasm. Explain how comments on your work like "It's good," or "I like it" are a waste of time. Describe a comment or two made during a critique of your work or ideas that were helpful to you.
- List important vocabulary words you learned by working on system fame and lighting components.
- Describe and explain important components of aquaponic, coral propagation, and broodstock larval rearing system frames design and lighting. Explain why these components are important.
- Describe three important skills you developed and ideas you learned while working on development of system frames and lighting components.
- Assembly of frames and installation of lighting in project two required students to apply many forms of measurement, and to work with a variety of materials, tools, and aquatic ecosystem equipment. Make a list of the tools, materials, and aquatic ecosystem equipment that you worked with in project two. Write a description of how the tools, materials and equipment have been used to create your aquatic ecosystem frames and lights. Explain ways you used STEM knowledge and skills while using materials and tools.
- Describe and explain the procedures or steps you would follow to assemble a frame for an aquaponics, coral propagation, and broodstock larval rearing system.
- Do the system frames and lighting components function properly? What could have been done differently to improve these system components?
- Describe valuable information you obtained during the field trip. What was most helpful about the support the field trip tradespersons guides provided? Explain any new understanding about safety and working with power tools that you have discovered.
- Describe the ways that different light bulbs operate. For fluorescent bulbs, explain the function of the lighting system's ballast and starter

Additional Project Two Assessment Activities and Questions

Project a light through a prism onto a screen or wall. Ask students to write what they know about light. Ask students to write about the colors they observe in the order they see them in the spectrum. Ask students to explain how absorption and reflection relate to the colors. Tape a small piece of blue plastic over the light source that is being sent through the prism. Ask students to explain why the color that now appears is blue.

According to properties of absorption and reflection, what order are colors are filtered from the water?
Ask students to explain which colors would be best for camouflage if they were fish in the mesopelagic twilight zone. Or, ask if you were to become a scuba diver, what color wetsuit would you wear to become less visible to fish (like sharks)? Your wetsuit color choices include yellow, orange, red, green, and blue.
Describe optimal levels of light intensities for each of the three systems. Compare and contrast the light intensities and system purposes. How does knowledge of light help people establish successful systems?

Create a chart, diagram, poster presentation of waves between 400nm and 700nm (labeled) and showing the waves in scale to one another and using colors of light that correspond to each color's wavelength in the spectrum, using the colors of the rainbow (ROYGBIV) as a guide.

Explain two functions that light enables organisms to perform.

List and summarize the three properties of light how each is measured.

Give two reasons why marine scientists find knowledge of light wavelengths important.

Explain why wavelengths have more energy if they are short and how energy from short wave light energy is used in aquatic ecosystems.

Explain black body radiation. Create a chart that illustrates that low frequency radiation produces long wavelengths etc.

Explain CRI and how it is used to help select lighting for aquatic ecosystems.

Explain light intensity and the units of measurement used.

Explain three factors that must be considered in order to determine lighting necessary to grow coral. Give an example of the calculations to make for growing coral at the bottom of a 12-inch deep body of water.

Using information provided in *Beyond the Classroom Aquarium*, such as the information by Carner (1997) or some other reliable source as a model, create a poster and diagram that documents the steps you followed in designing and assembling an aquatic ecosystem frame and lighting component (include photos of your progress when possible). Prepare this for class presentation and critique.

Review activities and information that interested you in this project and determine if this information may fit on your "Research-ability" Potential Topics Form on page 59. If so, add it to the list.

Create a "technical data on light bulbs" table that follows the format used by Weast (1987), Lundegaard (1985), and Osborne, (1983) (and as shown in Moe, 1992b) including information about LED and T-5 light bulbs. Develop a presentation of your updated chart and describe the influence this information may have on decisions students make about light bulbs they place in the aquatic ecosystems they design.

BULB TYPE	WATTS	INITIAL LUMENS	INITIAL EFFICIENCY	COLOR TEMP (K)	HOURS OF LIFE	CRI	LENGTH INCHES
INCANDESCENT							
FLUORESCENT							
GREEN							
HIGH INTENSIT DISCHARGE – MERCURY VAPOR							
				NA			
				NA			
				NA			
HIGH INTENSITY DISCARGE – METAL HALIDE							
				NA			
				NA			
LIGHT EMITTING DIODES (LED)							
T-5 BULBS							

CRI (Color Rendering Index) signifies the spectral distribution of light sources. The CRI of sunlight, as a standard, is 100. The higher the CRI of a bulb, the more all colors appear natural to the human eye.

Beyond the Classroom Aquarium STEM Self or Peer Assessment Rubric

This self-assessment is valid only to the extent to which you can document and display evidence that supports it. For each category in the left hand column, write a STEM-Oriented example in the box beneath 3 if you can display evidence that your STEM-Oriented example represents work that exceeds expectations. If your evidence shows that your STEM-Oriented statement meets but does not exceed expectations, write the statement in the box beneath 2. Similarly, for STEM-Oriented examples that demonstrate some progress toward expectations.

Name: **Project Number:** **Date Project Work Began:** **Date of Project Completion:** **List Math Topics and/or Objectives Currently Under Study:**		**3** **Work Exceeds Expectations**	**2** **Work Meets Expectations**	**1** **Work Is Progressing Toward Expectations**
Write examples of questions and defining problems related to your STEM work.				
Describe how you have developed and used models in your STEM work.				
Write examples of STEM investigations you planned and carried out.				
Describe STEM related data that you analyzed or interpreted.				
Give examples of how you used mathematics and computational thinking related to the five standards listed to the right.	Problem Solving (Math Process Standard One)			
	Reasoning & Proof (Math Process Standard Two)			
	Communication (Math Process Standard Three)			
	Connections (Math Process Standard Four)			
	Representation (Math Process Standard Five)			
Describe STEM-Oriented explanations and solutions you designed				
Explain evidence you used while engaging in a STEM project-based argument.				
Give examples of STEM information you obtained, evaluated, and/or communicated				
TOTAL POINTS				

In project three, students continue assembling classroom aquaponics, coral propagation and, (one or two) broodstock and larval rearing aquatic ecosystems by designing and installing water recirculation and containment fields in the three system frames developed in project two. This project is divided into three major sections for the purpose of enabling student teams to continue working simultaneously and independently on one of the three aquatic ecosystems.

As mentioned in project two, aquatic ecosystem dimensions provided in *Beyond the Classroom Aquarium* are expected to vary between teachers, schools, classes, and student groups. Therefore, it is expected that student groups will likely need to modify the given design configurations in order to fit spaces available in their own classrooms. The purpose of the second field trip is similar to the first. However, with aquatic ecosystem development well under way, more (if not all) of this field experience can be dedicated to providing supportive and professional guidance to students as they develop and assemble water containment and recirculation components of their aquatic ecosystems.

In project three, student teams design and present modified aquatic ecosystem water containment and flow proposals based on information reviewed (from this text and other sources). As in project two, presentations include recommendations for specific water containment and recirculation features they would like to include in the system they are developing, describing how their water containment and recirculation system would impact the organisms in the system. Presentations include information about materials, tools, aquatic ecosystem equipment, knowledge, and skills important for development of water containment and flow components of aquatic ecosystems. When reports have all been given, classes are guided to compile report information, critique it, and adopt designs and design modifications suggested in student team reports. Once a critique of student water containment and recirculation reports is complete, classes will hold elections to arrive at a consensus regarding the adoption of each component of each system's water containment and recirculation components. By making these decisions prior to the second field trip, it is possible to provide this information to the supportive tradespeople who will be guiding the students through their field trip activities.

In the field trip feedback form (p. 21), students are asked to include contact information about supportive local tradespersons with whom they are familiar and whom they may wish to invite to give a class demonstration. Many students find their motivation to assemble high quality aquatic ecosystems increase when skilled tradespersons and parents contribute their time, knowledge, and support.

As in project two, project three activities again require students to work with a variety of tools, materials, and aquatic ecosystem equipment—much of which, for many of them, may initially be unfamiliar. It is possible that many students have never before worked with such tools, materials, and aquatic ecosystem equipment. It is also most probable that, following the second field trip, with additional guidance and support, there will be at least three primary systems in the designated classroom space and that all three will contain water and operational lights and will be ready to support aquatic life by the time students have completed work on project three. The instructor shall already have acquired many of the organisms to inhabit the new systems. After the systems mature, the organisms will be transferred to the new systems from the classroom's four model systems during projects four, five, and six.

To get the three systems up and running as soon as possible, students are introduced to essential plumbing skills and water management equipment such as: pumps, PVC pipe and connectors, flexible hoses, siphons, submersible heaters and chillers, filters, water processors, and storage tanks. After water containment and flow components have been incorporated into system frames, titration techniques used to determine water quality are demonstrated and mastered. Use of digital water quality equipment is introduced and skills developed. Students perform comparisons of different water quality measurement technologies.

Although construction is initially in the forefront of each project in project three, important protocols for students to follow in the collection and recording of aquatic ecosystem water quality data are also introduced during the concluding section of each project in this chapter. After each of the three (or four) student teams have filled their aquatic ecosystems with water, a culminating project three activity challenges students on each team to lead students in each of the other two groups through the water quality assessment protocols they were guided to perform at the conclusion of their successful aquatic ecosystem assembly process. Once students from each group can efficiently perform a comprehensive series of water quality tests on any of the three classroom aquatic ecosystems, students are directed to set up a computer file in which they enter and save water quality data for future analysis.

At some point after the following water electrolysis activity is complete (perhaps during field trip two), and prior to the discussion that follows, students should be shown how to operate a reverse osmosis water purification unit and a water de-ionization unit. A storage space for water (and related water treatment chemicals) should be reserved in the classroom. Preferably the containers will be in an out of the way space and located so that water may be drained from the water storage containers via a spigot. Also, these containers should be easily accessible so that students may refill the storage containers without much difficulty. Lifting and transporting large volumes of water can be difficult and dangerous.

In order to measure the relative gas volumes from the electrolysis of water and study the gaseous electrolysis products, water containing a dissolved salt is electrolyzed in an apparatus where the volumes of gas produced can be collected and measured. This particular activity is from Brooks (2012).

Materials

- Hoffman apparatus with 2 platinum electrodes
- approximately 150 mL distilled water mixed with 1 g sodium sulfate and dyed with food coloring
- 2 medium size test tubes (25 x 150 mm, wrapped in plastic tape)
- 2 wooden splints
- matches
- DC power supply
- 2 wires with alligator clips
- distilled water
- Bunsen burner
- 1-L beaker
- stirring rod

Electric currents transport electric energy through matter. To have an electric current there must be charge carriers in the matter and there must be a force that moves those carriers. The charge carriers in the electrolytic solution are positive and negative ions.

- Pure water conducts electricity poorly because it contains very few positive and negative ions. Sodium sulfate is added to provide charge carriers in the water.
- When a direct electric current is passed through water to which a small amount of sodium sulfate has been added, a chemical reaction takes place, and two different gases are produced.
- Elemental hydrogen is formed at one electrode and elemental oxygen at the other.

Procedure

1. Dissolve 1 g sodium sulfate in about 150 mL of distilled water. Add food coloring.
2. Open the stopcock on the Hoffman apparatus, and fill the apparatus with the water-salt solution.
3. Connect the platinum electrodes to the DC power supply.
4. Close the stopcocks on the apparatus.
5. Plug in the power supply and slowly turn up voltage until you see bubbles forming at the electrodes.
6. Run for approximately 5-10 minutes or until you have about 20 mL of oxygen gas collected.
7. Collect hydrogen gas in a plastic-tape-wrapped test tube (keep inverted) and test with a lighted splint. You should hear a "pop."
8. Collect oxygen in the other plastic-tape-wrapped test tube and test with glowing splint. The splint should reignite.

Lab Hints

- Larger test tubes will produce a "pop" or "whoosh" when testing the H_2 gas.
- Be sure to allow enough oxygen gas to form before testing (approximately 20 mL).
- Carbon electrodes also work in the Hoffman apparatus.

- Have student volunteers test the gases.
- Make sure that students observe the gas levels for H_2 and O_2 before any gas is removed. Before testing, ask them to predict which tube contains H_2 and which tube contains O_2.

Teacher preparation: 20 minutes

Presentation: 25 minutes

- There is an electric shock hazard possible with the power supply. An explosion might shatter test tubes filled with gas.
- Use caution when handling electrical apparatus. A tingling sensation implies an electrical shock hazard. Use an outlet with a ground fault interrupter as a source.
- Do not allow all of the water to run out of the apparatus while producing hydrogen and oxygen gas.
- Do not displace so much water that the level drops below the cross tube and allows mixing of the gases.
- Wrap the test tubes used for testing gases with plastic tape.

Disposal

Dispose of the sodium sulfate solution at the sink. Rinse Hoffman apparatus and platinum electrodes with distilled water. Burnt splints and matches are ordinary trash.

Presentation Question:

- Should the test tube from the anode compartment, which has the smaller volume of gas collected, be held with its mouth pointing up or down?
    - If the density of a gas is denser than air, the mouth should be held up. The gas in this test tube is oxygen, which is denser than air, so it should be held mouth-up. Gases are fluids, so the less dense fluids float on the more dense air, and the more dense fluids sink. Therefore, oxygen sinks in air. Hydrogen, however, rises in air and its test tube should be held mouth-down.

Discussion

- When a direct electric current is passed through water to which a small amount of sodium sulfate has been added, a chemical reaction takes place, and two different gases are produced.
- Elemental hydrogen is formed at one electrode and elemental oxygen at the other.
- The balanced equation for this reaction is:
    - $2\ H_2O(l) \rightarrow 2\ H_2(g) + O_2(g)$
- Note the relationship between the coefficients of the balanced equation and the volumes of hydrogen and oxygen produced during the electrolysis.

According to (Williamson and Smoak, 1998), "water itself is very simple. Biologically, water is the essence of life. Physically, water can exist in three forms: as a liquid, as a solid (ice), and as a gaseous vapor (steam). Pure water is neither an acid nor a base (alkaline). Hydrology is the study of water, and hydrologists are scientists who study water." Buchheim's (1998c) exposition on water and the following bulleted list of water facts is provided below.

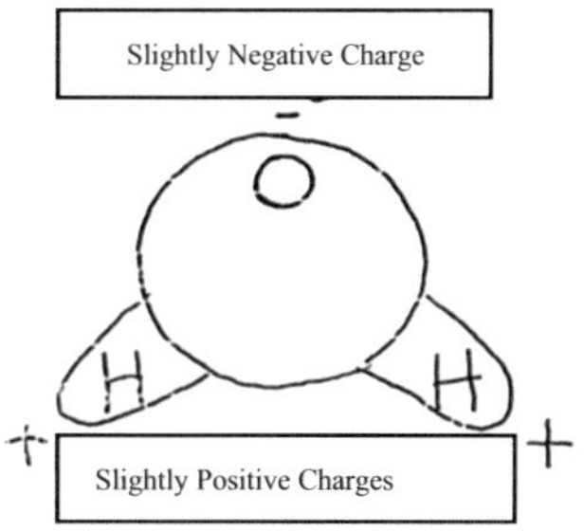

Each molecule of water is composed of two hydrogen atoms and one oxygen atom. The hydrogen atoms bond to the oxygen atom asymmetrically by sharing electrons (each hydrogen atom shares its only electron with the oxygen atom. The oxygen atom receives the two electrons needed to complete its outer shell, making it a stable molecule). Important interactions occur because of the electron sharing. The oxygen atoms, in adjacent molecules of water, tend to draw the electrons furnished by the hydrogen atoms, closer to its nucleus, creating an electrical separation and a polar molecule. The polar nature results in the hydrogen end of the molecule having a positive charge, attracting the oxygen end (with a negative charge) of adjacent water molecules. This forms hydrogen bonds between adjacent water molecules. These bonds are weak compared to the electron sharing bonds (6% as strong) and are easily broken and reformed. The hydrogen bonding and polarity of water molecules is responsible for many of the unique characteristics and physical properties of water.

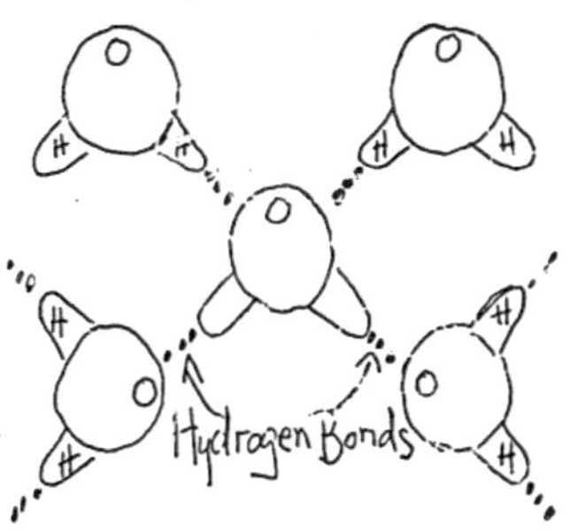

- If not polar, water would be a gas at room temperature and have an extremely low freezing point making life impossible.
- At the air-water interface, the sticky polar nature of water allows it to form a skin over the water surface, strong enough to support small objects. This phenomenon is known as surface tension, and water has the highest surface tension of all common liquids.
- Water has a great capacity to hold heat energy with the highest heat of vaporization of most common substances (thus a high boiling point – allowing it to be a liquid on the surface of the relatively warm Earth). When water evaporates, it absorbs considerable amounts of heat.
- Water has a high latent heat of fusion; when ice is formed, considerable amounts of heat energy is released. Water therefore acts as a buffer against temperature changes and keeps the Earth's climate from fluctuating rapidly.
- When water freezes, it becomes less dense – hence, ice floats (a lucky thing as if it were not so, the oceans would be frozen solid.
- Possibly most important for the chemical processes of life, water is a universal solvent. It has the ability to dissolve more substances than any other liquid (due, once again, to its polar nature and hydrogen bonding). When in water, salts turn into their ions. This allows many free radicals to be available to the chemistry of life.
- Water is very dense, some 800times denser than air. The density allows large and small organisms to float along effortlessly for long periods of time (compared to land, where terrestrial organisms must fight gravity with each step in order to move around).
- Water absorbs light differently. The red end of the light spectrum is absorbed in shallow water while the blues and greens penetrate the deepest (important for plants because different plants use different part of the light spectrum for photosynthesis, and the differential absorption can the vertical distribution of marine plants).

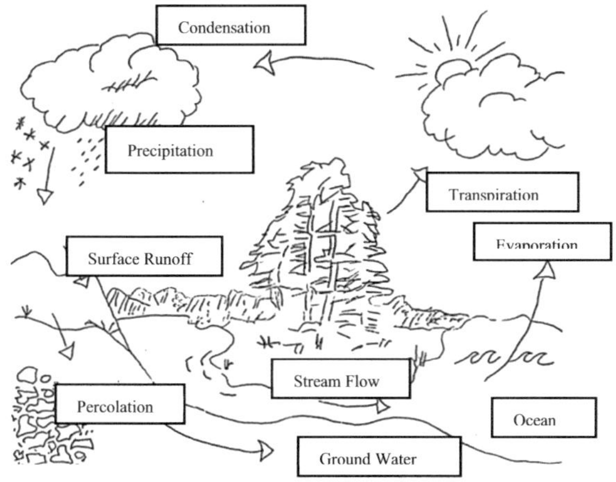

"The water cycle is an endless process by which Earth's water is collected, purified, and circulated from the atmosphere to plants and animals, and back to the atmosphere" (Williamson and Smoak, 1998). Williamson and Smoak indicate that this cycle began billions of years ago, when the Earth cooled and that it is driven by the Earth's gravity and the heat of the sun. They inform us that the sun's heat causes the water to vaporize and then the Earth's gravitational pull causes condensation and eventually precipitation and that the water cycle will never end as long as there are plants and sunlight. The following information about water is from their book *Curious about Water?*

> Water evaporates from oceans and surface water into the atmosphere, where it condenses to clouds. From the clouds, it falls back to Earth as precipitation, and eventually returns to oceans through a drainage system of streams and rivers. The total amount of water circulating through the water cycle never changes, as it continually shifts from gaseous to liquid and solid states.
>
> After precipitation falls to Earth as rain, hail, sleet, or snow, it meets one of several fates. Our ocean gets about 8 percent of all precipitation. However, some water which falls on land may run across the surface of the soil (runoff) into lakes, ponds, rivers and streams. Water that remains at the Earth's surface is surface water. Precipitation may also soak in or infiltrate into soil. Plants absorb some of this water, and some of it seeps down through soil until it reaches a saturated zone. Water that reaches this saturated zone is groundwater.
>
> Vaporization occurs when water molecules evaporate into water vapor from liquid water. Some water may also change directly from ice to vapor through evaporation. This is a unique evaporation process called sublimation Evaporation also occurs when water enter the air as a gas. Once the vapor cools, condensation occurs and water fall to the Earth as precipitation – rain, snow, sleet, hail. As water evaporates, it leaves behind any dissolved or suspended solids, and is cleansed.

Mills (1986), reports that hardness of water is only relevant to freshwater systems "as seawater has so much mineral content that hardness analysis would be immaterial." In freshwater broodstock and larval rearing systems in *Beyond the Classroom Aquarium*, students will explore ways to identify and modify water hardness parameters. In general, water hardness is "a measure of the quantity of certain metallic ions present in the water, notably, calcium and magnesium. . . . To a much lesser extent, barium, strontium, iron, copper, zinc and other metallic ions have an influence on water hardness. These ions are generally present in three main forms: hydroxides, carbonates and bicarbonates, while smaller quantities of sulfates, chlorides, silicates, phosphates, and borates also exist" (Andrews, C; Exell, A; Carrington, N;, 1988). Water with high concentrations of these salts is referred to as "hard." Water with low levels of these salts is called "soft." The degree of hardness scale "dH" ranges between zero and 30. Water between four and eight degrees of hardness is called soft. Water between 18 and 30 on the hardness scale is considered hard. Most freshwater fish do best in water between four and 15 degrees of hardness.

- There is 20 – 30 times more freshwater under the Earth than in all the surface lakes, streams, and reservoirs combined Williamson & Smoak (1998).
- In gallons, the volume of groundwater is estimated to be 33 to 59 quadrillion gallons Williamson & Smoak (1998).
- Together there is a total of 369 quintillion, 820 quadrillion, 250 trillion gallons of water on Earth (369,820,250,000,000,000,000) Williamson & Smoak (1998).

In *Beyond the Classroom Aquarium*, students design, assemble, and maintain both fresh and saltwater aquatic ecosystems. Preparing aqueous solutions of proper salinities and reviewing literature to gain a fuller appreciation of the vastness of the saltwater world are important activities in many of the projects described here. A brief description of seawater is provided below as described by Buchheim (1998c).

> Seawater is pure water plus dissolved solids and gasses. The dissolved solids come from "weathering" processes of the continental land masses rocks being dissolved by rain water and flowing out to sea with the rivers. The gasses come from the atmosphere. As water is a universal solvent, many different compounds are dissolved in it. A one kilogram sample of saltwater contains 35 grams of dissolved compounds, including inorganic salts, organic compounds from living organisms, and dissolved gasses. The solid substances are known as salts and their total amount in the water is referred to by a term known as salinity (expressed as parts per thousand). Oceanic salinities generally range from 34 to 37 parts per thousand (ppt). Variations from place to place are due to factors such as rainfall, evaporation, biological activity and radioactive decay. Salinities are higher in the tropics due to high evaporation rates. Fresh supplies of salts are now

being added to the oceans from the rivers at roughly the same rate that they are being removed by various physical, chemical, and biological processes.

Inorganic salts compose most of the solid matter of the "salts" (99.2%). The percentages of seawater's inorganic salt compounds remain constant regardless of the waters salinity; therefore, salinity can be measured by just measuring the concentration of one of the salts, such as chlorine.

The remaining 0.72% "salts" are inorganic and crucial to life. These include phosphates, nitrates (required for photosynthesis) and silicon dioxide (required by diatoms to construct their glass skeletons). In contrast to other salts, the nitrates and phosphates vary in concentration due to biological activity. In surface waters, where plants are actively in the process of photosynthesis, the nitrates and phosphates can be in short supply, limiting the amount of biological activity that can take place.

Regardless of whether an aquatic ecosystem contains freshwater or some level of salinity, Moran, Goudy, and Hankin (2001) explain that "recirculating systems are a unique and environmentally friendly way to culture aquatic animals. This is because all environmental effects and impacts to the animals being reared are controlled. One must be aware of the importance of keeping the culture water clean to allow the fish to thrive. Many biological and chemical reactions are occurring in a recirculating system that can affect the fish." In view of this idea and the many processes students must understand to maintain an aquatic ecosystem once they have made it operational, it is important to consider their additional comments on maintenance of water quality in recirculating aquatic ecosystems.

Seawater's Inorganic Salt Compounds	Chemical Symbol	Percentage
Chloride	Cl^-	55.04%
Sodium	Na^+	30.61%
Sulfate	SO_4^{--}	7.68%
Magnesium	Mg^{++}	3.69%
Calcium	Ca^{++}	1.16%
Potassium	K^+	1.16%
Carbonic Acid	HCO_3^-	0.41%
Bromine	Br^-	0.19%
Boric Acid	H_3BO_3	0.07%
Strontium	Sr^{++}	0.04%
Total		**99.28%**

Buccheim (1988c)

When fish eat and metabolize, they produce waste as well as excrete ammonia, urine, and carbon dioxide through their gills. In order for the fish to thrive the waste needs to be removed from their water. Carbon dioxide can be removed by aeration or by plants through photosynthesis. Nitrifying bacteria break down nitrogenous waste. Plants can use the final form of nitrates present at the end of the nitrogen cycle.

In a recirculating system, the biological filter is the main component that allows for the culture water to be recycled; it is the home for nitrifying bacteria. Nitrifying bacteria are the good bacteria you actually **WANT** to grow in your system because it uses the waste products from the fish as an energy source. Nitrosomonas first breaks down ammonia to nitrite. Then, Nictrobacter breaks down the nitrite to form nitrate. Ammonia and nitrite are dangerous forms of nitrogenous wastes because they interfere with the body's normal functions and exchanges. Nitrate, which is the end product of these reactions, is the least toxic form of nitrogenous water and has a minimal effect on the fish at moderate concentrations. Water exchange is needed (approximately 10% per day) in a well-seeded recirculating system to rid it of excessive amounts of nitrate. Alternatively, a polyculture that include plants or shellfish can be used to remove nitrates.

A variety of water filtration techniques and media are available. Filtration media is categorized by the amount of surface area that it has. Surface area of the media determines the space available for nitrifying bacteria to colonize and grow upon. "Types of media are often determined by the space available for the system and the amount of fish being grown. Once a system is functioning for a few months, bacteria will begin to colonize the biofilter media and on the inside of pipes of the system" (Moran et.al, 2001).

Bacteria require oxygen to breakdown the ammonia and nitrites. Oxygen demands of the system rise as fish and bacteria grow. You do not want algae growing inside your tanks because these algae will consume the oxygen within your system. The dissolved oxygen (DO) level needs to be monitored within your system to be sure that there is enough oxygen available for your fish and nitrifying bacteria. You should maintain a minimum of 80% saturation of dissolved oxygen in your system at all times. To prevent algae from growing in your system, do not let direct sunlight come in contact with any part of the system. Low dissolved oxygen levels in your system can stress the fish and inhibit their survival and growth rates.

. . . There are several measurements that need to be monitored along with dissolved oxygen (**DO**) to ensure a healthy environment for your fish. Here is a list of the parameters that need to be tested in a saltwater system. There are several types of chemical test kits and probes that are available to monitor the water quality parameters.

Much of the following water quality information about pH, ammonia, nitrite, nitrate, dissolved oxygen, alkalinity, carbon dioxide, salinity, and temperature is available from a range of common sources. (Moran, et. al., 2001) provide it as follows:

Safe pH levels: 6.0 to 9.0; pH measures the intensity of the acidity of the water. A normal reading for saltwater is between 7.5 and 8.0: pH is affected by other water quality parameters such as alkalinity and CO_2 levels. When pH is high, it can limit the ability of gills to transport ions essential to the fish, which can lead to osmoregulatory failure or death.

Ammonia: Safe levels: 0.0 to 0.1 mg/L. Ammonia is the result of fish excreting waste and metabolizing feed. It is toxic to fish. Levels that remain high (+5ppm) for a long or short period will stress the fish and make them more susceptible to other pathogenic bacteria and viruses present in the system. Low levels of ammonia (1ppm) that persist for a long time can have an overall negative effect on the fish as well. Fish will become stresses form the toxic ammonia in their water, which will allow them to be more susceptible to acquiring a disease.

Nitrite: Safe levels: 0.0 to 1.0 mg/L. Nitrite is a toxic form of nitrogen. It is the result of the nitrifying bacteria Nitrosomonas breaking down ammonia. When it is present in high concentrations, it will stress the fish and allow them to be more susceptible to pathogenic bacteria and viruses present in the system. Also, when nitrites oxidize the iron in the blood, it enables the blood to bind oxygen and carry it to the tissues.

Nitrate: Safe levels: Up to 10 mg/L. Nitrate is the end result of the conversion by the nitrifying bacteria Nitrobacter and is considered to be non-toxic to fish. If nitrates are continually present in an overabundant amount (+150 mg/L), osmoregulatory failure can occur in the fish.

Dissolved Oxygen: Safe levels: 7.0 to 9.0 mg/L. This is the amount of oxygen available to the fish in the water. This is an important parameter. Low **DO** concentration levels can have adverse effects of fish health including respiratory stress, tissue hypoxia and eventually death. The saturation point of the water is dependent upon the temperature of the water.

Alkalinity: Safe Levels: 90 to 150 mg/L. This is the measurement of the total concentration of basic substances dissolved in the culture water. Alkalinity serves as a buffer to the culture system and does not allow wide swings in the pH to occur.

Carbon Dioxide: Safe Levels: 0.0 to 8.0 mg/L. CO_2 in tanks is present due to fish respiration and metabolism. Adverse effects of carbon dioxide are affected by temperature and **DO**. CO_2 toxicity will increase when **DO** is low. Increased water temperatures decrease CO_2 toxicity by decreasing its solubility. Also, if the fish become stressed, they respire at a higher rate which produces more CO_2. If the level of CO_2 in the blood rises, the oxygen carrying capacity will decrease causing respiratory distress because little oxygen is being transported to the tissues.

Salinity: See life history requirements for the culture of fish. Salinity should be within 2 ppt of the desired number.

Temperature: The temperature of your system should be maintained at $\pm 1^{\circ}$ of the temperature required for the fish that was found in the research about the chosen species. The use of a heater or chiller may be necessary to maintain the desired temperature. Fluctuations of temperature will stress the fish and allow them to be susceptible to pathogenic bacterial and viral infections. Having temperature control over the whole culture room is important as well. A temperature change of more than 5 degrees in a period of 2 hours can cause fish health problems as well as hinder their growth due to stress. Continuous temperature changes to the culture water should be avoided because the fluctuating temperature will continuously stress the fish.

Moran et.al (2001), describe procedures to follow when water quality tests are higher than recommended.

> The best thing to do is to replace more than the typical 10% water in the system. Depending on the magnitude of the of the high test results, up to 50% of the culture water can be changed at one time to lessen the effect of the outlying parameter. Be sure to have enough storage of water (at the correct temperature & pH) for emergency water exchange situations. They will happen! Common problems in new systems will be high initial ammonia loads and low DO. If parameters are high, continue to test the water daily until the parameters are acceptable for more than 3 days in a row. Do not replace more than 50% of the original water per day. If you do this, you are eliminating the source of ammonia for the bacteria and risk a change in your system's temperature if the storage and culture tanks are not maintained at the same temperature.

After describing protocols for setting up a school-based hatchery system, Moran et.al (2001) further explain that "recirculating systems are the most environmentally friendly method of aquaculture. This is because only a limited amount of water is needed and that water is recycled by the system's components. Only a small amount of water replacement is needed in a functioning system on a routine basis to control evaporation and nitrogen levels. Recirculating technology is a great technique to be used in any situation where there is a minimal source of available quality water and there is a need for minimal discharge from classrooms and demonstration facilities."

They report that the system components your students set up will be capable of helping them "focus on each of the bottlenecks that occur during larval rearing: egg holding, hatching, and rearing larvae through metamorphosis to juveniles. Raising food for the newly hatched fish, weaning fingerlings onto dry foods and raising the fingerlings to market size will also be learned." Moran et.al (2001), report that "recirculating systems need to have numerous components to allow for the maximum reuse of culture waters." Their discussion of the components of a recirculating system that provide for maximum reuse of culture water include a description of: the tank, weir box, mechanical particle filtration, pump, chiller, foam fractionator, biological filtration, UV treatment, degassing column, aeration for the system, and storage for saltwater. Their presentation is provided below.

> Tanks provide a living environment for the fish, eggs/larvae/juveniles. Tanks can be any shape and size depending on the space available. Circular tanks are best because they encourage good water circulation and waste removal. Tanks must be dark in color (preferably black). The tank walls need to be smooth for easy cleaning and minimize the amount of available area for bacterial growth. A central drain in the bottom of the tank is needed for the waste to be flushed out of the tank. The size of the drain screen should be small enough to keep the larvae or fish in the tank and large enough to let out waste and debris.
>
> A weir box or settling box is used to collect larger pieces of waste (excess feed, feces). The water flow slows down in this box which allows the larger particles to sink to the bottom. This box reduces the amount of particles to enter the filter bag. This is important in a large system with a lot of waste products.
>
> For larval rearing systems especially, mechanical particle filtration helps filter excess feed out of the tank water so the water can be reused. For hatchery systems, the filter must be capable of filtering the water down to 25 microns to ensure proper water quality in the sensitive larval stage. A felt or fabric bag that can filter is ideal for larvae that are feeding on live foods. Once the fish are feeding on dry pellets, the filtration can go up to 50 microns. It is important to design the system so that the filter is easily accessible because it will need to be cleaned on a routine basis.
>
> A pump is needed to drive water through the system. The pump is always located after the mechanical/ particle filtration so that the water passing through is free of particles and will not clog the pump.
>
> A chiller is used to control the temperature of the system. Chillers can be plumbed into the system or a titanium chiller coil can be placed in the bottom of the filter box.

"A foam fractionator removes dissolved organics, proteins and small particulates out of the water. Water flows down the column as air moves up through the water column. As the air bubbles rise to the top of the water column, the particles and proteins bound to the bubbles produce" concentrated foam that is collected and removed. "The foam can be directed to exit the foam fractionator at the top by creating a directed opening such as an elbow" (Moran et.al, 2001).

[*The Marine Aquarium Reference* (Moe 1992b, pp. 71 & 72) features a section that includes directions on making a homemade protein skimmer from PVC pipe (p. 71 & 72)]. Students may be guided to explore ways to document beneficial effects that variously sized and maintained protein skimmers have upon water within marine aquatic ecosystems.

In order to filter water most completely, it is intended that water leaving the foam fractionator be free of all particles. Moran et.al (2001) indicate,

> once all particles are removed from the water being filtered, the water can be pumped into a biological filter. This component of the system should be scaled to the amount of fish you are holding in your system because its purpose is to remove the ammonia and nitrite created by waste of the firsthand excess feed. A variety of plastic materials are available to be used as bio-media. The bio-media provides surface area for the nitrifying bacteria to populate. The more surface area available for the nitrifying bacteria, the more ammonia it can break down.

According to Moran et.al (2001), "ultraviolet light (UV) treatment assists in the control of 'bad' bacteria, and viruses in the system. Water passes beside a light emitting UV light which minimizes the amount of bacteria and viruses present in the water."

A degassing column "is used to strip excessive buildup of carbon dioxide and nitrogen that occur in a recirculating system out of the culture water. The water enters the top of the column and bounces down through the large plastic media that is inside the column. As the water bounces through, it comes in contact with the air and the excess carbon dioxide and nitrogen dissipate into the air. The water becomes properly oxygenated and can be distributed back into the system" (Moran et.al, 2001; p.16).

Moran et.al (2001; p. 16) also report: "each culture tank and biofilter as well, needs an airline with an air stone or perforated ring attached. Aeration is important to create an upwelling motion in the culture tank for the eggs and larvae. Also, it helps maintain the system's dissolved oxygen level and agitation in the biofilter media." They suggest and report that students check the local "water treatment facility to determine if they treat the water with chloramine. Aerating tap water for 24 hours will remove chlorine from treated water – but not chloramine. To rid water of chloramine a water additive must be added to the water. Chloramine removing additives can be purchased from most stores that sell water quality test kits."

After treating the water to remove chlorine or chloramine, Moran et.al (2001; p. 16) recommend that students "mix in sea salt to make up saltwater for your marine hatching system. One airline and air stone per storage tank is sufficient. Any plastic container such as a sterilized garbage can or vat can be used. Storage tanks should be able to hold at least 50 gallons of freshwater or mixed saltwater."

They also explain why a hydrometer (salinity meter) is a necessary piece of equipment; it "has the ability to read salinity to the nearest 1.5 part per thousand (ppt). A hydrometer measures the specific gravity of seawater and tells us what the salt concentration is in the water. Open ocean seawater is approximately 34 ppt (or 3.4%), but many fish can live in or tolerate higher and lower salinities" (Moran et.al, 2001; p. 16).

Other basic components include a thermostatically controlled heater, a thermometer, a lighting timer, and a gravel vacuum or gravel hose.

After visiting a municipal water treatment facility my students were able to differentiate between four types of water filtration. I worked to guide them to appreciate that applying the appropriate combination of filtration technologies can be as critical to the well-being of a classroom aquatic ecosystem as it is to the well-being of a municipalities residents. There are thousands of pages of literature that explain the principles that underlie the water filtration and the various techniques that have been developed to do so.

In general the four technologies may be listed as involving processes that fall in to one of these four categories; mechanical, chemical, sterilization, or biological.

The process of mechanical filtration is one in which objects that range widely in size (from tree trunks to microscopic particles) are progressively (by size from large to small) screened or sifted out of a body of water. The mechanical filtration process is similar to the basic idea of "physical changes in matter" to which chemistry students are introduced early on in their study of chemistry. For some bodies of water a thorough mechanical filtration process will be enough to make the water clean and safe enough to use. All too often however, this is not the case and mechanically well-filtered water that appears to be clean and safe enough to use is not. It is important to know that many toxic particles in water cannot be removed through the process of mechanical filtration. In many cases water may not be clean and safe enough to use until after it is filtered chemically (and biochemically) and in some cases also sterilized. In this curriculum students must be guided to appreciate the idea that when water is well-filtered mechanically, additional filtration technologies become much less difficult to employ for the purpose of filtering water so that it is in fact absolutely clean and safe to use.

A number of technologies have been developed to remove the majority of chemical and microbiological impurities from water. Chemical and microbiological impurities are difficult to remove because they are either molecularly bonded to molecules of water or too small to remove inexpensively by using mechanical filtration methods. In order to remove these aquatic contaminants, technologies that disrupt molecular bonds between a water molecule and a toxic chemical must be broken and/or the life processes of small microbiologic organisms must be disrupted. As students design their aquatic ecosystems and consider their options for filtration of system water they may be guided to explore ways that chemical filtration technologies like activated carbon, protein skimming, chlorination, and other such treatments may be used to make water cleaner and safer to use. Chemical filtration technologies can provide many students with good topics for their project nine research activity.

Once student designed aquatic ecosystems are operating in the classroom, the recycled water in these systems must be filtered continuously to counteract the toxins and organic waste products that are introduced to the classroom systems every day. Although the classroom aquatic ecosystems students design and assemble must provide for ways to gently and thoroughly filter recycling water – mechanically and chemically; after initial system set-ups are complete, as systems mature and become established, students must also ensure that biological filtration technologies are operational. Students must ensure that each classroom system also allows recycling water to be continuously filtered biologically through the use of bacterial/biochemical activity. Despite the fact that biological filtration of recycling system water is a necessity this filtration process does create unavoidable problems for ecosystems in which students do not regularly monitor system water quality data and carefully perform partial water changes based on standard water quality data criteria. Topics related to the nitrogen cycle can also provide many students with good topics for a project nine research activity. A discussion of the role to the nitrogen cycle and the necessity of biological filtration in aquatic ecosystems is probably described in every aquarium book written in the 21st century. I believe students should be guided to develop an explanation of the nitrogen cycle, how to monitor the biological filtration process in their systems, and how they may use partial water changes to sustain their systems. By guiding their reading and subsequent classroom discussions students should be capable of listing between five and ten factors that are commonly considered influential in limiting or reducing the carrying capacity of a system and describing – in detail – at least three symptoms commonly observed in systems that are overcrowded.

As students learn to assess water quality in project three, they will be encouraged to use the tools at their disposal to thoroughly document the nitrogen cycle in their newly established systems. For the project nine aquatic ecosystem research projects, some students may choose to study the impact of live rock, live sand, macroalgae and aquatic plants, and other substrates on the processes of nitrification and de-nitrification in aquatic systems. Research topics may also include applications of the processes of properly curing live rock and live sand. Additional forms of biological filters are described below and involve filtration techniques of which all students should be aware before they design water filtration systems for the three primary systems they will be developing throughout project three. These additional biological filtration techniques include but are not limited to under-gravel, wet-dry, compound, hang on the tank, trickle, bio-wheel, fluidized bed, and algae and de-nitrification filters.

In order to continuously filter recycling water mechanically, chemically and biologically, students will incorporate a combination of filtration technologies in each system they design and assemble.

Under close supervision, in a well-planned field trip workshop setting, some aquatic ecosystem design and assembly projects may involve students in the hands-on experience of drilling a hole in a glass-walled aquarium (to establish a reliable plumbing connection between the display section of the system and the "sump" filtration area of the system). However, I can imagine some educational settings in which it will be better to decide that drilling a hole in a tank to provide internal drainage for a trickle filter is not desirable. In these instances educators may go to any local aquarium shop and select from among a number of easy-to-use and inexpensive compound filters, that hang on the tank and include mechanical, chemical, and wet-dry biological filtration. Respective shop owners can usually provide all necessary directions to follow for operating such a store bought compound filter.

Another common and widely used biofiltration technology is called the biowheel filter. Biowheel filters are a type of trickle filter that actively rotate a circular media drum through a stream of water being pumped from the tank and over the rotating circular media. Many students may wish to study bio wheel technology and either design and assemble their own or purchase one of many good bio-wheel filters that are available at many local pet/aquarium shops.

And yet another type of filter to which students may be introduced for consideration as a design project might involve assembling and studying the filtration effectiveness of a "fluidized bed filter." These filters consist of a vertical tube that contains fine sand. Water is pumped into the bottom of the tube. The upward flow through the tube creates a rotating mass of water and sand particles. All the sand grains in the filter are exposed to oxygenated water and the grinding movement of the sand also breaks up detritus.

And still another biological filter for students to consider for study and use in a classroom system is called an algae filter, algae scrubber or by some a living machine. Information on this filtration methodology is extensive. Adey and Loveland (1991) describe many aspects of a system they developed for use in a Smithsonian Museum exhibit. Descriptions and explanations of algal filter technologies abound on the internet and students who decide to design and assemble such a filtration system will have opportunities to collect many interesting kinds of data and perform fun and interesting research for project nine. Some students who work with (or research) algae filters may wish to design, assemble and study the operational parameters and effectiveness of what is commonly known as a reverse algae filter. By setting up parallel model systems students may study differences in water quality parameters that may occur when the algal filter on one system is kept in the dark during the day (when the display tank is well lit) and illuminated at night, while the algae filter lighting on a second model go on/off when lights in the display tank are switched on/off. Or, for a project nine research project some students may wish to compare trickle filter performance with filtration performance of "live rock" or an algal filtration system.

De-nitrification filter technology is usually irresistible to a few students, and this technology provides rich opportunities for exploration and research. Moe (1992a) indicates the basic principles upon which de-nitrification filters are based. The information he provides should be enough to get students who want to design, assemble and analyze the performance of such a system working in the right direction.

Ion exchange and ozonation filtration methodologies are interesting however these methodologies may usually be reserved for advanced independent study projects and their use for most classroom aquatic ecosystem settings developed in this curriculum are not usually warranted.

The process of water sterilization by ultra-violet radiation or other technologies may become a project design for advanced systems and students but will generally be beyond the introductory one year scope of this curriculum.

With additional classroom aquatic ecosystem experience students may be guided to learn ways to design aquatic ecosystems in such a way that technologies to filter water biologically are always supplemented with mechanical, chemical and sterilization filtration technologies.

The Aquaponics System Water Containment and Flow Design
(Field Trip #2 Applications)

Photo by Robert Jakus

The work performed at this point in project three is an important lead-in to completion of the aquaponics system in project four. Without completing the project three aquaponics system plumbing work described here, a fully operational aquaponics system will remain incomplete. Aquaponics system design and assembly work in projects two and three make it possible for students to complete the assembly of an aquaponics system in project four and to learn about collection and analysis of water quality data that is necessary for the successful daily operation of an aquaponics ecosystem. In this section of project three, students work to set up aquatic ecosystem habitat spaces and create plumbing connections between the aquaponic system sump, growing pipes, and fish habitat.

The first step in the process of incorporating water containment and flow components into the aquaponics system frame—to which lighting system components have been attached—requires students to review aquaponic system water containment and flow information found in this text (and other sources). Following this review, and before going on the second field trip, students are given the assignment of developing a report that demonstrates their understanding of the knowledge, skills, materials, and equipment that development of this systems water containment and flow components require. Reports about aquaponics system materials should contain dimensional details about PVC pipe sizes, connectors, placement, and cuts (and safely using PVC cement). In addition, information about flexible hose dimensions and the use and placement of main and ancillary pumps, filtration equipment, and materials should be included in their reports. Reports should conclude with a team developed proposal that gives detailed dimensions, design layouts, and system modifications that students would follow if their water containment and flow design were to be implemented. Aquaponic system designs that students propose for development must also address water quality needs of the organisms that will inhabit the system. The proposals must provide explanations as to how the water containment and flow design is capable of sustaining high water quality standards. Finally, the most successful reports will also provide strategies that may be used so that assembly and setup of the water containment components of the system can be performed separately (and simultaneously when possible) with the assembly and setup of water flow components of the system.

The class is then guided through a critique of these reports. Following the critique, students vote for the aquaponic system design they think will best meet the needs of the class. Before going on the second field trip, it is recommended that students (or the teacher) share specific class design plans with the individuals who will be demonstrating design and assembly protocols to the students during the field trip. Students are asked to acquire contact information for tradespersons with whom they are familiar and whom they might invite to donate guidance, support, or possibly further demonstration of tools, equipment, and materials students must safely work with as they develop a water-containment and flow system. This information should be added to their reports. Many students find their motivation to assemble high quality aquatic ecosystems increase when skilled tradespersons and parents contribute their time, knowledge, and support.

Teachers may develop their own Aquaponics System Water Containment and Flow Report or use the one provided below to guide student efforts to identify information to include in their reports.

Photo by Robert Jakus

Aquaponics System Water Containment and Flow Report

	Materials Needed	Design and Assembly Steps AND Skilled Tradesperson	Function Performed by this Piece	Safety Concerns
Fish Habitat				
Sump				
PVC pipes				
Drain from fish habitat to sump				
Connect pump from sump to top PVC pipe				
Drainage from pipe to pipe				
Drain lowest PVC pipes to fish habitat				

From this point forward in project three, student activities transition to those that require direct hands-on work with equipment and materials to implement plans critiqued and approved by the class to install water containment and flow components in their classroom aquaponics system frame.

For the example system described here, an eight foot diameter pool 18 inches deep (564 gallons) was set up on two 4 x 8 foot tables with 4 x 8 foot pieces of plywood placed across their surface and which was then covered by a waterproof tarp. The height of the water surface in the pool rose to about 48 inches, so the heights of the two tables were 32 inches, and the water depth was allowed to rise to 16 inches; thus, the volume of the fish habitat was about 500 gallons. A hole was drilled in the side of the pool at 16.75 inches. When water in the pool rose to this level, it was channeled to the aquaponics sump via a 1.5-inch PVC pipe. A screen was placed at the location of the drain from the pool to the sump to prevent fish, fish food, and larger particles in the pool to enter the sump. The 58-gallon rectangular plastic bin "sump" was placed between two 6 x 3 foot tables that were 36 inches high. The sump was raised (by means of bricks) so that the upper lip of the bin was above the surface of the two tables. A hole drained in the lip of the sump permitted water to flow down the PVC drain and pipe from the fish habitat to the sump. In addition to providing a good spot for the sump, these two tables also provide space upon which to place the A-frame.

Photo by Robert Jakus

Five (4-inch diameter) 5-foot long PVC pipes were attached to the A-frame, and water from the sump was pumped to the highest pipe on the A-frame. Water then drained to the lower pipes finally exiting the 5-foot long growing tubes on the A-frame by joining a PVC pipe extending from the A-frame to a point above the center of the pool. PVC cement prevented many leaks. The length of the pipe that was suspended over the pool sprinkled water on the surface of the pool as it returned to the fish habitat. Additional aeration of water was provided in the sump and in the pool. The pump in the sump was on a timer, allowing water to cycle through the system for 15 minutes every hour. At times, pumps and various filters (including a UV filter) were used in the pool to provide additional water circulation and filtration.

Fill the aquaponic system with water, turn pumps on, and let them run for the day. Turn the pumps off at the end of the day. If no leaks are found by the next day, allow the pumps to continue operating and make final adjustments before moving ahead.

Before installing an appropriate water containment and flow system in the coral propagation system frame, students are now introduced to laboratory techniques for identifying levels of oxygen dissolved in the water and for determining an aquatic ecosystem's biological filtration needs. This knowledge is needed in order to provide organisms with the healthiest aquatic environments possible and to maintain the life supporting capacities of the water in aquatic ecosystems.

Photo by Robert Jakus

Aquatic plants and animals require certain minimal levels of oxygen in their water to survive and grow. Consequently procedures for monitoring and measuring dissolved oxygen (DO) in classroom aquatic ecosystems are essential for students to understand and perform. In this activity students learn proper protocols for collecting water samples from which they will determine the volume of DO in their samples. They will be introduced to two methods for determining the amount of DO in water (the Winkler method and through the use of a DO meter). It is important that all individuals participating in this activity are aware that the corrosive chemicals concentrated sulfuric acid and alkaline iodide-azide must be handled carefully. Protective eyewear is mandatory. Glassware such as BOD bottles must always be handled with caution. At the conclusion of the activity students should wash their hands to remove contaminants they may have come in contact with during the activity.

While introducing this activity, students should be made aware of the commonly understood idea that the water samples they collect (to be tested for DO) must not be shaken or mixed with air when collected. Shaking sample containers often aerates the sample. DO readings from water in aerated sample containers cannot be used to accurately represent the DO level of water in the system from which it was taken. In addition, water samples with air bubbles must be avoided because atmospheric concentrations of air in an air bubble can dramatically raise oxygen levels of water in a sample bottle.

The following DO protocol is a compilation of procedures presented by Rounds et.al. (2013), and in an APHA (2005) document in which a standard procedure to follow to evaluate DO in a water sample using the Winkler Method is explained.

The Iodometric (Winkler) method is recognized as a reliable standard for producing accurate results when correctly implemented.... It is widely accepted in the scientific community and is used to check the calibration of, and the measurements made with, electrometric DO instrument systems. (p. 36)

For students with minimal chemistry background it is preferable to have them use reagents available in pre-measured pillow packets from commercial suppliers as they conduct a DO water analysis. For students with some chemistry background the teacher may consider it valuable to have students themselves identify and perform the protocols necessary for preparing the reagents used in this Iodometric method of DO analysis, as described in American Public Health Association (2005).

Equipment and supplies needed for the Iodometric method are (p.37)

Beaker, 2,000 ml, glass or Teflon (to contain sample water that will be placed in BOD bottles)	Chemical reagents • Alkaline iodide-azide reagent • Manganous sulfate reagent • Sulfamic acid granules • Sodium Thiosulfate, 0.025 N titrant
Bottles for biochemical oxygen demand (BOD) analysis, glass stoppered, 300ml	Starch indicator solution
Stirrer, magnetic	Clippers for opening reagent pillows
Stirring bars, Teflon coated	Appropriate safety gloves, glasses, and apron
Cylinder, graduated, 250 ml	Waste disposal container
Flask, Erlenmeyer, 250 ml	White background sheet
Burette, 25-ml capacity with 0.05-ml graduations and Teflon stopcock	Deionized water (maximum conductivity of 1 u(cm)
Burette, support stand	Bottle, squeeze dispenser for deionized water
Burette, clamp, double	Thermometer, calibrated
Pocket altimeter-barometer, calibrated, or DO-measurement equipment that incudes barometer	

Collect the sample and perform the titration described below on at least two subsamples to provide the appropriate quality control. When filling the BOD bottles, a minimum of three bottle volumes of sample should pass through the bottle to collect the final volume (p. 38).

Using a hand-held method to fill a 2000ml beaker, fill the beaker from the water body by tilting it slightly to allow the beaker to slowly fill in a manner so as to avoid turbulence, bubbling, or otherwise entraining air. Keep the filled beaker submerged (in the surface water body) for about 30 seconds. Next, while the beaker is submerged, insert a stopper firmly in the beaker (taking care not to trap air bubbles).

Pour off excess water that is trapped on the lip of the stoppered beaker.

Carefully fill two clean BOD bottles with three or more bottle volumes of sample water from the beaker, taking care not to introduce any air bubbles by slowly overflowing the bottles adequately to remove any tapped air bubbles (p. 39).

Determine the DO concentration of the sample water in each BOD bottle as follows:

a. Add one each of the following dry reagent pillow packets
   - Alkaline iodide-azide (white powder)
   - Manganous sulfate (pinkish-colored powder)
b. Recap the bottle to prevent air bubbles from being trapped in the bottle
c. Invert the bottle 25 times or more to completely dissolve the reagents.
   - An orange-brown flocculent indicates the presence of DO
   - Allow the flocculent to settle halfway down the bottle (approximately 5 minutes)
   - Invert the bottle 25 times again; let the flocculent settle again until the upper half of the solution is clear
d. Add one reagent pillow of sulfamic acid (laboratory/classroom-prepared reagents might be prepared instead, depending on data-quality requirements – and student preparedness)
e. Recap the bottle without introducing air or air bubbles. Invert the bottle 25 times until all of the flocculent and granules are dissolved, leaving a yellow color.
f. Fill a clean 25-ml burette with 0.025 N (*Normal*), sodium thiosulfate titrant. Remove any air bubbles from the delivery tube beneath the stopcock and zero the meniscus.
g. Using either a clean 200-ml pipet or a 200-ml volumetric flask, measure 200 ml of the sample and dispense the sample into a clean wide-mouth Erlenmeyer flask.
h. Place the flask on a magnetic stirrer. Carefully insert a clean Teflon stirring bar and stir the sample at a moderate rate without aerating the sample.
i. Add increments (drops) of sodium thiosulfate titrant until the color turns pale straw-yellow.
j. Add 1 to 2 ml of starch indicator solution. (This causes the sample to turn dark blue.)
k. Very slowly add more sodium thiosulfate titrant until the sample just turns clear. (A white background behind or below the flask will help to see the color change.)
l. Record the volume of sodium thiosulfate titrant used, in milliliters.
   - For a 200 ml sample, the volume of titrant added is directly proportional to the amount of DO in milligrams per liter.
   - The formula to calculate DO for a sample volume greater or less than 200 ml is:
     DO (mg/L) = {(200) x (titrant added [in ml])}/(sample volume)
m. Record the DO value. Rinse the equipment thoroughly with deionized water.

Photo by Robert Jakus

Results of two Iodometric titrations should agree within 0.1 mg/L. If they do not agree, repeat the titration on one or more additional subsamples until this quality-assurance criterion is met.

Measurement – Digital Dissolved Oxygen Meter Method

Obtain a commercial DO meter read the directions, calibrate the equipment and collect DO samples from two or three samples of water from each classroom system. Average the respective result s of each system's DO meter readings. How do digital Dissolved Oxygen meter readings compare with the Winkler method D.O. results for each classroom system?

*** It is important that students performing the activity on this page have sufficient skill and background understanding of the water quality and analysis activities presented here (including calibrating a digital DO meter, to effectively lead students in other groups so that they can also understand the techniques and perform them (per the "Sharing of Knowledge and Accountability" activity on page 121).**

In this activity, students are introduced to the procedures they must follow to quantify concentrations of the nitrogenous compounds, ammonia and nitrite in their classroom aquatic ecosystems. One of the many important skills that all successful aquatic ecosystem designers must acquire is the ability to determine surface area requirements of an ecosystems biofilter so it will be sized to accommodate the desired fish biomass. In this activity, students are guided to calculate the surface area requirements for a biofilter.

Once the aquatic class/lab is properly set-up and all students demonstrate that they conform to the aquatic class/lab safety criteria regarding clothing, jewelry, and hair and are also prepared to properly dispose of potentially dangerous equipment and materials as necessary student teams (or pairs) may perform this activity – under supervision and guidance.

To begin, students in each group must read the directions (with guidance when necessary). By doing this they get an idea of how many steps they will be required to take and also more easily inventory the materials with which they will need to work. Each group should also obtain a data sheet and collect the necessary class/lab equipment from the designated supply station. In addition to a classroom aquatic ecosystem, each group should be sure to have a water quality test kit, nutritional information from the food(s) that is given to organisms in the aquatic ecosystem (this is provided on the label of each package), a Celsius thermometer (or C = 5/9[F – 32]), a balance, and in some cases (entirely up to the teacher… a calculator).that reports the to measure concentrations of ammonia and nitrite in water.

The steps students must follow to complete this activity require them to collect the following data.

	Write which system water in this report was taken from	
Step 1	**Ammonia** in the system (in parts per thousand)	
	Nitrite in the system (in parts per thousand)	
Step 2	**Percentage of crude protein** in the feed being given to the organisms in the system	
Step 3	**Temperature** of system water in degrees **Celsius**	
Step 4	Estimate the **biomass in kilograms** of organisms in the classroom aquatic ecosystem (an average sample weight may be used if the system contains many organisms)	
Step 5	**Feeding Rate** (calculate this by dividing the amount of feed put in the system water each day – in kilograms, by the biomass of the system)	
Step 6	Calculate the **Ammonia Production Rate** (use the ratio that says 1 Kg of protein generates 56 grams of total ammonia nitrogen TAN) by determining the proportional amount of total ammonia nitrogen that is produced by the amount of crude protein that you put into your system each day (in grams of ammonia per day).	
Step 7	Use this formula to determine the **total ammonia nitrogen** in the system – based on your daily feeding rate by multiplying - **Ammonia** in the system (step 1) by the **feeding rate** (step 5) and then multiply this product by the biomass of the system (step 4).	
Step 8	Use the formula: 0.0527 x (Temperature – step 3) – 0.0391 [in g/m^2] To determine the ammonia removal (or nitrification rate) (This formula is from Hanes et. al., 2002)	
Step 9	Multiply nitrification rate (step 8) by Total Ammonia Nitrogen (step 7) to determine the surface area requirements for your systems biofilter in m^2.	

Specific surface area comparisons for different substrates

Particle Size			**Specific Surface Area**	
Media Type	Inches	mm	ft^2/ft^3	m^2/m^3
Medium sand	0.12	3	270	886
Stone	0.33	-	19	-
Pea gravel	0.57	14.5	85	280
Rock	1	25	21	69
Large Rock	4	102	12	39
Plastic biofilter media	1	25	85	280
Plastic biofilter media	2	50	48	157
Plastic biofilter media	3.5	89	38	125
Revolving Biocontactor (**RBC**)	-	-	200	-

* estimated to be approximately that of small diameter plastic biofilter media

What media does your classroom aquatic ecosystem design require? How will you incorporate this into your design?

____

What does the calculation in step nine indicate about the filtration requirements of the classroom aquatic ecosystem being studied?

____

Does this activity suggest modifications to the design of the classroom aquatic ecosystem filtration system used in the activity presented here? If so, what kinds of design modifications might be good to include? (Hint: consider steps 1 – 8)?

____

*** It is important that students performing the activity on this page have sufficient skill and background understanding of the water quality and analysis activities presented here to effectively lead students in other groups so that they can also understand the techniques and perform them (per the "Sharing of Knowledge and Accountability" activity on page 121).**

**** This activity as presented here is a modified version of the "Calculating Bio-Filter Surface Area Requirements" activity as developed by Hanes et.al. (2002).**

***** The table contains information from Hanes et.al. (2002) and Storey (2012)**

The work performed at this point in project three is an important lead-in to completion of the coral propagation system in project five. Without completing the project three coral propagation system plumbing work described here, a fully operational coral propagation system will remain incomplete. Coral propagation system design and assembly work completed in projects two and three make it possible for students to complete the assembly of a coral propagation system in project five and to continue learning about collection and analysis of water quality data that is necessary for the successful daily operation of a coral propagation system. In this section of project three, students work to design and assemble two coral propagation system tanks (using plywood and acrylic) and a coral propagation system sump (using plywood and acrylic) and create plumbing connections between the coral propagation system sump and propagation tanks.

Photo by Robert Jakus

The first step in the process of incorporating water containment and flow components into the coral propagation system frame to which lighting system components have been attached requires students to review aquaponic system frame water containment and flow information found in this text (and other sources). Following this review, and before going on the second field trip, students are assigned to develop a report that demonstrates their understanding of the knowledge, skills, materials, and equipment that development of this systems water containment and flow components require. In addition to describing plywood dimensions and cuts as well as epoxies and bulkhead installations, reports about coral propagation system materials should contain dimensional details about pumps used, PVC pipe sizes, connectors, placement, and cuts (and safely using PVC cement). In addition, flexible hose dimensions and use and placement of ancillary pumps and filtration equipment and materials should be described. Reports should conclude with a team developed proposal that gives detailed dimensions, design layouts, and system modifications that students would follow if their water containment and flow design were to be implemented. Coral propagation system designs that students propose for development will also address water quality needs of the organisms that will inhabit the system. The proposals must provide explanations as to how the water containment and flow design is capable of sustaining high water quality standards. Finally, the most successful reports will also provide strategies that may be used so that assembly and setup of the water containment components of the system can be performed separately (and simultaneously when possible) with the assembly and setup of water flow components of the system.

The class is then guided through a critique of these reports. Following the critique, students vote for the coral propagation system design they think will best meet the needs of the class. It is recommended that when students go on the second field trip, they share this design plan with their tradespersons guides prior to the field trip. Students are asked to include contact information for tradespersons with whom they are familiar and whom they might invite to donate guidance, support, or possibly further demonstration of tools, equipment, and materials students must safely work with as they develop a water-containment and flow system. Many students find their motivation to assemble high quality aquatic ecosystems increase when skilled tradespersons and parents contribute their time, knowledge, and support.

Teachers may develop their own Coral Propagation System Water Containment and Flow Report or use the one provided below to guide student efforts to identify information to include in their reports.

Coral Propagation System Water Containment and Flow Report

	Materials Needed	Design and Assembly Steps AND Skilled Tradesperson	Function Performed by this Piece	Safety Concerns
Sump tank				
Upper tank				
Middle tank				
PVC drains from tanks to sump				
Protein Skimmer				

From this point forward in project three, student activities transition to those that require direct hands-on work with equipment and materials to implement plans critiqued and approved by the class to install water containment and flow components in the classroom coral propagation system frame.

In this system, water is pumped independently from the lowest tank (the sump) to each of the coral propagation system shelves above the sump. And, PVC outlets allow water to drain from each coral propagation tank to a common drain pipe and back to the sump for filtration, oxygenation, and recirculation.

Photo by Robert Jakus

Creating the three water containment "tanks" for the coral propagation system is the largest part of this project. In general, plywood is cut to fit the shelves and also cut to form two sides and the back wall of the tank. Pieces of wooden framing placed horizontally along the internal base of the bottom piece of plywood and vertically at the left and right rear of the tank are all screwed into the plywood base of the tank. For the sump a plywood front can also be screwed into wooden frames. However, to allow students to view the contents of the coral propagation system from the front, pieces of plexiglass are cut to fit into grooves in the frames at the front of the tank. In the back wall of each tank (not the sump though) 1.5-inch circular bulkhead holes are drilled and installed to allow water to drain from the tank back to the sump.

After the wooden components of the two tanks are completely assembled, but before the front plexiglass is place in the groves, the wood must be thoroughly painted with a non-toxic and waterproof epoxy. The front plexiglass walls are then inserted into the grooves and the edges secured for leak prevention. The sump, too, must be painted with the epoxy. After the epoxy has dried, the shelving may be assembled, tanks and sump placed upon them, PVC pipes connected to a common drain pipe to the sump, lights fixed to the frame, water placed in the tanks, and two pumps and other filtration equipment connected to the sump so that water will circulate freely throughout the system. Egg crate shelves may be created.

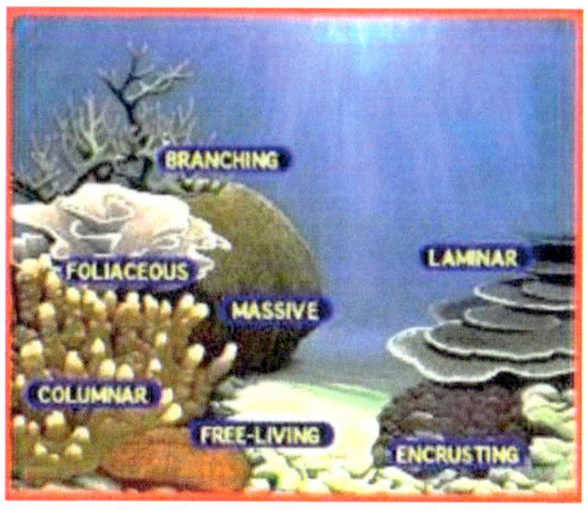

In George Parsons Shedd Aquarium PPT

Fill the system with water, turn on the pumps and let it run for the day and check for leaks. If no leaks are found, mix a good salt mix into the system. Students will gather and record salinity data on this system every day. Students will learn to determine salinity levels by using a hydrometer and a refractometer. Salinity levels will also be found digitally and chemically through the method described below.

Before installing a water containment and flow system in the broodstock and larval rearing system frame, students are now introduced to laboratory techniques for mixing saltwater and determining the salinity of the water and also for identifying the level of nitrogenous wastes that water contains. This knowledge is needed in order to provide corals with the healthiest saltwater possible and to maintain the life supporting capacities of the water in all aquatic ecosystems (freshwater and saltwater).

Once the aquatic class/lab is properly set-up and all students demonstrate that they conform to the aquatic class/lab safety criteria regarding clothing, jewelry, and hair and are also prepared to properly dispose of potentially dangerous equipment and materials as necessary student teams (or pairs) may perform this activity - under supervision and guidance.

Student teams should gather this equipment in their lab workspace

- 25 ml burette
- 10 ml volumetric pipettes
- 1 ml volumetric pipette
- 1 L volumetric flask
- Beakers, reagents, bottles, etc.

Student teams should gather these reagents in their lab workspace

- Potassium chromate solution: Dissolve 5 g of potassium chromate (K_2CrO_4) in one liter of distilled water.
- Silver nitrate solution: Weigh out exactly 2.400 g of silver nitrate ($AgNO_3$) and dissolve in one liter of distilled water. Store this solution in the dark.

Student teams should follow this procedure:

- Measure exactly 10.0 ml of seawater into a beaker via a volumetric pipette. Add exactly 10.0 ml of distilled water via volumetric pipette and mix thoroughly.
- Transfer exactly 1.00 ml of the accurately diluted seawater to a beaker.

Student teams should prepare three samples and

.

- Add 10 ml of potassium chromate to each beaker.
- Titrate each sample; add drops of silver nitrate solution until the yellow chromate color turns (just-permanently) to a faint reddish brown or coffee tint. Record the number of ml (estimate to the nearest one-hundredth ml) required for each of three samples. Average the results.
- The average titration in ml is equivalent to chlorinity in parts per thousand.

Student teams should use the following formula to calculate salinity
("parts per thousand" - ppt)

Salinity = 0.03 + 1.805 X chlorinity (ppt)

When available, students should also use a refractometer and a specific gravity meter (from an aquarium shop) to compare salinity readings of water samples from classroom aquatic ecosystems.

*** It is important that students performing the activity on this page have sufficient skill and background understanding of the water quality and analysis activities presented here to effectively lead students in other groups so that they can also understand the techniques and perform them (per the "Sharing of Knowledge and Accountability" activity on page 121).**

The process of monitoring and managing nitrogenous wasters in classroom aquatic ecosystems can involve highly technical analytical methods. In this activity, for introductory and practical purposes of monitoring and controlling the three nitrogenous waste products (ammonia, nitrite and nitrate) in their classroom aquatic ecosystems, students will be introduced to the use of commercial (nitrogenous wastes) water test kits and the interpretation of the results of these water tests. Students will be guided to better understand how concentrations of these compounds in ecosystem water indicate whether the biological filter is functioning properly. When this procedure is presented with systems that are newly set-up the raw data often fluctuates in ways that yield dramatic graphs of the nitrogen cycle. However, it is important also to note that the nitrogen data (ammonia, nitrite and nitrate) levels may need to be taken and recorded for as many as 8 weeks before the nitrogen cycle is fully active in newly setup classroom aquatic ecosystems. Students will be able to develop graphs and use them to help explain when ammonia, nitrite and nitrate first appear in the system and what their maximum concentrations were.

Once the aquatic class/lab is properly set-up and all students demonstrate that they conform to the aquatic class/lab safety criteria regarding clothing, jewelry, and hair and are also prepared to properly dispose of potentially dangerous equipment and materials as necessary student teams (or pairs) may perform this activity – under supervision and guidance.

Students should be guided to develop clearly labeled aquatic ecosystem data sheets for each classroom aquatic ecosystem. Keeping a copy of this sheet on a clipboard close to each respective system is a convenient way to prevent data from different systems from getting mixed up with data from a different system.

To test their classroom aquatic ecosystems for ammonia, nitrite and nitrate students should

1. Collect water in the appropriate water sample vials (provided with the test kit) to the proper fill lines.
2. Add the properly labeled tablet from the kit to each vial (ammonia, nitrite, and nitrate).
3. Shake the vials according to the manufacturer instructions.
4. Wait until the color of the water in the vials stop changing.
5. Compare the color of the water in the vials to the colors on the charts provided with the kit. The color on the charts that match the color of the samples in the vials show the concentrations of ammonia, nitrite and nitrate in your sample.

Photo by Robert Jakus

Once students have learned to correctly perform these water tests using commercial test kits they should be assigned to work on the preparation of a report about a titration technique that may be used to determine levels of ammonia, nitrite or nitrate in classroom aquatic ecosystem waters. Students should prepare these reports in such a way that they may use their reports to help them demonstrate every step of a titration technique to identify a nitrogenous waste in water. Students may compare titration levels of nitrogenous wastes with levels identified by commercial test kits. Students should also be guided to keep a digital record of aquatic ecosystem data and use software such as Excel to graph their system data. Finally, these reports should also explain safe and healthy emergency procedures that must be performed when concentrations ammonia or nitrite in a system are dangerously high.

Students who wish to study about the relation between ammonia and pH may use the equation provided to fill in the given table and go to the EPA (PDF file) to verify their results. Hanes et.al (2002), have an interesting activity about this that students may wish to explore independently.

*** It is important that students performing the activity on this page have sufficient skill and background understanding of the water quality and analysis activities presented here to effectively lead students in other groups so that they can also understand the techniques and perform them – including testing water, developing a spreadsheet, entering data, graphing data on a computer in the math lab, and performing calculations correctly (per the "Sharing of Knowledge and Accountability" activity on page 121).**

Percentage unionized ammonia (NH_3) in aqueous solution at different pH values and temperature. Use the equations in each row to complete this table. Review the EPA (1979) document to curve-fit additional percentages of un-ionized ammonia to pH values by varying temperature levels. To convert percentage values to decimals (proportions), divide tabulated numbers by 100 (e.g. pH 6.4 and 24° C, the proportion of **TAN** in the toxic form is 0.133/100 = 0.00133)

	Let temperature = X. Substitute the temperature " x" in each equation and enter your result (rounded to at least three decimal places) in each cell. The top row has been completed for you (The equations used for calculations of percent un-ionized ammonia in each row, were derived from Thurston et.al. (1979).								
pH and equation for determining percent un-ionized ammonia	**16° C**	**18° C**	**20° C**	**22° C**	**24° C**	**26° C**	**28° C**	**30° C**	**32° C**
pH 6.4 5.1978114 E^{-6} X^3 + -6.634199 E^{-5} X^2 + 0.0036772607 X + 0.019712843	**0.074**	**0.086**	**0.099**	**0.115**	**0.133**	**0.154**	**0.176**	**0.202**	**0.231**
pH 6.6 8.9436027 E^{-6} X^3 + - 1.536797 E^{-4} X^2 + 0.0069166065 X + 0.0091443001									
pH 6.8 1.4835858585909 E^{-5} X^3 + - 3.0032467532869 E^{-4} X^2 + 0.01237896825407 X + 0.00399134199046									
pH 7.0 2.7316433566071 E^{-6} X^4 + -1.938455063421 E^{-4} X^3 + 0.00591418997657 X^2 + -0.06360981610807 X + 0.4172183372									
pH 7.2 6.0096153845678 E^{-6} X^4 + -4.5802577052153 E^{-4} X^3 + 0.01430288461524 X^2 + 0.17065624190431 X + 1.0206682206582									
pH 7.4 1.0562354312183 E^{-5} X^4 + -8.20383320366748 E^{-4} X^3 + 0.02570658508099 X^2 + - 0.31165436414518 X + 1.8144832944311									
pH 7.6 1.8939394 E^{-5} X^4 + -0.0015340909 X^3 + 0.0496590909 X^2 + - 0.6409090909 X + 3.754848485									
pH 7.8 2.458479 E^{-5} X^4 + - 0.001959256 X^3 + 0.0633602855 X^2 + -0.7985615773 X + 4.811328671									
pH 8.0 1.2747668998479 E^{-5} X^4 + -5.1249676257486 E^{-4} X^3 + 0.00435460373237 X^2 + 0.29860593106331 X + - 1.8246309243893									
pH 8.2 4.4798951 E -5 X^4 + -0.0034820966 X^3 + 0.1110912005 X^2 + -1.29965553 X + 8.171289821									
pH 8.4 -8.322407 E^{-5} X^4 + 0.0097961182 X^3 + - 0.3924672203 X^2 + 7.210920422 X + -43.15994561									
pH 8.6 5.6271853 E^{-5} X^4 + 0.0042857258 X^3 + 0.138661859 X^2 + -1.363836118 X + 10.74982906									
pH 8.8 0.00108375420876 X^3 + -0.04919372294456 X^2 + 1.7932515632714 X + -4.797691197841									

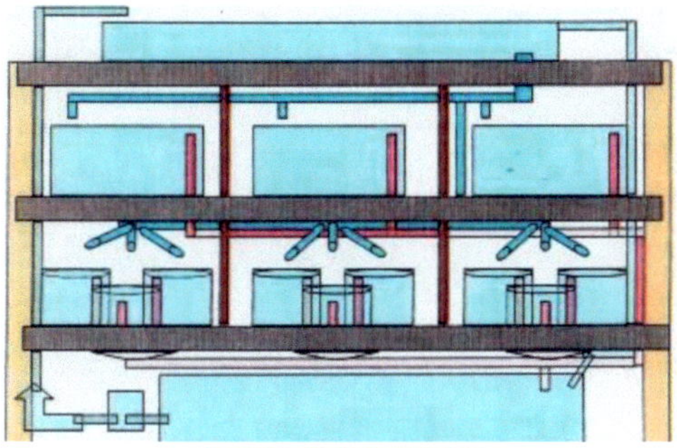

Drawing by Robert Jakus

The work performed in this section of project three is an important lead-in to work that students will complete in project six. Without completing the project three broodstock and larval rearing system plumbing work described here, a fully operational broodstock and larval rearing system will remain incomplete. Broodstock and larval rearing work in projects two and three make it possible for students to complete the assembly of a broodstock and larval rearing system in project six while continuing to refine their abilities to collect and analyze water quality data that is necessary for the successful daily operation of a broodstock and larval rearing ecosystem. I find it valuable to have students develop two such systems (one for freshwater and the other for saltwater). If it is not possible to create two systems, I recommend that a saltwater broodstock larval rearing system be developed. By creating a saltwater system, a wider range of opportunities for access to aquatic larvae become possible. And, the existence of saltwater larvae will allow students to make full use of the algae, rotifers, and other micro-organisms that are produced as a result of project seven. For classes that create a freshwater system, activities related to softening water with peat may be useful with certain aquatic species.

The first step in the process of incorporating water containment and flow components into the broodstock and larval rearing system frame to which system lighting components have been attached requires students to review broodstock and larval rearing system frame water containment and flow information found in this text (and other sources). Following this review, and before going on the second field trip, students are assigned to develop a report that demonstrates their understanding of the knowledge, skills, materials, and equipment that development of this systems water containment and flow components require. In addition to describing bulkhead cuts and installation, reports about broodstock and larval rearing system materials should contain dimensional details about the pump used, PVC pipe sizes, connectors, placement and cuts (and safely using PVC cement). In addition, flexible hose dimensions and use and placement of ancillary pumps and filtration equipment and materials should be described. Reports should conclude with a team developed proposal that gives detailed dimensions, design layouts, and system modifications that students would follow if their water containment and flow design were to be implemented. Broodstock and larval rearing system designs that students propose for development will also address water quality needs of the organisms that will inhabit the system. The proposals must provide explanations as to how the water containment and flow design is capable of sustaining high water quality standards. Finally, the most successful reports will also provide strategies that may be used so that assembly and setup of the water containment components of the system can be performed separately (and simultaneously when possible) with the assembly and setup of water flow components of the system.

The class is then guided through a critique of these reports. Following the critique, students vote for the broodstock larval rearing system design they think will best meet the needs of the class. Before going on the second field trip, it is recommended that students (or the teacher) share specific class design plans with the individuals who will be demonstrating design and assembly protocols during the field trip Students are asked to acquire contact information for tradespersons with whom they are familiar and whom they might invite to donate guidance, support, or possibly further demonstration of tools, equipment, and materials students must safely work with as they develop a water-containment and flow system. This contact information should be added to their reports. Many students find their motivation to assemble high quality aquatic ecosystems increase when skilled tradespersons and parents contribute their time, knowledge, and support.

Teachers may develop their own Broodstock and Larval Rearing System Water Containment and Flow Report or use the one provided below to guide student efforts to identify information to include in their reports.

Broodstock and Larval Rearing System Water Containment and Flow Report

	Materials Needed	Design and Assembly Steps AND Skilled Tradesperson	Function Performed by this Piece	Safety Concerns
Sump				
Top Shelf Reservoir				
3 Broodstock Tanks				
3 Circular Hatchery Tanks				
Pump from sump to reservoir				
Overflow drain from reservoir to sump				
Drain from reservoir to water distribution manifold				
3 Pipes from water distribution manifold to 6 faucets.				

Photo by Robert Jakus

From this point forward in project three, student activities transition to those that require direct hands-on work with equipment and materials to implement plans critiqued and approved by the class to install water containment and flow components in their classroom broodstock and larval rearing system frame(s). As this work is completed, students will have developed a move-in ready broodstock and larval rearing system to which they will introduce living organisms in project six.

In this project, students are guided to plumb the system so that the flow of water, from the upper shelf, may be directed independently to any one of six or to all six life support sections in the system. Water delivered to the upper level from the system sump is drained from the upper level in one of two ways. First, water may drain through a pipe that distributes water to one of three vertical half-inch pipes that supply water to the six life support sections of the system. Second, as water accumulates in the upper level, it may drain to the sump via an overflow pipe that is designed to prevent water in the upper reservoir from overflowing. As students work with aquatic ecosystems through the year, they will come to appreciate system designs that serve to prevent accidental and unwanted floods. The overflow drain channels excess water in the reservoir to the sump for additional filtration and recirculation.

Photo by Robert Jakus

As water exits the upper reservoir, it is directed to one of the three upper shelf life support systems in which broodstock are held. The rate of water flow into the broodstock containers is controlled by a ball valve. As water levels in these life support systems rise to maximum levels, water drains through bulkheads that students have installed (by carefully drilling 1.5-inch diameter holes in the glass tank) in the glass tanks to a common drainpipe which then channels all water to the sump for filtration and recirculation.

Photo by Robert Jakus

Water draining from the upper reservoir that does not flow into the broodstock tanks is directed to flow into circular hatchery containers in the shelf below the broodstock tanks. These containers are designed to contain eggs and larvae. When holding eggs and larvae, these tanks do not require high flow rates and the ball valves that control flow to these containers allow students to limit flow rates to drops per minute if necessary. Supplemental water flow to eggs and larvae are often better provided with an air pump and air stone. It is commonly believed that in most situations, larval tanks do not need an external biofiltration system. Larval tanks do not carry a large biomass and are in operation for only a few weeks, thus wastes do not quickly build up to toxic levels. Therefore, many fish and invertebrate larvae can be reared in a relatively small tank with a few fairly large water changes during a four-week larval period. Students will design the circular egg and larval hatchery containers on the second shelf to drain via small overflow bulkheads that drain to the common drainpipe which again deposits all water to the sump for filtration and recirculation.

Broodstock and larval rearing tasks for students to perform in project three include broodstock tank size selection, the drilling of circular bulkhead holes in glass broodstock tanks, drilling drain holes in upper shelf reservoir, drilling input hole in upper shelf reservoir, designing egg and larval flow through systems to fit on shelves beneath broodstock tanks, using properly-sized PVC pipes and ball valves to manage flow of water to and from each of the six life support containers and then to the common drain pipe, attaching drainage pipes to the frame and directing water to proper placement within the sump, and finally, connecting pump in the sump to the upper reservoir with a check valve in the line to prevent back flow.

Fill the system with water, turn on the pumps, and let it run for the day and check for leaks. If no leaks are found student groups are assigned the responsibility of collecting daily water quality data.

Before introducing life to the three “move-in-ready” aquatic ecosystems in projects four, five, and six respectively, students must first be introduced to laboratory techniques for determining the pH of water. This is knowledge is also needed in order to provide organisms with the healthiest water possible and to maintain the life supporting capacities of the water.

According to information on websites such as: http://www.wikihow.com/Measure-the-pH-of-Water, three common techniques for measuring the pH of water include ("p" represents the word parts and the letter "H" represents the idea of hydrogen ions). It is possible to measure the concentration of hydrogen ions in water by using either:

- a pH meter (by project ten students may be challenged to develop digital pH probe technologies to collect pH data of classroom aquatic ecosystem water)
- a commercially prepared field kit and
- litmus paper

In this simple activity students will learn to measure pH of water in classroom aquatic ecosystems using these three measurement techniques. They will be guided to compare the accuracy of each technique. Once the aquatic class/lab is properly set-up and all students demonstrate that they conform to the aquatic class/lab safety criteria regarding clothing, jewelry, and hair and are also prepared to properly dispose of potentially dangerous equipment and materials as necessary student teams (or pairs) may perform this activity – under supervision and guidance.

The pH meter method for measuring pH in classroom aquatic ecosystems

1. Calibrate the probe and meter following the manufacturer specifications.
2. Collect a sample of the water in a clean container.
   - The water sample must be deep enough to cover the tip of the probe.
   - Do not splash or agitate the water when taking samples.
3. Rinse the probe with water from the sample before placing it in the sample container.
4. Set the probe into the sample.
5. Adjust the meter, according to the manufacturer specifications for temperature if the meter requires a temperature adjustment.
6. Wait for the meter to come to equilibrium.
7. Record the pH measurement of the sample from the probe.

Photo by Robert Jakus

The commercially prepared field kit method for measuring pH in classroom aquatic ecosystems

1. Collect water in the appropriate water sample vial (provided with the test kit) to the fill line. Do not splash or agitate the water when taking samples.
2. Add a tablet from the kit.
3. Shake the vial according to the manufacturer instructions.
4. Wait until the color of the water in the vial stops changing.
5. Compare the color of the water in the vial to the colors on the chart provided with the kit. The color on the chart that matches the color of the sample in the vial shows the pH of your sample.

The litmus paper method for measuring pH in classroom aquatic ecosystems

1. Collect a sample of the water in a clean container. Do not splash or agitate the water when taking samples.
2. Dip a piece of litmus paper in your sample.
3. Compare the end of the litmus paper that you dipped in the water with the color chart that came with the litmus paper. The color on the chart that matches the color of your litmus paper shows will show you the pH of your water sample.

Each student should be guided to collect pH data from the same classroom aquatic ecosystem using each technology. Students should then be guided to pool their pH readings from each technique with pH data collected by others in the class. Once the class's data from each technique is pooled students may be guided to be aware of subtle differences in the accuracy, precision and sensitivity of each method. Designated teams of students must use their knowledge of pH to collect and record pH data from every system every day. Students should then write a pH report as described in the assessment section for this project (at the bottom of page 124).

*** It is important that students performing the Effect of Buffers on pH activity have sufficient skill and background understanding of the water quality and analysis activities presented here to effectively lead students in other groups so that they can also understand the techniques and perform them (per the "Sharing of Knowledge and Accountability" activity on page 121).**

Working independently of one another, three student groups (or more) have each completed the chapter II design and assembly phase of one classroom aquatic ecosystem. After filling each system with water, students in each team/group were given two water quality-related activities to participate in. The water quality-related activities that the different groups of students participated in were performed only by the one group. Students in the other two groups completed entirely different water quality-related activities. However, a complete set of data on each of the classroom aquatic systems will include a battery of tests on all the above measures of water quality.

The following sharing activity is designed to prepare every student to be accountable for the water quality data that is collected and analyzed throughout the year. It is assumed that as the assembly of each group's aquatic ecosystem was completed and water was placed in the system (by each respective group of students), the teacher had enough time to guide the students in this group through the water quality-related activities as described in the project before then leading students in the second and third classroom aquatic ecosystem groups through their respective water quality-related activities. It is important that students in each respective group have sufficient skill and background understanding of their water testing activities in order to effectively share their specific water quality analysis information and techniques with students in other groups.

When all systems are operating and students in each "home group" are familiar with their own water quality activities, the cycle of information sharing and accountability begins by having students in each group count off by twos. The following chart may be much easier to use than this verbal description. It is helpful to provide each student group with the paperwork and forms that are necessary for this activity (from the text above).

According to the chart, the "ones," who assembled the aquaponics system will work to guide students ("number ones") who assembled the coral propagation system through the DO and filter surface area water quality-related activities. According to the chart, as the X1 students lead the Y1's, the Y2 students will lead the Z1's through salinity and nitrogen water quality-related activities and the Z2 students will lead the X2's through the pH and buffer-related water quality activities. As this activity concludes, students shall have completed the first cycle of water quality-related activities. In cycle two, students in the X1 group lead students in the Z1 group. Students in the Y2 group lead the X2 group, and students in the Z2 group lead students in the Y1 group. As the chart below indicates, at the end of cycle two, the ones and twos in each group exchange (leader/learner) roles and pass through cycles three and four.

Leaders Cycles 1 & 2	**Learners Cycle 1**	**Learners Cycle 2**	**Leaders Cycle 3 & 4**	**Learners Cycle 3**	**Learners Cycle 4**
X1 DO/Filter Surface Area	**Y1**	**Z1**	**X2** DO/Surface Area	**Y2**	**Z2**
Y2 Salinity/Nitrogen	**Z1**	**X2**	**Y1** Salinity/Nitrogen	**Z2**	**X1**
Z2 pH	**X2**	**Y1**	**Z1** pH	**X1**	**Y2**

Assessment questions in the following section of project three help students process their understanding of these water quality-related activities. The forms that students complete as they work through these activities are helpful for students to critique their understanding of water quality protocols. Related paperwork should be kept for inclusion in a portfolio of *Beyond the Classroom Aquarium* project work.

Write a list of **science-based** activities you have participated in during this project and briefly describe evidence that verifies your interaction with and knowledge of scientific concepts for each item on your project three list.

Write a list of **technologically-oriented** activities you have participated in during this project and briefly describe evidence that verifies your interaction with and knowledge of technologically-oriented concepts for each item on your project three list.

Write a list of **engineering-oriented** activities you have participated in during this project and briefly describe evidence that verifies your interaction with and knowledge of engineering-oriented concepts for each item on your project three list.

Write a list of **mathematical** activities you have participated in during this project and briefly describe evidence that verifies your interaction with and knowledge of mathematical concepts for each item on your project three list.

Write a brief list of **Non-STEM-related** activities you have participated in during this project and briefly describe evidence that verifies your interaction with and knowledge of Non-STEM related concepts for each item on your project three list.

Rate your level of involvement in STEM-related activities in this project (1=lowest, 10=highest).

1—2—3—4—5—6—7—8—9—10

Beyond the Classroom Aquarium Daily Journal Guidelines

Entries students place in their daily journals should be:

1. Dated accurately and logged chronologically.
2. Kept in a notebook and written neatly in ink or typed.
3. Written or typed in complete sentences using correct capitalization, spelling, and punctuation.

In their journals, students should keep a daily record of:

1. Who they worked with each day by including the names of school staff, fellow students, community members, and volunteers.
2. What they did each day, such as lecture topic, activities or project design, assembly, or maintenance.
3. What they learned each day by asking questions, following instructions, and/or observing activities.
4. How *Beyond the Classroom Aquarium* experiences relate to past, present or future life experiences

For bonus points, students may complete these journal options, including illustrations, diagrams, or creative expression relating to the journal entry. For project one, journal writing bonus topics may include answers to the following questions or descriptions of a project activity or assignment. Reflection on these topics may also help students prepare for exams, quizzes, and other assessments.

- What do you want to understand better in order to contribute more to your team's productivity as it independently designs, assembles, and then manages an aquatic ecosystem?
- Create a new drawing and story about an organism or system that is most interesting to you.
- Select a book or magazine to read that is related to a topic in project three. Write a report on this book and include your report in your *Beyond the Classroom Aquarium* portfolio.
- Create a journal entry that tells what you learned during project three. What you would like to know more about? Do you know someone who might help you learn more if you asked them?
- Create additional posters and signs so that people who visit your aquatic ecosystems will be able to read about new and important system components and the organisms in them.

- Explain the proper technique to follow when using PVC cement. Describe ways PVC cement was used in the three systems your class assembled.
- Develop a brief report on coral fragging and ways to use superglue as part of the coral propagation technique. Also explain ways to use calcium carbonate substrates, like aragocrete to form base materials and shapes to which coral frags may be attached.
- What skills and knowledge did you develop and use as you created system water containment and flow components? Do the system water containment and flow components function properly? What could have been done differently to improve these system components?
- List important vocabulary words you learned by working on system water containment and flow components. Describe three important skills you developed and ideas you learned while working on development of system frames and lighting components.
- What is the importance of water quality testing, and how does water quality affect the growth of aquatic organisms?
- What are sources of poor water quality conditions, and how do we fix water quality problems?
- Read and summarize the article *Why is the Ocean Salty* and explain how information in this article justifies the use of the titration method for measuring salinity by chloride titration.
- Research and report upon important elements of undergravel filtration and explain why each factor is important and how to be sure that any undergravel filter they design would operate efficiently.
- Explain ways to determine the acidity of aquaponic system water and describe optimal pH requirements of organisms in the aquaponics system. In what ways do plants influence the choice of organisms that can be kept in the aquaponics fish habitat? Explain how water pH in the aquaponic system can be adjusted.
- Explain ways to determine the pH of water in a coral propagation system and describe optimal pH requirements of organisms in coral propagation systems. Explain techniques and equipment that may be used to adjust water pH in a coral propagation system.
- Explain the importance of each method of filtration (biological, physical, and chemical) and how each is included in the design of one of the student developed classroom aquatic ecosystems.
- Explain why it is possible to determine the salinity of water by identifying the concentration of chloride in the water.
- Create and present a water quality database in which students will enter water quality data from each system every day.
- Without being personally critical, what was most difficult about leading other students through the water quality related activities? Without being personally critical, what was most difficult about understanding the water quality related activities led by students from another group?

Additional Project Three Assessment Activities and Questions

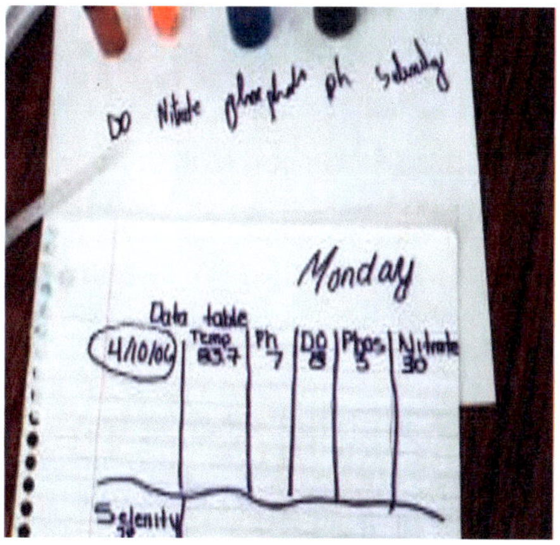

Photo by Robert Jakus

Create a presentation about the nitrogen cycle (including charts and graphs) include an explanation of how it relates to a recirculating system. Prepare this report for presentation and display.

List various water test kits and what they measure on a matching quiz, and have students match the kit to the parameter in a pre and post-test format. List the source of the problems that occur for each parameter (i.e. DO = dissolved oxygen low due to too much decomposition in the water or too little aeration).

Information about fish illness is available from many sources. Review this information and develop a brief report on four common fish (or other aquatic organism) illnesses. Explain what is known about the cause of each illness. Describe the expected outcome of each illness. Explain important protocols to follow to reduce the possibility that fish in the aquatic ecosystems in this classroom will become ill. Finally, design a quarantine aquatic ecosystem that would be adequate for the organisms in each of the three systems.

Soaps, perfumes, deodorants, and many other chemicals that people use every day are toxic to many aquatic organisms. Create a poster that illustrates how to avoid the use of toxic materials before, during, and after working with aquatic ecosystems. Include steps to ensure that the environment surrounding the aquatic ecosystem is nontoxic. Finally, explain the proper use of soaps for hand-washing, etc. when work with organisms and lab equipment is completed.

Create a report that demonstrates your knowledge about the critical importance of water movement in successfully operating aquatic ecosystems. Include an explanation of the role of pumps, protein skimmers, constant-level siphon cups, sponge filters, and air pumps useful for maintaining stable water conditions.

Students work with siphons and surface skimming water filtration mechanisms to demonstrate their knowledge about operational principles of siphons and surface skimming.

Students demonstrate their knowledge of appropriate water containment and flow processes by designing a wave and/or surge generator.

Students design and assemble a prototype bell-siphon, and create a presentation that proposes a way to include an additional aquaponic grow bed in the aquaponics system through which water on its way from the fish habitat to the sump may be diverted and in which a bell-siphon is utilized.

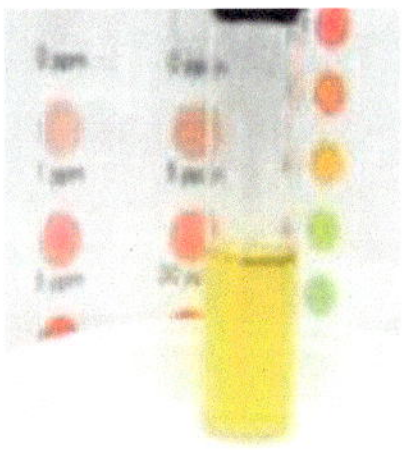
Photo by Robert Jakus

Students design and assemble a prototype "refugium" for a saltwater system and propose a way to include it on one of the classroom demonstration systems, incorporating the use of a siphon overflow through which water in the demonstration tank may be diverted before being channeled to the sump.

Students design and assemble a prototype piece of equipment described in project three and create a presentation that proposes a way to incorporate it into one of the student developed aquatic ecosystems.

Collect water quality data, from each student developed classroom system every day after they are first filled with water. Create and present a chart that illustrates nitrogen containing compounds in the water during the first month of operation. Be sure to explain fluctuations between ammonia, nitrite, and nitrate.

Develop a chart that compares and contrasts chemical versus digital techniques for analysis of water quality.

What did the electrolysis of water activity demonstrate?

What is water surface tension? What makes it possible?

How does the density of water change as it transforms from a liquid to a solid? In what ways is this important in natural aquatic ecosystems?

Create reservoirs of saltwater for future use. Describe the process and how you knew the mixture was the proper salinity.

Explain the importance of bacteria in maintaining good water quality in an aquatic ecosystem.

What are the four major kinds of water filtration? Describe a piece of equipment that performs this method of filtration.

What function does a bulkhead perform in an aquatic ecosystem?

Explain how temperature, pH, protein concentrations of fish food and the mass of organisms in a system influence the rate at which nitrogen can be removed from system water. Explain how an understanding of these factors can be used by system designers in the interest of optimizing a system's life stabilizing functions.

Review activities and information that interested you in this project, and determine if this information may fit on your "Research-ability" Potential Topics Form on page 59. If so, add it to the list.

Explain why, according to the dissolved oxygen meter's manufacturer's instructions (that accompany the instrument), what does it mean to "zero" and calibrate an oxygen meter?

Write a brief comparison of the measurement of dissolved oxygen concentration obtained with the oxygen meter with the oxygen concentration determined from the Winkler method?

Write a report that explains why management of pH is vital in aquaculture, common causes of pH changes in aquatic ecosystems, acceptable pH ranges for organisms in the classroom aquatic ecosystems, and how students might monitor pH and the use of buffers in classroom systems to prevent extreme pH changes in classroom systems.

Beyond the Classroom Aquarium STEM Self or Peer Assessment Rubric

This self-assessment is valid only to the extent to which you can document and display evidence that supports it. For each category in the left hand column, write a STEM-Oriented example in the box beneath 3 if you can display evidence that your STEM-Oriented example represents work that exceeds expectations. If your evidence shows that your STEM-Oriented statement meets but does not exceed expectations, write the statement in the box beneath 2. Similarly, for STEM-Oriented examples that demonstrate some progress toward expectations.

Name: **Project Number:** **Date Project Work Began:** **Date of Project Completion:** **List Math Topics and/or Objectives Currently Under Study:**		**3** **Work Exceeds Expectations**	**2** **Work Meets Expectations**	**1** **Work Is Progressing Toward Expectations**
Write examples of questions and defining problems related to your STEM work.				
Describe how you have developed and used models in your STEM work.				
Write examples of STEM investigations you planned and carried out.				
Describe STEM related data that you analyzed or interpreted.				
Give examples of how you used mathematics and computational thinking related to the five standards listed to the right.	Problem Solving (Math Process Standard One)			
	Reasoning & Proof (Math Process Standard Two)			
	Communication (Math Process Standard Three)			
	Connections (Math Process Standard Four)			
	Representation (Math Process Standard Five)			
Describe STEM-Oriented explanations and solutions you designed				
Explain evidence you used while engaging in a STEM project-based argument.				
Give examples of STEM information you obtained, evaluated, and/or communicated				
TOTAL POINTS				

CHAPTER III: Students Design, Assemble, & Maintain Primary Aquatic Ecosys-STEMs (Projects 4 - 6)

Of the 14 projects in *Beyond the Classroom Aquarium*, projects four through six guide students through the final stages in creating classroom aquaponic, coral propagation, and broodstock/larval rearing systems. These three projects generate many enriching experiences, some during the construction phase, others, over time, as a result of aquatic ecosystem biological productivity. Throughout the year, subsequent project work extends from these three operational classroom aquatic ecosystems. The more time students now have to optimize performance of the aquatic ecosystems, the fuller the aims of going *Beyond the Classroom Aquarium* shall have been met.

In order to move the development of each system along most expeditiously, a lot of system work has been completed in projects two and three. In many cases, this is in no small part due to the support of parents and local tradespersons and the essential skills and contributions they made in the workshop sessions in the first two field trips. In chapter III, some teachers may choose to accelerate the pace yet again. This decision is entirely left to the individual teacher and must depend upon setting and level of student readiness. Teachers who would prefer to guide students through projects four, five, and six—one at a time—can disregard "field trip three" (as optional) and guide their entire class in a linear sequence through the respective projects four, five , and then six introductory activities followed by other related pacing related suggestions. Regardless of which path the teacher selects, students should take part in respective introductory activities described below before moving into projects four, five and six. Teachers who wish to expedite the development of three classroom aquatic ecosystems will pre-arrange the three following separate and simultaneous introductory field experiences. Students in the:

- Broodstock and larval rearing system teams will participate in activities in a field trip three workshop in which they may be introduced to and/or guided through the aquatic ecosystem activities described in project six.
- Coral propagation system team will participate in activities in a field trip three workshop in which they may be introduced to and/or guided through the aquatic ecosystem activities described in project five.
- Aquaponics system team will participate in activities in a field trip three workshop in which they may be introduced to and/or guided through the aquatic ecosystem activities described in project four.

By adapting the "division of labor" activity introduced in chapter II, the AM workshop portion of the third field trip can similarly be used to introduce each group of students to skills and techniques needed to complete their own specific system. This arrangement and subsequent follow up provides for successful and timely completion of the three classroom aquatic ecosystems.

Initially, construction is in the forefront of each project in chapter III. Once construction is complete and the systems are cycling, the emphasis in projects four, five, and six switches from construction to that of developing student awareness of ecosystem dynamics, essential protocols to follow, and information to understand. In each field trip, students are guided through construction activities and introduced to activities unique to their project. Students in the aquaponics group are introduced to activities related to the germination and transfer of aquaponics system seedlings, as well as photosynthesis and respiration. Students in the coral propagation group are introduced to activities related to creation of aragocrete calcium carbonate plugs, coral fragging, determination of concentrations of salt and calcium in water, and also the design and assembly of a calcium reactor. Students in the broodstock and larval rearing system groups are introduced to activities related to developing a freshwater black worm culture system or a saltwater brine shrimp raceway/hatchery and participating in fish anatomy, and health analysis activities.

Regardless of the approach that a teacher adopts, in projects four, five, and six, students are guided to wrap up aquaponics, coral propagation, and broodstock/larval rearing ecosystem design and assembly work respectively. As students put finishing touches on each of these systems, they perform routine system maintenance and daily system data collection and recording, thereby monitoring, documenting, and analyzing system performance every day. An Activity and Consent Form related to ethical treatment of life in classroom aquatic ecosystems is provided on page 126. Teachers may consider requiring all students who are to work directly with livestock in the classroom to complete this form.

Project four focuses student attention on aquaponics system setup and operation. It includes a lengthy historical narrative of hydroponics/aquaponics (used by permission of the creator of the file 'History of Hydroponics' [Gary V. Deutschmann, 1988]) that underscores many important hydroponic principles and skills that students will be guided to apply in classroom systems.

Project five is designed to focus student attention on coral reefs and the coral reef oceanic environment. Coral reefs are some of the most biologically rich and economically valuable ecosystems on Earth. They are also in serious jeopardy, threatened by an increasing array of impacts from overexploitation, pollution, habitat loss, invasive species, diseases, and climate change. The rapid decline and loss of these valuable marine ecosystems has significant social, economic, and environmental consequences in the U.S. and around the world. Action is needed on a wide variety of fronts to address the coral reef crisis, especially on issues of global proportions such as the impacts of climate change, increasing coastal development, and persistent over-fishing of reef systems.

In project five, students are guided through coral fragging activities. Coral fragging is a propagation technique that students will use to stock the coral propagation system. Selling healthily growing coral fragments from the classroom system provides additional incentive for success. The development of a classroom coral propagation system challenges students to construct a system that provides living coral fragments and cuttings with all environmental conditions that will stimulate coral growth and development.

A National Coral Reef Action Strategy was produced by the National Oceanic and Atmospheric Administration (Evans, 2002)—in cooperation with the U.S. Coral Reef Task Force—to fulfill the requirements of the Coral Reef Conservation Act of 2000 (CRCA) (P.L. 106-562; 16 U.S.C. 6401 et seq.) and help track implementation of The National Action Plan to Conserve Coral Reefs (U.S. Coral Reef Task Force 2000). The Task Force National Action Plan was the first national blueprint for U.S. action to address the loss and degradation of U.S. and international coral reef ecosystems. Based on extensive input from: government and non-government organizations, scientists, resource managers, stakeholders and the public; the National Action Plan: (1) identified key threats and issues driving the loss and degradation of coral reefs; (2) established thirteen major goals to address these threats, and (3) outlined objectives and priority actions needed to achieve each goal.

Project six is designed to expand student awareness beyond specific applications of aquatic ecosystem principles in aquaponic and coral propagation settings to ways these principles may be applied more generally in relation to the aquaculture of a wide selection of organisms. Although motivation to use aquaculture successfully has a large basis in securing access to reliable food production, it encompasses many other valuable enterprises. Many who practice aquaculture use it to: help restore ecological diversity; provide unique materials for medical blood testing and pharmaceutical development; develop deeper insight into the nature of biochemical processes; and as a functional activity that benefits from well-used skill and knowledge by endowing the careful practitioner with ever expanding knowledge and skill. The purposes to which students may apply skills and knowledge acquired by participation in aquaculture activities are described in a U.S. Department of Commerce Aquaculture Policy (2011), indicates a strong national commitment to aquacultural activities.

> The purpose of this policy is to support the development of sustainable aquaculture within the context of the Department of Commerce's (DOC) goals of encouraging economic growth and employment opportunities in the United States and of enhancing United States competitiveness in, and exports to, global markets. This policy applies to a broad range of responsibilities at DOC relating to trade, technology, innovation and entrepreneurship, economic development, and environmental stewardship. (p. 1)

The National Oceanic and Atmospheric Administration, which is the primary bureau within DOC with programmatic aquaculture responsibilities, has a more detailed marine aquaculture policy (NOAA Marine Aquaculture Policy; 2011) that complements this overarching policy "reaffirming that aquaculture is an important component of NOAA's efforts to maintain healthy and productive marine and coastal ecosystems, protect special marine areas, rebuild overfished wild stocks, restore populations of endangered species, restore and conserve marine and coastal habitat, balance competing uses of the marine environment, create employment and business opportunities in coastal communities, and enable the production of safe and sustainable seafood." As stated in the D.O.C. document (p.2) "it is the policy of the D.O.C., consistent with the overarching National Aquaculture Act of 1980," to:

1) Create a business climate and technological base that encourages and fosters the development of sustainable aquaculture in the United States that provides domestic jobs, products, and services while conserving aquatic resources.
2) Encourage and foster environmentally sound and sustainable aquaculture innovation that increases the value of domestic aquaculture production and creates American business, jobs, and trade opportunities.
3) Ensure agency aquaculture decisions protect wild species and healthy, productive, and resilient coastal and ocean ecosystems, including the protecting of sensitive marine areas.
4) Advance scientific knowledge to develop and refine aquaculture technologies and methods to improve production, safeguard the environment, and sustain local food and cultural benefits, in cooperation with federal, state, local and tribal and academic partners.
5) Encourage and foster the development and application of aquaculture technologies that provide economic and/or ecological value by enhancing or restoring depleted, threatened, and endangered wild fish stocks and restoring habitat (e.g., oyster reefs).
6) Support the National Export Initiative by promoting a level playing field for U.S. aquaculture businesses to engage in international trade, working to remove foreign trade barriers and enforcing our rights under U.S. trade agreements.
7) Work collaboratively with our federal, state, regional, local, academic, and business partners to support the development of sustainable aquaculture in locations compatible with other uses.
8) Exchange scientific insights with other nations and promote joint participation in cooperative research that is of potential multinational value, including addressing impacts of aquaculture that breach international boundaries.
9) Advance public understanding of environmental, social, and economic benefits and challenges of sustainable aquaculture.

Ideally student groups will have produced one freshwater system and one saltwater system. Two different introductory activities are provided at the start of project six. In project six, students also learn about freshwater black worm culture and saltwater brine shrimp hatcheries and practice the process of brine shrimp cyst decapsulation.

By chapter's end, students are prepared to explore aquatic ecosystem topics and organisms more fully and explore the possibilities of using their knowledge and skills to develop fund-raising events which may include the sale of growing coral frags and plant sales in the fall (pumpkins), winter (poinsettias), and spring (garden starters) as another incentive to celebrate the success of their knowledge and skills. Many remaining projects depend on the productivity of the three classroom aquatic ecosystems. Projects four, five, and six are designed to ensure that students are developing an aquatic ecosystem background to call upon as they are guided to apply aquatic ecosystem skills and knowledge as they work on activities in projects eleven and twelve. With three operational classroom systems as chapter III begins, it is reasonable to expect that there should soon be enough plant and larval production so that all students will have opportunities to experience and reflect upon many manifestations of the cycle of life.

Bioethics Research Form

Ethical Treatment of Organisms and the Environment

REQUEST FOR NON-HUMAN VERTEBRATE ENDORSEMENT

Modified with Permission of the Chicago Public Schools Science Fair Inc. Form (2014)

These guidelines may be enforced as deemed proper by each educator (school or organization) responsible for instructing/supervising students as they participate in *Beyond the Classroom Aquarium* Projects involving vertebrate organisms.

Students and sponsors doing non-human vertebrate projects must complete this form. The signature of the student or students and the sponsor indicates that the project was done within these rules and regulations. Failure to comply with these rules will mean the disqualification of the project.

1. The student and the sponsor have the responsibility to see that all animals have proper care in well-ventilated, properly lighted locations with proper nutrition, proper temperature, adequate water, and sanitary surroundings. Care must be taken to see that the organisms are properly cared for during weekends and vacation periods.
2. NO primary or secondary cultures involving warm-blooded animals taken directly (mouth, throat, skin, bone, etc.) or indirectly (cage debris, droppings, etc.) will be allowed. However, cultures purchased from reputable biological supply houses or research facilities are suitable for student use.
3. No intrusive or pain producing techniques may be used. Included in these techniques would be such things as surgery, injections, taking of blood, burning, electrical stimulation or giving over the counter, prescription, illegal drugs or alcohol to measure their effect.
4. No changes may be made in an organism's normal environment that could result in undue stress, an injury, or death to the animal.
5. No vertebrate can be used as an independent or dependent variable in an experiment that could result in undue stress, injury or death to the animal.
6. For maze running and other learning or conditioning activities, food or water cannot be used or withheld for more than 24 hours. If the animal has a high metabolic rate, then food or water cannot be withheld for a length of time that would produce undue stress on the animal.
7. Chicken or other bird embryo projects are allowed but the treatment must be discontinued at or before 72 hours prior to the scheduled hatch day.
8. Projects that involve behavioral studies of newly-hatched chickens or other birds will be allowed if no change has been made in the normal incubation and hatching of the organism and all vertebrate rules are followed.
9. Vertebrate research involving the manipulation of any aspect of a vertebrate animal's environment MUST be supervised by a research scientist who is a veterinarian, medical doctor or biological research scientist. The project will be disqualified without the required evidence of this supervision.
10. No non-human vertebrate project will be allowed that is in violation of any of these rules. No person may perform any experiment for the students that violate any of these rules.

EXCEPTIONS to the rules above will not be allowed unless the student is working with a university, hospital or research facility on an ongoing research project being conducted by the institution and the following conditions are met. 1) The endorsement request is made before the established deadline. 2) The consulting scientist completes and signs the appropriate section of this request for and includes the required documentation and 3) The endorsement is granted before the student begins the project.

Student Names: ______________________ Project title: ______________________

In this space, briefly describe the use of non-human vertebrates in your project (include types of vertebrate animals used, the hypothesis or problem to be studied [independent and dependent variables], a brief procedure description, description of housing and care provided the animal(s) – such as cage size number of animals per cage, environment, bedding, type of food, frequency of food and water, how frequently animals will be observed, etc. And what will happen to animals after you have finished your project. (Use the back page if necessary.)

The signatures of the student or students and sponsor below indicate that the project conforms to the above rules of the Illinois Junior Academy of Science.

(Sponsor)
(Date) ______________________

(Student)
(Students) ______________________

TO BE COMPLETED BY THE CONSULTING SCIENTIST (per determination of supervising educator)

1. **Consultant's contact information**: Name of consulting professional (veterinarian, doctor or biological researcher)
   ______________________ Title: ______________________
   Profession ______________________ Position and name of institution: ______________________
   Phone number (extension) ______________________ E-mail address: ______________________
   "By signing this document I certify that I have reviewed the project described by this student (team) and have determined that this project does not violate any rules listed above. "
   Signature of Consulting Scientist: ______________________ Date: ____________

This Sheet Must Be Typed. It is recommended that a copy of this form be displayed on the exhibitor's display board. It may be reduced to a half sheet of paper if necessary.

The idea of an aquaponics system in a classroom setting, as presented by Nicol (1990), immediately captured my attention as a valuable and timeless learning experience for students. Aquaponic systems provide dramatic illustrations of nature's recycling processes. In *Beyond the Classroom Aquarium*, students are guided to use their aquaponic system to illustrate the integration of important aquatic ecosystem concepts. In addition, applications of aquaponics (Allen, 2012) can become powerful agents to promote social change. On the Wikipedia web site, aquaponics is described as a

> sustainable food production system that combines traditional aquaculture (raising aquatic animals such as snails, fish, crayfish or prawns in tanks) with hydroponics (cultivating plants in water) in a symbiotic environment. In aquaculture, effluents accumulate in the water, increasing toxicity for the fish. This water is led to a hydroponic system where the by-products from the aquaculture are filtered out by the plants as vital nutrients, after which the cleansed water is recirculated back to the animals. The term aquaponics is a portmanteau of the terms aquaculture and hydroponic. Aquaponic systems vary in size from small indoor or outdoor units to large commercial units, using the same technology. The systems usually contain freshwater, but saltwater systems are plausible depending on the type of aquatic animal and which plants.[citation needed] Aquaponic science may still be considered to be at an early stage, relative to other sciences." [Wikipedia, 2014] Aquaponic science may still be considered to be at an early stage, relative to other sciences.

Photo by Robert Jakus

In project number four, students make final adjustments to the aquaponics system prior to introducing aquatic livestock to the system. Once these adjustments have been made and the livestock have been procured (possibly already in one of the teacher's model systems), students then release the fish that were described in the plan that the class adopted. Final aquaponic system adjustments include preparation and placement of grow tubes to hold seedling plants, germination of seedlings, verification of stable water and ambient air temperatures, development of a daily water quality data log and related spreadsheet, development of a corresponding aquaponic system aquatic organism log book and related spreadsheet, and student submission of signed Consent Forms related to requirements about ethical treatment of the organisms.

Seed Germination - Project Four Opening Activity

Photo by Robert Jakus

Nicholls (1990) explains that "seeds are plants in potential. The embryonic plant encased in a hard coat has already been differentiated into several parts – a stem, one or two leaves, and a miniscule root. Each seed also contains a supply of sugars, proteins, fats, and starch to fuel the initial growth of the plant once it has burst free of its protective covering." In his book *Beginning Hydroponics Soilless Gardening*, Nichols (1990) indicates that plants do not emerge until "certain favorable conditions exist. If sufficient water, warmth, and oxygen are not available, the plant will not shed its coat. In this manner, seeds can survive for long periods of time until conditions turn in their favor. Once these initial conditions are met, the plant will germinate - that is it will shake off its protective coat and push up through the growing medium. Once this has occurred, an adequate supply of light will be essential if the plant is to keep on growing." He describes the function that water performs in the germination process.

> Although a seed will not germinate until all three elements are present, water is at first the most important, for it softens the hard outer layer of the seed, allowing the plant to emerge more easily, while at the same time causing the embryo to swell outwards from within. To effectively protect the seed, the outer layer must be very hard, and it requires this combination of the embryo swelling from within while the cat is softened by water for this layer to be cast off. In addition, water serves as the means by which oxygen is carried to the embryo. . . . If you supply

adequate amounts of warmth, moisture, and air to seeds you have placed in a growing medium, you should be able to consistently produce a high rate of successful germination (p.89).

Nicholls provides the following useful tips about the process of growing plants from seeds.

- Because seeds and seedlings have very specific needs which must be met, it would be impractical and awkward to mix mature plants and seedlings in a unit.
- A word of caution; when working with seeds, make certain that you do not bring your hands up to your mouth. Many kinds of seeds are now pretreated with a fungicide, and the chemicals could be harmful if you swallow them.
- Whether you start seeds in soil or in an aggregate, the procedure is largely the same. After the aggregate has been poured into the container, small holes are made in the surface.... These holes should be no more than a quarter of an inch deep. The seeds are tucked into place and aggregate gently brushed over them to fill in the holes entirely. Very small seeds need only a thin layer of aggregate spread over them. When dealing with small seeds, it is simpler to arrange them on the surface and then spread a thin layer of aggregate on top.
- Arrange the seeds in parallel rows. It is a general practice to sow more seed than you need. Before planting seeds, water the aggregate enough to make it moist but not waterlogged. After seeds are in place give them a light watering. Thereafter, the growing medium must be kept constantly moist, but never so moist as to be sodden.
- The day after the seeds have been sown, substitute nutrient solution for plain water and apply solution rather than water. The solution should be the same dosage as used for mature plants.
- The first two weeks are the most critical for seeds. To germinate properly they need warmth as well as steadily moist conditions. Most kinds of seeds require a temperature in excess of 65° F to germinate. A waterproof insulated heating cable is the best, surest way to guarantee sufficient warmth for the seeds.
- Hydroponic germination units should be shielded from the direct rays of the sun until the seeds have sprouted and sent shoots above the aggregate. While direct sunlight is harmful, bright but diffuse light is not only acceptable, but essential. After the seedlings have appeared you can give them full exposure to the direct rays of the sun, protecting them only during the midday hours when sunlight is most intense. If the unit is of manageable proportions, it would be helpful to rotate it each day to prevent the seedlings from growing too far in one direction, since they are inclined to bend toward the dominant light source.
- After the 2nd pair of leaves appear on a seedling, it is ready to be transplanted. (pp. 89 -91)

Materials, Equipment, and Instructions for Germinating Seeds

Materials needed for germinating seeds include: rectangular plastic germination trays with an insertable porous plastic frame with 1-inch diameter round slots; 1-inch diameter pieces of Rockwool media (to fit into the plastic frame); transparent plastic top for each germination tray; seeds to germinate; water from aquaponics system; light to suspend above transparent plastic top on germination trays in which seeds have been placed; electrically powered seed germination heating pads to set beneath the germination trays; 1-inch diameter cups in which to transplant seedlings from germination unit to 1-inch diameter holes in aquaponic grow tubes.

Photo by Robert Jakus

Place the electric germination heating pad in the place where the germination tray will be placed. Place germination trays upon the heating pads and fill them halfway with water from the aquaponics system. Place the porous plastic germination frame into the tray. Place a piece of Rockwool into each circular opening in the porous frame; water from the tray should saturate the Rockwool inserts. Once a piece of Rockwool media has been placed in each slot, in the porous frame, put enough water in the tray so that it is halfway full. Spread a seed to be germinated on the surface of each piece of damp Rockwool media in the tray. Tamp each seed into the media so that it is slightly embedded in it. Fit the transparent plastic top to the top of the tray. Position the light above the tray and turn it on. Plug in the heating pad. When the seedlings have at least two leaves on them, place the seedlings (embedded in the Rockwool media) into 1-inch diameter plastic cups (porous) and transfer them to the aquaponics system.

*** It is important that students performing the activity on this page have sufficient skill and background understanding for the Germinating Seeds activity presented here to effectively lead students in other groups so that they can also understand the techniques and perform them (per the "Sharing of Knowledge and Accountability" activity on page 141, 165, and 186).**

Aquaponics consists of two main parts: an aquaculture part for raising aquatic animals and a hydroponics part for growing plants (Rakocy, Masser, Losordo, 2006; and Diver, 2006). Aquatic effluents, resulting from uneaten feed or raising animals like fish, accumulate in water due to the closed-system recirculation of most aquaculture systems. The effluent-rich water becomes toxic to aquatic animals in high concentrations. But these effluents are nutrients essential for plant growth (Rakocy; Masser; Losordo, 2006). Although consisting primarily of these two parts, aquaponics systems are usually grouped into several components or subsystems responsible for the effective removal of solid wastes, for adding bases to neutralize acids, or for maintaining water oxygenation (Rakocy; Masser; Losordo, 2006). Typical components include:

- *Rearing tank*: the tanks for raising and feeding the fish.
- *Solids removal*: a unit for catching uneaten food and detached biofilms and for settling out fine particulates.
- *Biofilter*: a place where the nitrification bacteria can grow and convert ammonia into nitrates which are usable by the plants (Rakocy; Masser; Losordo, 2006).
- *Hydroponics subsystem*: the portion of the system where plants are grown by absorbing excess nutrients from the water.
- *Sump*: the lowest point in the system where the water flows to and from which it is pumped back to the rearing tanks.

Depending on the sophistication and cost of the aquaponics system, the units for solids removal, biofiltration, and/or the hydroponics subsystem may be combined into one unit or subsystem (Rakocy; Masser; Losordo, 2006), which prevents the water from flowing directly from the aquaculture part of the system to the hydroponics part.

Nitrification - Nitrification, the aerobic conversion of ammonia into nitrates, is one of the most important functions in an aquaponics system as it reduces the toxicity of the water for fish and allows the resulting nitrate compounds to be removed by the plants for nourishment (Rakocy; Masser; Losordo, 2006). Ammonia is steadily released into the water through the excreta and gills of fish as a product of their metabolism but must be filtered out of the water since ammonia is highly toxic to fish even at very low concentrations (Brennan, 2013). Brennan reports "levels of 0.2 to 0.5 parts per million are enough to kill some aquarium species. The precise level of ammonia that will suffice to kill a fish depends on the temperature and pH of the water. Given the serious threat posed by ammonia to your fish, however, the best general rule is that it shouldn't be present at levels you can detect. If you can detect ammonia in the aquarium with a test kit, the level is probably too high." Although plants can absorb ammonia from the water to some degree, nitrates are assimilated more easily (Diver, 2006), thereby efficiently reducing the toxicity of the water for fish (Rakocy; Masser; Losordo, 2006). Ammonia can be converted into other nitrogenous compounds through healthy populations of:

- *Nitrosomonas*: bacteria that convert ammonia into nitrites.
- *Nitrobacter*: bacteria that convert nitrites into nitrates.

In an aquaponics system, the bacteria responsible for this process form a biofilm on all solid surfaces throughout the system that are in constant contact with the water. The submerged roots of the vegetables combined have a large surface area so that many bacteria can accumulate there. Together with the saliency of ammonia and nitrites in the water, the surface area determines the speed with which nitrification takes place. Caring for these bacterial colonies is important as to regulate the full assimilation of ammonia and nitrite. This is why most aquaponics systems include a bio-filtering unit which helps facilitate growth of these microorganisms. During system startup, spikes may occur in the levels of ammonia and nitrite, with nitrate levels peaking later in the startup phase. Typically, after a system has stabilized, levels of ammonia and nitrite (measured in parts per million) diminish. Nitrate levels increase inversely to decreased levels of ammonia and nitrate. Since the nitrification process acidifies the water, non-sodium bases such as potassium hydroxide or calcium hydroxide can be added for neutralizing the water's pH (Rakocy; Masser & Losordo, 2006) if insufficient quantities are naturally present in the water to provide a buffer against acidification. In addition, selected minerals or nutrients such as iron can be added in addition to the fish waste that serves as the main source of nutrients to plants (Rakocy; Masser; Losordo, 2006).

A good way to deal with solids buildup in aquaponics is the use of worms which liquefy the solid organic matter so that it can be utilized by the plants and/or animals.

Hydroponics subsystem - Plants are grown in hydroponics systems, with their roots immersed in the nutrient-rich effluent water. This enables them to filter out the ammonia that is toxic to the aquatic animals or its metabolites. After the water has passed through the hydroponic subsystem, it is cleaned and oxygenated and can return to the aquaculture vessels. This cycle is continuous. Common aquaponic applications of hydroponic systems include:

- *Deep-water raft aquaponics*: This method utilizes rafts of Styrofoam floating in relatively deep aquaculture troughs.
- *Recirculating aquaponics*: solid media such as gravel or clay beads, held in a container that is flooded with water from the aquaculture. This type of aquaponics is also known as *closed-loop aquaponics*.
- *Reciprocating aquaponics*: solid media in a container that is alternately flooded and drained utilizing different types of siphon drains. This type of aquaponics is also known as *flood-and-drain aquaponics* or *ebb-and-flow aquaponics*.
- Other systems use towers that are trickle-fed from the top, nutrient film technique channels, horizontal PCV pipes with holes for the pots, plastic barrels cut in half with gravel or rafts in them. Each approach has its own benefits (Lennard and Leonard, 2006).

Most green leafy vegetables grow well in hydroponic systems. Profitable crops include varieties of Chinese cabbage, lettuce, basil, roses, tomatoes, okra, cantaloupe, and bell peppers (Diver, 2006). Other species of vegetables that grow well in an aquaponic system include beans, peas, kohlrabi, watercress, taro, radishes, strawberries, melons, onions, turnips, parsnips, sweet potatoes, and herbs. Since plants at different growth stages require different amounts of minerals and nutrients, plant harvesting is staggered with seedlings growing at the same time as mature plants. This ensures stable nutrient content in the water because of continuous symbiotic cleansing of toxins from the water (Nichols, ed. 2004).

Aquaculture subsystem - Freshwater fish are the most common aquatic animal raised using aquaponics, although freshwater crayfish and prawns may also be used (Backyard Aquaponics, 2006). In practice, "tilapia are the most popular fish for home and commercial projects that are intended to raise edible fish, although barramundi, silver perch, eel-tailed catfish or tandanus catfish, jade perch and murray cod are also used" (Diver 2006). And according to information provided in *Aquaponic Community* (2011), "for temperate climates when there isn't ability or desire to maintain water temperature, bluegill and catfish are suitable fish species for home systems; and Koi and goldfish may also be used, if the fish in the system need not be edible." Many people agree with Bernstein (2010) that "fish are clearly the heart and soul of an **aquaponics** system. They don't ask much of us – oxygen, water in the right pH and temperature, and food – and they give our plants life. And perhaps the best part is how excited they get when I approach their system at feeding time. The mature tilapia all crowd around where I usually throw in the food and wait at the surface. Then the frenzy ensues as I feed them."

Normal operations - Aquaponic systems do not typically discharge or exchange water under normal operation but instead recirculate and reuse water very effectively. The system relies on the relationship between the animals and the plants to maintain a stable aquatic environment that experience a minimum of fluctuation in ambient nutrient and oxygen levels. Water is only added to replace water loss from absorption and transpiration by plants, evaporation into the air from surface water, overflow from the system from rainfall, and removal of biomass such as settled solid wastes from the system. It is commonly believed that aquaponics uses approximately two percent of the water that a conventionally irrigated farm requires for the same vegetable production. As Viladomat (2011) indicates, this allows for aquaponic production of both crops and fish in areas where water or fertile land is scarce.

There seems to be some belief that aquaponic systems can also be useful for water treatment by reclaiming potable water from typical household sewage (Aquaponics Gardner, 2013). The nutrient-filled overflow water can be accumulated in catchment tanks and reused to accelerate growth of crops planted in soil, or it may be pumped back into the aquaponic system to top up the water level.

The three main inputs to the system are water, feed given to the aquatic animals, and electricity to pump water between the aquaculture subsystem and the hydroponics subsystem. Spawn or fry may be added to replace grown fish that are taken out from the system to retain a stable system. In terms of outputs, an aquaponics system may continually yield plants such as vegetables grown in the hydroponics section and edible aquatic species raised in the aquaculture section of the system.

Hydroponic and Aquaponic History (Deutschmann, 1998 - with permission) - Hydroponics and aquaponics has ancient roots, although there is some debate on its first occurrence. Hydroponics is at least as ancient as the pyramids. Many believe hydroponics, as a farming technique, started in the ancient city of Babylon with its famous hanging gardens which are considered one of the seven wonders of the ancient world. Some writers have suggested that the Babylonian gardens were an elaborate hydroponic system into which freshwater, rich in oxygen and nutrients, was regularly pumped. Aztecs cultivated agricultural islands known as chinampas which are considered by some the earliest form of aquaponics practice (Boutwell, 2007 and Rogosa, 2010). According to Crossley (2004) and Boutwell (2007), plants were raised on stationary (and sometime movable) islands—called chinampas—in lake shallows, and waste materials dredged from the Chinampa canals and surrounding cities were used to manually irrigate the plants.

The nomadic Aztec tribe built rafts of rushes of rushes and reeds by securing the stalks together with tough roots. They then dredged up soil from the bottom of shallow lakes and piled it on the rafts. The roots vegetables, flowers, and trees planted on the rafts grew through the floor of the floating rafts and down into the water. The rafts, which never sank, were sometimes joined together to form floating islands that were as much as 200 feet long. On market days, the gardener might pole his raft close to a market place, picking and handing over produce as shoppers purchased them. William Prescott, the historian who chronicled the destruction of the Aztec empire by the Spaniards, described the chinampas as "Wandering islands of verdure, teeming with flowers and vegetables and moving like rafts over the water." Use of chinampas continued well into the nineteenth century.

The world's rice crops have been grown hydroponically from time immemorial. The floating gardens of the Chinese, as described by Marco Polo in his famous journal, are examples of hydroponic culture. Ancient Egyptian hieroglyphic records data back to several hundred years BC describe the growing of plants in water along the Nile without soil. Before the time of Aristotle, Theophrastus (327-287 BC) undertook various experiments in crop nutrition. Botanical studies by Dioscorides (writer of the Materia Medica) date back to the first century AD. Polycultural farming systems existed in many far eastern countries, such as China and Thailand, where the cultivation and farming of rice was practiced. Waters necessary for the growth of rice in the paddies were also used for the raising of fish such as the oriental loach, swamp eel, common and crucian carp, and pond snails in the paddies (McMurtry, Nelson, & Sanders, 1988; and Bocek, 2010). According to Deutchmann (1998) early scientific explorations in hydroponics (in the 1600s) were performed to determine what substances plants are composed of and what substances make them grow.

Aquaponics in the United States - While the development of aquaponics is often attributed to the various works of the New Alchemy Institute and the works of Dr. Mark McMurtry et al. at the North Carolina State University, many papers of initial development of aquaponics concepts pre-date both institutions by nearly a decade.

Tom and Paula Speraneo, owners of a small greenhouse operation near West Plains, Missouri, modified the North Carolina State method and raised tilapia in above-ground tanks inside a solar greenhouse. The effluent from the tanks was used to fertilize gravel-cultured vegetables in raised benches. In addition, the Speraneos manipulated the watering cycle, and this methodology of the Speraneos forms the basis for the style of "flood & drain" media grow bed aquaponics systems that have been widely adopted in Australia based on models promoted by Joel Malcolm and Murray Hullam and are now gaining increasing popularity in the United States.

Inspired by the aquaponic success at the New Alchemy Institute and North Carolina State University (NCSU), other institutes followed suit. Besides the reciprocating aquaponics based on the techniques developed by Dr. Mark McMurtry et al. at NCSU, Dr. James Rakocy and his colleagues at the University of the Virgin Islands (2010) researched and developed the "Deep Water" or "Raft Culture" aquaponics. The system combines tilapia with various vegetables.

In 1997, Rebecca L. Nelson and John S. Pade began publishing the Aquaponics Journal, a quarterly scientific journal that brings together research and various applications of aquaponics from around the globe.

Recent years have seen a shift toward community integration of aquaponics, such as the nonprofit foundation Growing Power that offers Milwaukee youth job opportunities and training while growing food for their community. The model has spawned several satellite projects in other cities, such as New Orleans where the Vietnamese fisherman community has suffered from the Horizon Disaster and in the South Bronx in New York City (Harris, 2011). In addition, aquaponic gardeners from all around the world have gathered in online community sites and forums to openly share their experiences and promote the development of this form of gardening.

Aquaponics in Canada – Lethbridge College (Lethbridge, 2013) reports its Aquaculture Centre, in operation since 1997, added a small aquaponics system onto an existing aquaculture research station at the college in 2002. Canada saw a rise in aquaponics setups throughout the 90s, predominantly as commercial installations, that, for example, combine trout with floating lettuce production (Nelson, 2007) or water fruiting vegetable crops that warm up the water too much to be recirculated back into the fish ponds. A setup based on the deep water system developed at the University of Virgin Islands was built in a greenhouse at Brooks, Alberta where Dr. Nick Savidov and colleagues researched aquaponics from a background of plant science. The team's findings (on rapid root growth in aquaponic systems and on closing the solid waste loop) demonstrated that aquaponics systems have certain advantages over traditional aquaculture, including the ability of the system to run well at a low pH levels, favored by plants but not fish. The Edmonton Aquaponics Society in Northern Alberta is adapting Dr. Savidov's commercially sized system to a smaller scale prototype that can be operated by families, small groups, or restaurants. They intend to further develop the closed solid waste loop.

Aquaponics in Barbados: According to Bishop, et.al. (2009), "Barbados is a densely populated island that deals with water scarcity." In 40 years' time, "focus has shifted from domestic fruit and vegetable production on small farms to importing 80% of all fruits and vegetables, for cost reasons" (Závodská & Dolly, 2009). Aquaponics would make Barbados and other Caribbean islands less dependent on the world food market and reduce stress on the dwindling fish supplies. "An inter-organizational project that started in late 2009 sets out to encourage and enable Barbadians to start aquaponics at home, with revenue generated by selling produce to tourists" (Bishop, Bourke, Connolly, & Trebic, 2009).

Aquaponics in Bangladesh - Bangladesh is the world's most densely populated country. Most of its farmers use/misuse agrochemicals to enhance food production/storage life. However, the country has yet to establish a good monitoring system to ensure chemical residue-free safe food in the market (Daily Sangbad, 2011). Capable city dwellers are looking for a better way to grow their own safe foods for domestic consumption. To help support mainly this client group and people living in adverse climatic conditions such as salinity-prone southern part and flood-prone haor area in the eastern region, a low-cost model of backyard aquaponics system (The Independent, 2012) has been developed by Professor Dr. M.A. Salam at the Department of Aquaculture of Bangladesh Agricultural University, Mymensingh (Daily Janakantha, 2011). The initial model has been developed in 2011 and demonstrated and reported to the press in January 2012 (Daily Janakantha, 2011). Further research may be necessary to develop appropriate models for different situations among client groups.

Aquaponics in Taiwan - Taiwan is a densely-populated island that is faced with freshwater scarcity. Dispensation of water is government controlled. The closed-loop system of aquaponics is used by agricultural farmers to save water by also rearing fish while fish farmers grow plants that filter the water from the fish tanks. Taiwan Aquaponics Association is the not-for profit federation to promote Aquaponics at Taiwan (Daily Janakantha, 2011).

Aquaponics in Japan - Japan is another highly densely-populated country that has a long history of agricultural innovation and technological improvement. Aquaponics, however, has only recently been introduced to this country, and there have been some projects to integrate aquaponics into communities in Tohoku, northern Japan, that suffered after the 2011Tohoku earthquake and tsunami. Japan Aquaponics, founded by Aragon St-Charles, is a leading social enterprise that is pioneering the use of aquaponics in this country.

Aquaponics in Australia - Due to a ban on tilapia in all states, native freshwater fish—including silver perch, jade perch, sleepy cod, murray cod, and barramundi—are popular in aquaponics and aquaculture systems (Diver, 2006), along with non-native rainbow trout, brown trout, and crayfish such as common yabby and redclaw. The law changed in Western Australia June 2012, banning tilapia.

Aquaponic Pros and Cons

The unique advantages of aquaponic systems are:

- Conservation through constant water reuse and recycling
- Organic fertilization of plants with natural fish emulsion
- Elimination of solid waste disposal from intensive aquaculture
- Reduction of needed cropland to produce crops
- Overall reduction of the environmental footprint of crop production
- Building small efficient commercial installations near markets reduces food miles
- Reduction of pathogens that often plague aquaculture production systems
- Reduction of erosion by eliminating the need to plough the soil

Photo by Robert Jakus

Some conceivable drawbacks of aquaponics are:

- Initial expenses for housing, tank, plumbing, pumps, and grow beds
- The infinite number of ways in which a system can be configured lends itself to equally varying results, conflicting research, and successes or failures
- Some aquaponic installations rely heavily on man-made energy, technological solutions, and environmental control to achieve recirculation and water/ambient temperatures.

However, if a system is designed with energy conservation in mind, using alternative energy and a reduced number of pumps by letting the water flow downwards as much as possible can be highly energy efficient

- While careful design can minimize risk, aquaponics systems have multiple 'single points of failure' where problems such as electrical failure or pipe blockage can lead to a complete loss of fish stock
- Like all aquaculture based systems, stock feed usually consists of fish meal derived from lower value species. Ongoing depletion of wild fish stocks makes this practice unsustainable. Organic fish feeds may prove to be a viable alternative that negates this concern. Other alternatives include growing duckweed with an aquaponics system that feeds the same fish grown on the system (Rogosa, 2010), excess worms grown from vermiculture composting, using prepared kitchen scraps (Amadori, 2012), as well as growing black soldier fly larvae to feed to the fish using composting grub growers (Royte, 2009)

Completing the Aquaponics System

After the initial testing of the system for leaks, in project three, it is now time to prepare it for plant seedlings and fish. After turning off the pump and disconnecting temporary hose and pipe connections on the A-frame that were needed to test the system, detach the 4-inch diameter pipes from the A-frame and allow them to dry out. Then measure and mark the pipes in preparation for the drilling of carefully spaced 1-inch diameter holes. This can be done by drawing a line from one end of the 5-foot pipe to the other end and marking equally spaced points on the line. The points will be used as guides for drilling the 1-inch diameter circular holes in the pipe. Seedlings roots will soon be immersed in recirculating aquaponic system water as the plants are set in 1-inch diameter plastic aquaponic cups and fitted into the drilled holes. Seedlings grow into plants that require additional space so the space between holes should be no less than four inches. Drill the holes and clean out the pipes. To prepare for the transfer of germinating seedlings to the A-frame pipes, re-install the pipes on the A-frame (with newly drilled holes facing up), and secure all water connections and eliminate all drips and leaks; PVC cement and hose clamps help keep the system watertight throughout the school year.

Photo by Robert Jakus

While all of this new assembly is occurring, it is good to keep in mind that a day for disassembly and storage of materials will come, and it is good to reserve some thought to that time. And, when the time does come to disassemble the system, students will not be able to separate pieces that are cemented together. I encourage students to fit as many hoses and pipes together well enough with the use of clamps (when possible) so that as little cement as possible is needed.

Photo by Robert Jakus

Determine temperature ranges that are necessary for system operation and ensure that water heaters are safely installed in the pool and the sump. Connect the pump in the sump to a timer that will allow the pump to run for 15 minutes of every hour. Establish pH and DO of system water (and protocols for maintenance of it). Provide a place for storage (and dechlorination and pH adjustment) of water to be used for regular water exchanges or to replace evaporated water. Ensure access to aquaponics system and species logbooks (enter data). Ensure system lighting is operational. Install an air pump to aerate water in the sump and pool. Reserve a place in which water filtration equipment may later be installed.

A final but necessary step before working directly with fish and other organisms is to review information related to proper and ethical treatment of all classroom organisms. Among other things, students must be aware that bug sprays, sun block, perfumes, and many other substances on our skin kill aquatic organisms in an instant. Working with aquatic ecosystems requires a hygiene protocol that many students are not familiar with. It is important that all students are guided to make every effort to work with aquatic ecosystems in ways that prevent exposure to substances that are toxic. In the effort to promote proactive aquatic ecosystem work, each student who is to work directly with the system must submit a signed form of agreement; the signature of a parent or guardian is required on this **Non-Human Vertebrate Endorsement Form** (see page 129). Review protocols for healthy transfer of organisms to a new system (from the **Introductory Aquatic Ecosystem Inhabitant Form**, page 57).

Prepare fish for introduction to the system by slowly acclimating them to the temperature and chemical properties of the water in the pool. Hopefully, differences between water conditions in new and old environments will not be too extreme. Fish may be acclimated for transfer to the new system by placing them in a half-full bucket of water from their old system, and then gently adding two quarts of water from the new system into the bucket; wait for 15 minutes and then again add two quarts of new system water to the bucket. After repeating this exposure to the new water three times, gently release the fish to the system. Transfer germinating plants to the system and allow them to grow. Harvest them when they become large enough. Harvested plants can be consumed or replanted in a school garden. When a plant is harvested, the system space can then be fitted with a newly germinated plant. Remember to keep data about the way that plants in the system are cycled through it, moving grown plants out of the system and new ones into it. Prepare to conduct yearlong system production and data collection activities and document your work and quantify your results.

Photo by Robert Jakus

Do not over feed the fish in the system, especially when the nitrogen cycle is just beginning. If necessary, place some kind of fencing around the pool to prevent fish from jumping out.

Congratulations for completing this stage of your aquaponics system development. I hope your harvest is abundant.

Photo by Robert Jakus

While studying Chesapeake Bay related topics with colleagues at the Horn Point Laboratory in Maryland, we were introduced to a photosynthesis and respiration activity structured similarly to the activity presented below. By performing this activity students can become more fully aware of the dynamic interaction between these two biological processes and how these interacting processes may influence the availability of oxygen and other substances in the classroom aquatic ecosystem environments they design, assemble and maintain. On the day preceding this activity, water used in this activity may be collected and stored from freshwater classroom aquatic ecosystems as part of the routine 10% to 20% water change system maintenance processes students perform on a regular basis.

Once the aquatic class/lab is properly set-up and all students demonstrate that they conform to the aquatic class/lab safety criteria regarding treatment of living organisms, clothing, jewelry, and hair and are also prepared to properly dispose of potentially dangerous equipment and materials as necessary student teams (or pairs) may perform this activity – under supervision and guidance. Students should ensure that they have access to a well calibrated DO meter. (Some educators prefer to have their students repeatedly perform the Winkler method of DO analysis instead of allowing them to use a DO meter – twice for each sample to determine an average.) Students using the Winkler method to assess DO of water in each jar will also need to assemble lab equipment and materials to perform this analysis technique (see page 108).

To help students work through this activity smoothly they should prepare themselves for the work they will be performing by creating a data table like the following one. From the table students may infer they will need eight jars (A through H), a place in which to expose some jars (A, C, E, G) to light and a second place in which to prevent jars (B, D, F, H) from being exposed to light. Before filling their eight jars with water students should measure the DO and temperature (to the nearest 0.1 degree Celsius) of the water in the container they will be using to fill their jars and record this information in column A of their table.

Once this data is collected, students should be able to obtain enough fresh water from the water collected from freshwater classroom systems on the previous day to completely fill all eight jars with water (students should be reminded that when DO must be measured it is important that the samples contain no air bubbles). Once jars A through H are filled with water students should place Anacharis in jars E through H, a living aquatic organism (guppy) in jars C, D, G, and H (remember – no air bubbles). All eight completely filled (and labeled) jars should then be tightly covered and placed in darkness or light.

After at least 90 minutes have elapsed students should work methodically to collect data from each jar, carefully release the living organisms back to a healthy holding tank and fill in the rest of the table. The table should provide enough space for students to record all of their data and calculations directly in it. Once all data has been collected students may be guided to review the data in order to identify patterns that seem to exist. As student teams are guided in the process of pooling of their data students may become more certain about conclusions they may make about the impact that any of the life forms in this activity had on levels of dissolved oxygen in the water.

Each team of students participating in this activity should be able to use the class's pooled data to formulate at least one conclusion about the influence of life forms in water on concentrations of dissolved oxygen in water.

*** It is important that students performing the activity on this page have sufficient skill and background understanding of the completed aquaponics system and the Photosynthesis and Respiration skills and knowledge presented here to effectively lead students in other groups so that they can also understand the techniques and perform them (per the "Sharing of Knowledge and Accountability" activity on page 141, 165, and 186).**

Photosynthesis and Respiration in Classroom Aquatic Ecosystems

	Jar	A		B		C	D	E	F	G	H
	Light or Dark	Light		Dark		Light	Dark	Light	Dark	Light	Dark
	Jar Treatment	Water Only		Water Only		Fish Only (in water)	Fish Only (in water)	Plant Only (in water)	Plant Only (in water)	Plant and Fish (in water)	Plant and Fish (in water)
1	Temperature of Water (to the nearest 0.1 degree Celsius)	Time 1	Time 2	Time 1	Time 2						
2	Dissolved Oxygen (mg/L)	Time 1	Time 2	Time 1	Time 2						
3	Minutes kept in light/darkness/light (90 minutes minimum)										
4	Change in DO concentration From DO of water in the original sample	A2i – A2ii		B2i – B2ii		Change = A2i – C2	Change = B2i – D2	Change = A2i – E2	Change = B2i – F2	Change = A2i – G2	Change = B2i – H2
5	Calculate the rate of oxygen change per hour	A4/hours		B4/hours.		C4/hours	D4/hours	E4/hours	F4/hours	G4/hours	H4/hours
6	Subtract the change in DO mg/h (observed in the control jars from the change in DO observed in jars with a fish or plant (mg/h)					C5 – A5	D5 – B5	E5 – A5	F5 – B5	G5 – A5	H5 – B5
7	Before replacing each fish and plant into a holding tank briefly and gently place each organism on a paper towel, remove excess water from their bodies an transfer them to a beaker with water that has been weighed previously on a balance. Record each plant and animals weight in grams										
8	Divide the calculated change in dissolved oxygen (in last step) (mg/h) by the mass of organisms (grams of fish or aquatic plant) giving the oxygen consumption or generation as mg O2/g organisms/h.					C6 / mass	D6 / mass	E6 / mass	F6 / mass	G6 / mass	H6 / mass

Coordinated Sharing of Aquaponic System Development and Operation Information with Classmates

For classes that followed the division of labor approach to assembling their classroom aquatic ecosystems, it should now be apparent that, working independently of one another, three student groups (or more) have each completed the design and assembly of one classroom aquatic ecosystem. The project four concluding activity described here is designed to be synchronized with project five and six concluding activities (see pages 165 and 186). As with the chapter two concluding activity on page 121, it is again assumed that students from each respective project team possess sufficient skill and background understanding of their ecosystem and activities presented in their respective projects to effectively guide students from other ecosystem teams to also understand their ecosystem and to capably perform the respective techniques featured in their project

Once all systems are operating and students in each "home group" are familiar with their system related information, the students are prepared to engage in a coordinated cycle of information sharing and accountability. This activity begins by having students in each of the three groups (aquaponics, coral propagation, and broodstock larval rearing) count off by twos. The following chart may be much easier to use than this verbal description. It is important to provide each student group with the paperwork and forms (contained in the text of each project) that were necessary for each group of students to complete.

According to the chart, the "ones," who assembled the aquaponics system will work to guide students ("number ones") who assembled the coral propagation system through the seed germination and respiration-related activities. According to the chart, as the X1 students lead the Y1's, the Y2 students will lead the Z1's through calcium carbonate egg plug production and coral fragging related activities and the Z2 students will lead the X2's through fish setting up black worm culture systems and decapsulating brine shrimp cysts. As this activity concludes, students shall have completed the first cycle of sharing specific aquatic ecosystem-related knowledge and skills. In cycle two, students in the X1 group lead students in the Z1 group. Students in the Y2 group lead the X2 group, and students in the Z2 group lead students in the Y1 group. As the chart below indicates, at the end of cycle two, the ones and twos in each group exchange leader/learner roles and pass through cycles three and four.

Leaders Cycles 1 & 2	**Learners Cycle 1**	**Learners Cycle 2**	**Leaders Cycle 3 & 4**	**Learners Cycle 3**	**Learners Cycle 4**
X1 seed germination photosynthesis and respiration	**Y1**	**Z1**	**X2** seed germination photosynthesis and respiration	**Y2**	**Z2**
Y2 calcium carbonate egg plug production and coral fragging	**Z1**	**X2**	**Y1** calcium carbonate egg plug production and coral fragging	**Z2**	**X1**
Z2 setting up black worm culture systems and decapsulating brine shrimp cysts	**X2**	**Y1**	**Z1** setting up black worm culture systems and decapsulating brine shrimp cysts	**X1**	**Y2**

This coordinated sharing activity is designed so that at its conclusion, students from each team will be familiar with all topics and issues that are relevant to the design, assembly, and operation of each of the three student-designed classroom aquatic ecosystems. Students may be asked to use details of each classroom system to develop a comparison and contrast poster.

Write a list of **science-based** activities you have participated in during this project and briefly describe evidence that verifies your interaction with and knowledge of scientific concepts for each item on your project four list.

Write a list of **technologically-oriented** activities you have participated in during this project and briefly describe evidence that verifies your interaction with and knowledge of technologically-oriented concepts for each item on your project four list.

Write a list of **engineering-oriented** activities you have participated in during this project and briefly describe evidence that verifies your interaction with and knowledge of engineering-oriented concepts for each item on your project four list.

Write a list of **mathematical** activities you have participated in during this project and briefly describe evidence that verifies your interaction with and knowledge of mathematical concepts for each item on your project four list.

Write a brief list of **Non-STEM-related** activities you have participated in during this project and briefly describe evidence that verifies your interaction with and knowledge of Non-STEM related concepts for each item on your project four list.

Rate your level of involvement in STEM-related activities in this project (1=lowest, 10=highest).

1—2—3—4—5—6—7—8—9—10

***Beyond the Classroom Aquarium* Daily Journal Guidelines**

Entries students place in their daily journals should be:

1. Dated accurately and logged chronologically.
2. Kept in a notebook and written neatly in ink or typed.
3. Written or typed in complete sentences using correct capitalization, spelling, and punctuation.

In their journals, students should keep a daily record of:

1. Who they worked with each day by including the names of school staff, fellow students, community members, and volunteers.
2. What they did each day, such as lecture topic, activities or project design, assembly, or maintenance.
3. What they learned each day by asking questions, following instructions, and/or observing activities.
4. How *Beyond the Classroom Aquarium* experiences relate to past, present, or future life experiences.

For bonus points, students may complete these journal options, including illustrations, diagrams, or creative expression relating to the journal entry. For project one, journal writing bonus topics may include answers to the following questions or descriptions of a project activity or assignment. Reflection on these topics may also help students prepare for exams, quizzes, and other assessments.

- What do you want to understand better in order to contribute more to your team's productivity as it independently designs, assembles, and then manages an aquatic ecosystem?
- Create additional posters and signs so that people who visit your aquaponics system will be able to read about new and important system components and organisms.
- Create a new drawing and story about an organism or system that is most interesting to you.
- Create a journal entry that tells what information you learned during project four. Also write down what you would like to know more about and if you know someone else who might help you learn more if you asked them.

- Select a book or magazine to read that is related to a topic in project four. Write a report on this book and include your report in your *Beyond the Classroom Aquarium* portfolio.
- Explain three important ideas and/or skills you gained by participating in the aquaponics field trip workshop.
- Explain how and why governmental agencies are concerned about aquaponics. Explain progress they are making and what measures need to be adopted in the future to improve the effectiveness of their policies and actions.

Journal entries will be evaluated on a point system, with four points being the maximum number a student can earn. A half-point will be awarded for each of the first eight journal requirements above. Bonus points will be awarded when journal options are completed. Journal points will be totaled and divided by the number of journal dates. A possible journal grade based on points is included in the project one assessment section.

Additional Project Four Assessment Activities and Questions

Create illustrations that depict applications of mathematical topics from your individualized math curriculum (possibly schematic drawings of systems and related calculations; interpretations of system data – descriptive statistics; data technologies; etc.)

Research and report upon how bell siphons and other surge devices may be used in aquatic ecosystem designs to simulate fluctuations of tides and the cyclic and/or periodic flow and oxygenation produced by incoming waves as they crash on the shores.

Work with a team of students to develop a design/assembly proposal for attaching additional grow beds to the aquaponic system, through which water may be diverted and in which bell-siphons are utilized. Include an explanation of the principles upon which the operation of a bell siphon relies. Also include a detailed recommendation of protocols for monitoring and maintaining proper water pH and E.C./TDS levels of water soluble plant nutrients in attached hydroponic grow beds (for class critique).

Aquaponic system assessment items from Nicol (1990) include:

Define hydroponics. What are some advantages it has over conventional agriculture? What are some drawbacks? To what extent is this technology being used in the United States today?

Observe the plants growing in the hydroponic garden. Note any indications of using the descriptions of plant nutrient deficiency symptoms. Diagnose any presumed deficiencies and recommend corrective measures.

Define aquaculture. What are its advantages over traditional techniques? What are its disadvantages? How does the U.S. compare in its use of this technology to other countries?

What is the white film found at various places in the system? Why is it found in these locations?

Take a sample of the green film from the side of the system, make a wet mount of it, observe under our microscope, and make a sketch of the organisms you see. What impact would each of these have on the ecosystem?

What is the principal nitrogen containing waste product of the fish in the system? What is the principle nitrogen containing waste product of a human? Why are they different? Why would the absence of plants endanger the fish? How do plants secure their nitrogen and what do they do with it?

What is the fishes' source of nitrogen? Trace a possible path of the nitrogen from the fish to plant?

What is the fishes' carbon source? Trace a possible path of carbon from fish to plant.

How do the fish get oxygen? How do the plants get their oxygen?

What chemicals are constantly being lost from this system and need to be replenished?

Describe for ways in which water is being lost from the system.

Describe a simple method of obtaining an estimate of the amount of water lost by the hydroponic plants through transpiration.

What is the source of nutrients like calcium and iron for the system?

State Liebig's law of the minimum and give examples of how it relates to system productivity.

What might cause calcium or iron to be the limiting factor for plant growth in this system?

What is chelated iron? If an iron deficiency is apparent why is chelated iron gluconate used rather than something like iron oxide (rust)?

Develop a brief report on regulation and control of temperatures in aquatic ecosystems; be sure to include factors that influence ambient temperatures and precautions that may be taken to prevent drastic and unexpected temperature changes.

Make a chart or diagram of operational system water quality parameters, ways these parameters may be disturbed, and ways to ensure system parameters may be maintained even in cases of power outage.

Develop a presentation about germination techniques that the class should adopt for unique plants that are not in their system.

Once plants become large enough, they may be removed from the aquaponics system. Newly germinated seeds may be used to replace plants that are overgrowing the system. Devise ways to sustain or utilize plants that are removed from the aquaponics system.

Identify places where agricultural/hydroponic topics make use of mangroves and sea grasses. Describe these programs and provide details for how systems for vegetation of this type could be developed. Work with a team of students to prepare a design for the class to critique.

Use other sources to outline and report upon ten historically significant developments in aquaponics since the 18th century. Develop a poster board to accompany the class presentation/critique.

Create an aquaponics system daily checklist and upon which observation protocols, system and organisms data may be logged each day.

Create a poster presentation describing your work on the photosynthesis and respiration activity. Explain your interpretation of the data that was collected. In addition, include data results that other students collected and compare the pooled results with your own.

Work with a group of students in your team and design a web page layout that you could use to feature the classroom aquaponics system. Explain the value of the information contained in your web design and describe who might find your web site valuable. Prepare this layout for classroom critique.

Review activities and information that interested you in this project and determine if this information may fit on the "Research-ability" Potential Topics Form on page 59. If so, add it to the list.

Beyond the Classroom Aquarium STEM Self or Peer Assessment Rubric

This self-assessment is valid only to the extent to which you can document and display evidence that supports it. For each category in the left hand column, write a STEM-Oriented example in the box beneath 3 if you can display evidence that your STEM-Oriented example represents work that exceeds expectations. If your evidence shows that your STEM-Oriented statement meets but does not exceed expectations, write the statement in the box beneath 2. Similarly, for STEM-Oriented examples that demonstrate some progress toward expectations.

Name: **Project Number:** **Date Project Work Began:** **Date of Project Completion:** **List Math Topics and/or Objectives Currently Under Study:**		**3** **Work Exceeds Expectations**	**2** **Work Meets Expectations**	**1** **Work Is Progressing Toward Expectations**
Write examples of questions and defining problems related to your STEM work.				
Describe how you have developed and used models in your STEM work.				
Write examples of STEM investigations you planned and carried out.				
Describe STEM related data that you analyzed or interpreted.				
Give examples of how you used mathematics and computational thinking related to the five standards listed to the right.	Problem Solving (Math Process Standard One)			
	Reasoning & Proof (Math Process Standard Two)			
	Communication (Math Process Standard Three)			
	Connections (Math Process Standard Four)			
	Representation (Math Process Standard Five)			
Describe STEM-Oriented explanations and solutions you designed				
Explain evidence you used while engaging in a STEM project-based argument.				
Give examples of STEM information you obtained, evaluated, and/or communicated				
TOTAL POINTS				

Photo by Robert Jakus

By developing an awareness of the delicate and dynamic balance of light, temperature, and chemistry upon which the survival of corals depend, students will be well prepared to improve their system design and develop ideas of ways to monitor and collect system data and respond appropriately to changes in parameters of system data (and avoid the expensive costs of system failure and coral death).

The primary student task for project five involves completion of the assembly of a student designed, coral propagation system that began at the end of project one. In project five, students will also: develop calcium carbonate egg crate plugs on which to mount coral "frags," propagate selected corals (of different species) by cutting coral frag from larger parent colonies, install a protein skimmer in the system, design and assemble a calcium reactor, and introduce live corals to an aquatic environment that will maximize coral growth. Students are introduced to a study of coral reefs, coral bleaching, and ecological awareness. Students study the potentials of profitable production of coral for local markets (business-related calculations) and establish ways to raise money from this project. Ongoing data collection occurs with the coral propagation system and many student teams will find it essential for their upcoming research projects.

Coral Frag Substrate Production - Project Five Opening Activity

This procedure is based on the work of Vendetti (2000). Assemble all tools and materials in a work location. Have a garden hose in place and ready – running at a slow trickle where the concrete is to be mixed. Materials include: a 60-pound bag of Portland concrete sand mix, an 18-pound bag of "CaribSea Aragonite Reef Sand," and 35 empty egg cartons (one dozen per carton) collected by the class. In addition, students will need a few gallons of inexpensive white vinegar and two 5-gallon buckets. Tools and equipment should include: a gardening hoe (for mixing the concrete), a garden hose (for adding water to the concrete mix), and a wheel barrow or similarly shaped container (in which concrete may be mixed). Ratios of six parts Aragonite to one part cement or four parts crushed oyster shell to one part cement may be used.

Concrete should be poured and mixed outside if possible.
Students should avoid breathing the concrete dust.
For hand protection, students may prefer to use latex gloves.
Make sure egg cartons are clean; rinse them with clean water.

Before adding water, mix the cement mix and the reef sand (calcium carbonate) in the wheel barrow. Try to get an even mix before adding the water. Then add water to the mix. Never use so much water that puddles form in the mix. If puddles occur, add leftover cement mix to dry out the mixture. Stir the mixture. The quantity of material listed above produces about 400 plugs.

Wikipedia Commons

The mixture is complete when the consistency of the mix allows students to get it to clump up in their fingers and not fall off as they roll it between their fingers. If the consistency is too wet, it will spill; if it is "just right," it will sit in a clump on your fingers. Look for clods of dry mix and be sure to break them up. Move the wheel barrow to the working area, preferably somewhere where students can be seated. Fill the egg cartons with the mixture (either lay the mixture in the egg carton mold and then press it in firmly or by picking up a reasonable amount of mixture and dropping it into the mold). Pressing down on filled molds helps prevent the formation of air pockets that cause the new plug to be weak and erode in the coral propagation system.

After using up all of the egg plug concrete that they mixed up, students should clean off all tools and wash their hands. Students must leave the egg carton molds in a space that will be warmer than 50° F for three or four days. This period of time is when the "plugs" cure. After the plugs have cured sufficiently they can usually be removed from the egg carton by popping them out from the bottom. If plugs are stuck together, students may break them apart using a chisel or by hand. Once the plugs have been removed from the egg cartons, they must be placed in a five-gallon bucket and be covered with white vinegar for 24 hours. Remove the pickled plugs from the bucket of white vinegar and place them into a second five-gallon bucket, and cover them with tap water for one week. Once the week has passed, the plugs may be removed from the water and allowed to dry until a coral fragment is attached to each plug and placed in the coral propagation system to grow.

*** It is important that students performing the activity on the previous page have sufficient skill and background understanding of calcium carbonate Coral Frag Substrate Production as presented here to effectively lead students in other groups so that they can also understand the techniques and perform them (per the "Sharing of Knowledge and Accountability" activity on page 141, 165, and 186).**

Introduction to Coral Related Topics

More than 50 million years ago, way before the first pyramid or even a crude hut, corals began construction of the largest structures ever built. Time passed, and the Earth was changed. Fighting waves and winds, dodging glaciers rising and falling with the level of the seas moving across the face of the planet with the continents these organisms, half plant and half animal, blind, unthinking but very persistent, toiled to the rhythms of the oceans, steadily building up and out. Capturing ions dissolved in the sea, they produced tiny and fragile crystals welded to older tissues day after day millennium after millennium. These accretions grew into massive boulders, and slowly, very slowly, into giant walls and buttresses that could withstand the mighty poundings of the ocean. Coral reefs are among the most ancient of ecosystem types, dating back to the Mesozoic era some 225 million years ago. Modern reefs can be as much as 2.5 million years old.

Although they cover only a tiny fraction (less than 0.2 percent) of the ocean's bottom, coral reefs capture about half of all the calcium flowing into the ocean every year, fixing it into calcium carbonate rock at very high rates. Coral reefs release carbon dioxide to the atmosphere due to the chemistry of calcium carbonate precipitation. The release of carbon dioxide from coral reefs is very small (probably less than 100 million tons of carbon per year) relative to emissions due to fossil fuel combustion (about 5.7 billion tons of carbon per year).

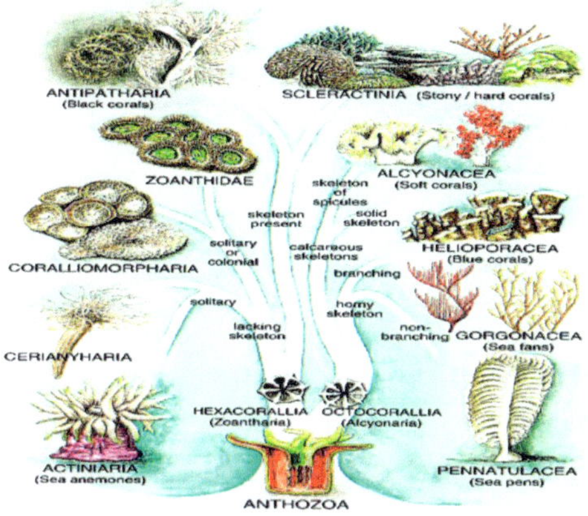

George Parsons Shedd Aquarium PPT image

Coral reefs store very little organic carbon and are not very effective "sinks" for carbon dioxide from the atmosphere. Forests are more effective sinks for atmospheric carbon because they convert carbon dioxide into long-lived structures (trees).

Coral reefs are among the most biologically diverse ecosystems on the planet. Although tropical rainforests contain more species than coral reefs, reefs contain more phyla than rainforests. Phyla are large groupings of organisms that are thought to be related. Covering less than 0.2 percent of the ocean floor, coral reefs contain perhaps 25 percent of all marine species. New studies indicate that biodiversity on coral reefs may be even higher.

Despite their limited area, coral reefs may be home to up to 25 percent of the fish catch of developing countries or 10 percent of the total amount of fish caught globally for human consumption as food. Reefs are also extremely valuable as breakwaters and draws for tourism.

Coral reefs are among the most endangered ecosystems on earth. Coral reefs in 93 of the 109 countries containing them have been damaged or destroyed by human activities. In addition, human impacts may have directly or indirectly caused the death of five to 10 percent of the world's living reefs, and if the pace of destruction is maintained, another 60 percent could be lost in the next 20 to 40 years.

The most important short-term threats to coral reefs are sedimentation (from poor land use such as clear-cutting on steep slopes and other activities such as dredging without silt curtains), eutrophication (over-fertilization caused by excessive fertilizer use and sewage pollution), and overfishing. Destructive fishing techniques such as fine mesh nets, cyanide poisoning, and dynamiting are common in coral reefs and have actually come to dominate fishing in parts of Indonesia and the Philippines. Physical damage to coral reefs by scuba divers and tourists would probably be a minor threat if the number of visitors to reefs were limited to moderate levels and if water quality was always high enough to support rapid recovery of corals. However, tourism often results in large numbers of visitors, which leads to extensive physical damage, sewage pollution, and other adverse water quality impacts which slow or eliminate recovery.

Coral reefs in every major tropical region of the world bleached white during the mass bleaching events of the 1980s. This bleaching depresses coral growth rates and in some cases results in mass coral mortality and enormous aquatic population loss and can even contribute to potential species extinctions. Bleaching is caused by a variety of factors, including siltation and changes in salinity resulting from poor land use, pollution, and slight increases in temperature. Coral reefs may bleach even more extensively if global warming continues unabated.

The United Nations Framework Convention on Climate Change and the United Nations Convention on Biological Diversity, both signed at the 1992 Rio Earth Summit, seek to protect sensitive ecosystems from the effects of climate change and to enhance protection of areas of high biological diversity such as coral reefs.

A new focus by pharmaceutical firms and scientists on coral reefs as a source of new medicines may generate a new form of sustainable development in coral reef areas. This will be contingent upon the continued development of a trend toward contracts designed to compensate less developed countries, indigenous peoples, and other effected parties for intellectual property and genetic resources used to develop medicines, thereby creating economic incentives to protect biological diversity.

Marine fisheries reserves have proven to be effective in restoring and maintaining fisheries resources. For example, the establishment of the Sumilon Island Reserve in the Philippines resulted in about twice the density of fish than in adjacent areas after nine years of protection. The catch-per-unit-effort was also about twice as high while protective measures were in effect compared to a period when they were not. Small scale waste treatment systems, employing both biological and chemical methods, are available for cost-effective treatment of sewage that will protect water quality in coral reef areas. Other measures can also reduce nutrient pollution. For example, an ordinance restricting phosphate levels in detergents was adopted in the Florida Keys. It will likely result in a 50 percent reduction in phosphate pollution from septic tanks. Phosphate pollution contributes to algal blooms and coral death in many coral reefs.

flowergarden.noaa.gov
Photo Schmahl/*FGBNMS*

Marine sanctuaries, such as Australia's Great Barrier Reef Marine Park and America's Florida Keys National Marine Sanctuary, can be effective management tools if properly implemented with the meaningful participation of resource users, citizens, scientists, and environmentalists. Sanctuaries need effective research, administration and enforcement to work (Buchheim, 1999b).

Coral bleaching, the whitening of diverse invertebrate taxa, results from the loss of symbiotic zooxantheallae and/or a reduction in photosynthetic pigment concentrations in zooxanthellae, residing within scleractinian corals. Coral reef bleaching is caused by various anthropogenic and natural variations in the reef environment including sea temperature, solar irradiance, sedimentation, xenobiotics, subaerial exposure, inorganic nutrients, freshwater dilution, and epizootics. Coral bleaching events have been increasing in both frequency and extent worldwide in the past 20 years. Global climate change may play a role in the increase in coral bleaching events and could cause the destruction of major reef tracts and the extinction of many coral species.

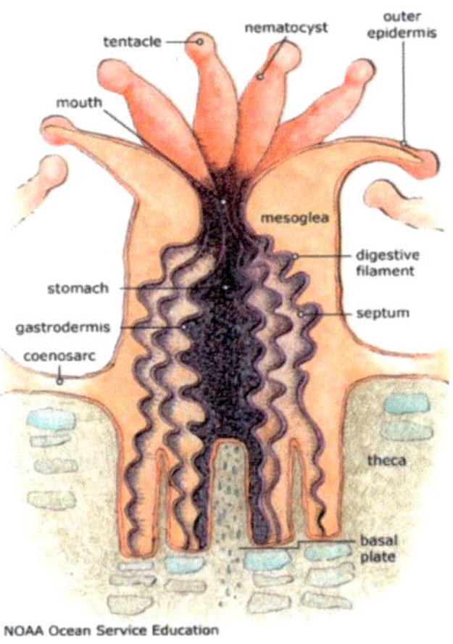

http://upload.wikimedia.org/wikipedia/commons/4/49/Coral_polyp.jpg

Coral reefs are found in circumtropical shallow tropical waters along the shores of islands and continents. The reef substrate is mainly composed of calcium carbonate from living and dead scleractinian corals. Many other invertebrates, vertebrates, and plants live in close association to the scleractinian corals, with tight resource coupling and recycling, allowing coral reefs to have extremely high productivity and biodiversity, such that they are referred to as "the Tropical Rainforests of the Oceans."

Corals live in very nutrient-poor waters and have certain zones of tolerance to water temperature, salinity, UV radiation, opacity, and nutrient quantities.

Scleractinian corals build skeletons of calcium carbonate sequestered from the water. When the coral polyp dies, this skeleton remains incorporated in the reef framework.

Scleractinian corals are in the Phylum Cnidaria, and they receive their nutrient and energy resources in two ways. They use the traditional cnidarian strategy of capturing tiny planktonic organisms with their nematocyst capped tentacles, as well as having an obligate symbiotic relationship with a single celled alga known as zooxanthellae. Zooxanthellae are autotrophic microalgae belonging to various taxa in the Phylum Dinoflagellata.

Zooxanthellae live symbiotically within the coral polyp tissues and assist the coral in nutrient production through its photosynthetic activities. These activities provide the coral with fixed carbon compounds for energy, enhance calcification, and mediate elemental nutrient flux. The host coral polyp in return provides its zooxanthellae with a protected environment to live within, and a steady supply of carbon dioxide for its photosynthetic processes. The symbiotic relationship allows the slow-growing corals to compete with the faster growing multicellular algae because the tight coupling of resources and the fact that the corals can feed by day through photosynthesis and by night through predation.

Wikimedia Commons

The tissues of corals themselves are actually not the beautiful colors of the coral reef but are instead clear. The corals receive their coloration from the zooxanthellae living within their tissues.

In *Tales of the South Pacific*, (Michener, 1947 p. 284) describes coral and the militaristic use of to which coral, surrounding the island, was put in building a military airstrip on Michener's fictionalized south Pacific island of" Konora" in 1942 (during World War II). He reported that coral grows in shapes that resemble plantlike branches and in "many colors, ranging from exquisite pastel greens to violent, bleeding reds. There is blue coral, orange, purple, gray, amethyst, and even now and then a bush of stark black coral." In describing the prominent anatomical features of the corals his crew crushed up and used to build a solid landing strip for warplanes, he reported that the coral "was composed of a stony base, already calcified. Next that was a pulpy, mineral segment, pale white in color. The extreme tip was, almost purely, vegetable. It exuded a sticky-milk; which smelled noxiously. Over all were suction cups like those on the tentacles of an octopus." Michener also observed potential tips on the coral branches, which had not matured.

Wikimedia Commons

After describing the coral, Michener exclaimed how incredible it was to realize that "this tiny organism had raised the island on which we stood and was at that moment raising thousands of new islands throughout the Pacific, most of which would never break the waves but would remain subterranean palaces of rare wonder." Michener's description of how bulldozers and steam shovels raked up submerged coral surrounding the island and transported the calcified coral material to the airstrip site where steamrollers would create a smooth surface for warplanes to land by crushing deep layers of the coral into a hard flattened surface is almost as incredible an idea for me to realize. In his description he says: "they sank a-half-dozen massive steel maws onto the coral bottom. The maws were slowly dragged onto the beach. A steam shovel scooped up the live coral and filled trucks waiting for the loads. Trees could not be used as filler for the airstrip itself. . . . Bulldozers slowly pushed the top layer of coral toward the ravine. Coral from the hillside rumbled to the airstrip 24 hours a day. At the same time, live coral from the sea was hauled to the two ends. Six team rollers worked back and forth constantly."

Wikipedia.org (Public Domain)

Buchheim (1998a) in his introductory description of coral reef bleaching indicates:

Coral reef ecosystems world-wide have been subject to unprecedented degradation over the past few decades; including being used as testing grounds for detonations of thermo-nuclear devices (such as atomic and hydrogen bombs). The Canadian Network to Abolish Nuclear Weapons provides the image at the left to suggest the destructive force of thermo nuclear devices. Disturbances affecting coral reefs also include anthropogenic and natural events. Recent accelerations in coral reef decline seem to be related mostly to anthropogenic impacts (overexploitation, overfishing, increased sedimentation and nutrient overloading). Natural disturbances which cause damage to coral reefs include violent storms, flooding, high and low temperature extremes, El Nino Southern Oscillation (ENSO) events, subaerial exposures, predatory outbreaks and epizootics.

Coral reef bleaching is a common stress response of corals to many of the various disturbances mentioned above. Beginning in the 1980s, the frequency and widespread distribution of reported coral reef bleaching events increased. Widespread bleaching, involving major coral reef regions and resulting in mass coral mortality has raised concerns about linkage of the events to global phenomena including global warming or climate change and increased **UV** radiation from ozone depletion. This paper examines the causes of coral reef bleaching and addresses the impact of global climate change on coral reefs.

Bleaching, or the paling of zooxanthellate invertebrates, occurs when (i) the densities of zooxanthellae decline and/or (ii) the concentration of photosynthetic pigments within the zooxanthellae fall (Kleppel et al. 1989). Most reef-building corals normally contain around 1-5 x 10^6 zooxanthellae cm^2 of live surface tissue and 2-10 phosphatidylglycerol (**PG**) of chlorophyll-A per zooxanthellae. When corals bleach, they commonly lose 60-90% of their zooxanthellae. Each zooxanthellae invertebrate that remains within a bleaching coral polyp may lose 50-80% of its photosynthetic pigments (Glynn 1996). The pale appearance of bleached scleractinian corals and hydrocorals is due to the cnidarian's calcareous skeleton showing through the translucent tissues (that are nearly devoid of pigmented zooxanthellae).

If the stress-causing bleaching is not too severe and if it decreases in time, the affected corals usually regain their symbiotic algae within several weeks or a few months. If zooxanthellae loss is prolonged, i.e. if the stress continues and depleted zooxanthellae populations do not recover, the coral host eventually dies.

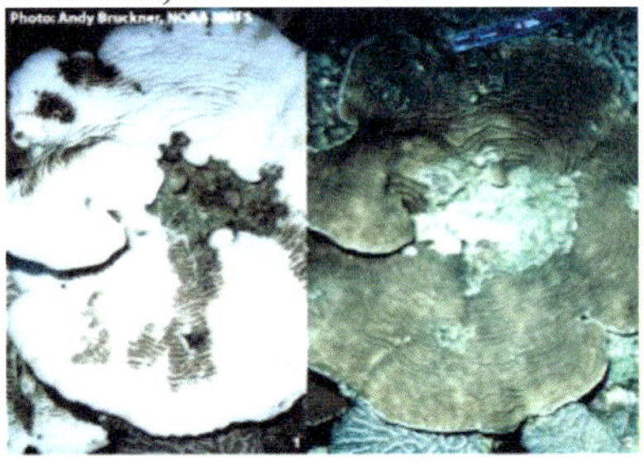
NOAA - Bruckner, 2001

Three hypotheses have been advanced to explain the cellular mechanism of bleaching, and all are based on extreme sea temperatures as one of the causative factors. High temperature and irradiance stressors have been implicated in the disruption of enzyme systems in zooxanthellae that offer protection against oxygen toxicity. Photosynthesis pathways in zooxanthellae are impaired at temperatures above 30˚C this effect could activate the disassociation of coral/algal symbiosis. Low- or high-temperature shocks results in zooxanthellae low as a result of cell adhesion dysfunction. This involves the detachment of cnidarian endodermal cells with their zooxanthellae and the eventual expulsion of both cell types.

It has been hypothesized that bleaching is an adaptive mechanism which allows the coral to be repopulated with a different type of zooxanthellae, possibly conferring greater stress resistance. Different strains of zooxanthellae exist both between and within different species of coral hosts, and the different strains of algae show varied physiological responses to both temperature and irradiance exposure. The coral/algal association may have the scope to adapt within a coral's lifetime. Such adaptations could be either genetic or phenotypic.

As coral reef bleaching is a general response to stress, it can be induced by a variety of factors, alone or in combination. It is therefore difficult to unequivocally identify the causes for bleaching events. The following stressors have been implicated in coral reef bleaching events.

Coral species live within a relatively narrow temperature margin and anomalously low and high sea temperatures can induce coral bleaching. Bleaching events occur during sudden temperature drops accompanying intense upwelling episodes, (-3˚C to –5˚C for 5-10 days), seasonal cold-air outbreaks. Bleaching is much more frequently reported from elevated se water temperature. A small positive anomaly of 1-2˚C for 5-10 weeks during the summer season will usually induce bleaching.

Bleaching during the summer months, during seasonal temperature and irradiance maxima often occurs disproportionately in shallow-living corals and on the exposed summits of colonies. Solar radiation has been suspected to play a role in coral bleaching. Both photosynthetically active radiation (PAR, 400-700nm) and ultraviolet radiation (UVR, 280-400nm) have been implicated in bleaching.

Sudden exposure of reef flat corals to the atmosphere during events such as extreme low tides, **ENSO**-related sea level drops or tectonic uplift can potentially induce bleaching. The consequent exposure to high or low temperatures, increased solar radiation, desiccation, and sea water dilution by heavy rains could all play a role in zooxanthellae loss, but could also very well lead to coral death.

Relatively few instances of coral bleaching have been linked solely to sediment. It is possible, but has not been demonstrated, that sediment loading could make zooxanthellate species more likely to bleach.

Rapid dilution of reef waters from storm-generated precipitation and runoff has been demonstrated to cause coral reef bleaching. Generally, such bleaching events are rare and confined to relatively small, nearshore areas.

Rather than causing coral reef bleaching, an increase in ambient elemental nutrient concentrations (e.g. ammonia and nitrate) actually increases zooxanthellae densities 2-3 times. Although eutrophication is not directly involved in zooxanthellae loss, it could cause secondary adverse effects such as lowering of coral resistance and greater susceptibility to diseases.

Zooxanthellae loss occurs during exposure of coral to elevated concentrations of various chemical contaminants, such as Cu, herbicides and oil. Because high concentrations of xenobiotics are required to induce zooxanthellae loss, bleaching from such sources is usually extremely localized and / or transitory.

Pathogen induced bleaching is different from other sorts of bleaching. Most coral diseases cause patchy or whole colony death and sloughing of soft tissues, resulting in a white skeleton (not to be confused with bleached corals). A few pathogens have been identified the cause translucent white tissues, a protozoan.

Mass coral moralities in coral reef ecosystems have been reported in all major reef provinces since the 1970's. The frequency and scale of bleaching disturbances has increased dramatically since the late 70's. This is possibly due to more observers and a greater interest in reporting in recent years. More than 60 coral reef bleaching events out of 105 mass coral moralities were reported between1979-1990, compared with only three bleaching events among 63 mass coral moralities recorded during the preceding 103 years.

Nearly all of the world's major coral reef regions (Caribbean/ western Atlantic, eastern Pacific, central and western Pacific, Indian Ocean, Arabian Gulf, Red Sea) experienced some degree of coral bleaching and mortality during the 1980s.

Wikipedia Commons

Prior to the 1980s, most mass coral moralities were related to non-thermal disturbances such as storms, aerial exposures during extreme low tides, and *Acanthaster* outbreaks. Coral bleaching accompanied some of the mortality events prior to the 1980s during periods of elevated sea water temperature, but these disturbances were geographically isolated and restricted to particular reefs zones. In contrast, many of the coral bleaching events observed in the 1980s occurred over large geographic regions and at all depths; and most of these bleaching events of occurred during years of large-scale ENSO activity.

Of the causing stressors of coral reef bleaching, many are related to local environmental degradation and reef overexploitation. Of the stressors mentioned above, only sea water temperature and solar irradiance have possible global factors driving changes and extremes. Global warming along with ENSO events changes sea water temperatures. Ozone depletion increases the amount of UVR reaching the Earth's surface, and possibly causing coral bleaching events.

Increased sea temperatures and solar radiation (especially UV radiation), either separately or in combination, have received consideration as plausible large-scale stressors. In most instances, wherever coral reef bleaching was reported, it occurred during the summer season or near the end of a protracted warming period.

Coral bleaching was reported to have occurred during periods of low wind velocity, clear skies, calm seas and low turbidity, when conditions favor localized heating and high penetration of short wave length (UV) radiation. Also less oxygen is held by water at higher temperatures. Potentially stressful high sea temperatures and UV radiation flux could conceivably cause coral reef bleaching on a global scale with suspected greenhouse warming and the thinning of the ozone layer.

As reef building corals live near their upper thermal tolerance limits, small increases in sea temperature (0.5–1.5°C) over several weeks or large increases (3-4°C) over a few days will lead to coral dysfunction and death. Anomalously high sea temperatures have often been reported in the Caribbean-wide series of bleaching events that occurred during 1986-88, leading to hypothesis that global warming was having an effect on the coral reefs in this region. A large-scale analysis of the relationship between coral reef bleaching and positive Sea Surface Temperature (SST) anomalies in the tropical western Atlantic, based on observations from ships and buoys, failed to disclose significant sea warming during major coral bleaching in 1987. In contrast, blended high resolution satellite **SST** data have shown that coral reef bleaching occurred during significant temperature increases at 5 of 7 Caribbean sites in the 1980s.

Solar ultraviolet radiation is potentially harmful to reef corals and their symbiotic. UV radiation can readily penetrate clear sea water, and reef–building corals contain UV-absorbing compounds capable of blocking potentially damaging UV radiation. These compounds are produced in response to ambient UV levels and the concentration in corals is usually an inverse function of depth, but it is not known if bleaching responses are related to variations in UV flux that exceed the protective capacity of UV-absorbing compounds. There is a possible interaction between temperature and UV, with temperature significantly reducing zooxanthellae densities and also the concentration of UV absorbing compounds in a reef zoanthid, thus potentially increasing the exposure of the symbionts to the direct effects of UV radiation.

If a global warming trend impacts on shallow tropical and subtropical seas, we may expect an increase in the frequency, severity and scale of coral reef bleaching. Coral mortality could exceed 95% regionally with species extirpation and extinctions. A conservative temperature increase of 1-2° C would cause regions between 20-30 degrees N to experience sustained warming that falls within the lethal limits of most reef-building coral species. In conjunction with sea temperature rise would be a sea level rise, and it has been suggested that sea level rise would suppress coral growth or kill many corals through drowning or lower light levels. Some coral populations and their endosymbiotic zooxanthellae may be able to adapt to the extreme conditions predicted during global climate change. Refuges in benign habitats, such as deep, sunlit reef substrates, oceanic shoals and relatively high latitude locations, might exist, but widespread coral mortality and reef decline would be expected in shallow reef zones in most low latitude. Even if significant sea warming and elevated irradiance levels do not occur, coral reef degradation from anthropogenic pollution and overexploitation will still continue, a result of unrelenting human population growth.

"The Reef's Tale" (Walton, 1980), contains important coral reef related information and provides a background that helps set a historical tone and context for the coral reef activities and projects in *Beyond the Classroom Aquarium*.

A living coral reef system is thriving in captivity at the Smithsonian Institution's Museum of Natural History – the first of its kind to be maintained away from its natural habitat. The reef however, is more than a novel exhibit. It represents a research breakthrough that could transform aquariums across the nation from simple fish tanks to valuable research tools in the quest to understand the mechanisms of the marine environment.

This is the first time an ongoing research project has been located in the natural history museum. Side by side, two windows give the museum visitors a glimpse of two seldom-seen populations. Behind one thick pane of glass, they can see exotic fish and sinuous plantlike animals swaying gently with the water's motion. Around a corner, behind another pane of glass, they can watch a small group of marine biologists as they go about the business of monitoring the reefs activity, accumulating data as imperceptibly as the reef accumulates new coral skeletons. It is for many persons, a sight no less unusual than damselfish and stony corals.

For the museum of natural history, this felicitous blending of exhibit and research is a welcome development. Museum director, Richard S. Fiske notes that visitors used to pass through the entire museum unaware of the 120 or so researchers two work in its labs or use it as a home base for their fieldwork. But this exhibit includes not only an explanation of the reef, but also explanations of the research projects in progress, which change with the research.

The whole arrangement is the product of nearly five years of thought, work, and planning by Walter H. Adey, a marine botanist and museum curator, with the staff of the Marine Systems Laboratory, which he directs. Adey is the researcher who cracked the coral reef's code. He is by no means the first to try to keep a coral reef alive – "there's hardly a marine biologist who hasn't," he says - but he is the first to succeed.

As in any living model, the reef system is separated from the natural contingent systems on which it is dependent. In place of the open ocean, an adjacent lagoon, the sun and the waves, the reef relies on human made machines to simulate their activity. The exhibit consists of two large tanks: one 9.5 kiloliter reef tank and an attached 1 kiloliter lagoon tank. An additional 0.75 kiloliter of water circulates throughout the system. Inside the larger tank, there is a two meter high, four meter long, and one meter wide reef, created from almost 3 tons of coral rubble hauled from the Caribbean Sea to Washington aboard the 27 meter ex-fishing boat *Marlys* Resolute. The rubble is supported by 40 acrylic rods, implanted in the tanks floor. Most of the surfaces in the tank are covered with a layer of fuzzy algae, kept short by grazing fish except where it is guarded by the vigilant damselfish. The light levels on the bottom of the tank – 2 meters down – are the same as light levels 18 meters down in a natural reef.

If one looks only at the tank, ignoring the equipment that surrounds it, the scene is similar to a patch reef viewed through a diving mask. But beyond the tank's boundaries, it again becomes clear that it is a model – a very good model, but a model nevertheless. Ten 400 watt and four 1000 watt metal halide lamps hang over the main tanks. Beginning at "dawn," the lights flick on one by one, until they attain the intensity of the tropical sun at high noon. As "evening" approaches, the lights blink off again, until only the illumination that would be provided by a tropical moon remains. The moon is always full in the Washington, **DC**, part of the Caribbean.

Tides play only a minor part in the reef's survival, but the motion of the water is crucial. As one watches the tank, a surge of water descends from the upper left hand side of it. This comes from an automatic wave generator, which consists of two large buckets that fill, become unbalanced and alternately dump their contents into the tank. This creates a current that surges to down to the bottom of the tank, then moves u and over the reef's crest. Every so often, a more dramatic surge occurs when the two buckets empty simultaneously, mimicking the big waves that occur in nature.

The water comes not from the Caribbean, but from Virginia's coast – Chincoteague water. Salinity is controlled by a device that sense even slight changes in the water level caused by evaporation. The change activates a pump that releases distilled water and returns the water to optimum salinity. The exhibit is also equipped to monitor oxygen, acidity, and nutrient levels, making certain that they correspond to the levels of a natural reef system. Normally, a small amount of the reef creatures' food comes from the surrounding ocean; in its absence, a machine automatically releases a "tiny amount' of live brine shrimp and dried krill into the tank. On an ocean reef, daytime creatures creep away at sunset and night denizens emerge to feed, some moving to the lagoon. In the exhibit reef, the channels into the lagoon are opened, and reef fish swim through to feed on the tiny crustaceans that live among the lagoon grasses.

In a natural setting, a reef this size would be built up over a period of hundreds of years. The Smithsonian scientists accelerated this pace considerably, creating their reef in several months. Adey and the crew sailed to the Caribbean in June of 1980 and during the voyage installed four small reef tanks, complete with wave makers, etc., to transport the plants and animals back to Washington. The reef on exhibit has existed for three months; Adey considers it a "colonial" reef. Currently, about 200 species make their home in the tank, although more may lurk in the cracks and crevices. It is, he says, healthier than many natural reefs.

The reef's tale begins about five years ago, when Adey began working on the problem of how to grow live algae for laboratory experiments. He did not set out to create a reef, but rather to recreate the conditions in which the algae lived in nature. "I knew that in addition to the algae, other organisms are nearly as important as temperature and light, so I tried to collect small blocks of reef with as many reef animals as possible. I brought them back and placed them in a small aquarium on my window, and for a long time, I thought the natural sunlight coming through the window would be all I needed," Adey says. But he discovered that his animals thrived during sunny weeks but during cloudy weeks "everything died."

Adey began experimenting with different kinds of light – fluorescent, quartz, and mercury vapor lights. The metal halide lamps, he discovered, gave the proper effect – "sufficient light energy to simulate the sun in a small area and yet a color spectrum close to natural sunlight so that the algae could gather the same light energy they use in the wild." Adey also began experimenting with different wave simulations before figuring out that that the violent two bucket wave surge was critical: "That surge keeps the sediments off the surfaces of the plants and animals and prevents the formation of a partially 'stagnant' water layer on the surface," he explains.

These experiments provided the answer to a very basic question: What are the components of a coral reef? The most striking characteristics, Adey noted are lots of light, wave action, and the presence of algae. In a natural reef system, the waves wash continually over the "algal lawn." The lawn is also relatively close to the surface and the sunlight. These three things – light, waves, and algae – allow the coral reef to be one of the most biologically productive places on earth. By including the algae, incorporated into a filter system through which water passes, Adey deviated drastically from filter systems traditionally used in aquarium science.

Having established the parameters for survival, Adey and the lab staff set up the first, 1.5-kiloliter, pilot tank in the basement of the Museum of Natural History. It was, the Smithsonian says, "the first time that a large, representative community of tropical reef organisms, including corals, had been kept alive and the natural ecological relationships of these organisms preserved." Many of the individual species now in the reef tank have lived in aquariums before, but no one had successfully transplanted the entire system. Corals had never survived the move to a tank.

After two years, in 1977, the small pilot tank was replaced with a bigger one, holding almost 7 kiloliters. This too proved successful, and, after several years, the National Science Foundation provided an $118,000 grant that allowed the researchers to construct the large exhibit tank. Except for one Pacific fish, all of the more than 200 species in the tank are Caribbean reef fish. "This really bothered people," Adey says, so the exhibit's tank inhabitants are geographically homogeneous.

There are Elkhorn and Staghorn corals, as well as several species of brain coral. Sea fans and anemones live there too as do Gorgonians (named for the mythological snake-haired gorgons). A spiny spider crab, the largest of the crabs that inhabit the Caribbean reefs, lives there; so does a spiny lobster, which prowls the night reef, feeding on small invertebrates. A Queen conch, dining on microalgae and plant debris, lives side by side with a horned feather worm, one of several filter feeding worms. And crucial to the survival of everything are the algae and other plants, on which fish and large invertebrates dine.

Some of the tank's algae are protected by the tank's two damselfish. These grazers stake out an algae covered territory and defend it from other grazers, a behavior that is useful to the algae as well as to the damsels since it prevents overgrazing. One of the damselfish was transferred from the pilot tank, in which she had already settled contentedly into guarding her territory. The move to the museum was like "a bad dream" for her, Adey says. "All of a sudden, all of her plants were gone." But she has now readjusted and found her new territory.

Another fish, pointed out by a research assistant, has not made the move as successfully; it noses nervously among the rocks and appears, an observer notices anthropomorphically, as if it were trying to act nonchalant about the people staring at it. They may transfer that fish back to the basement or, a staff member says; they may install mirrored glass allowing the visitors to see the fish but the fish to see only a reflection of their own surroundings.

The reef's most innovative aspect is the algal scrubber system. During the reef's night, when only the halide tropical moon is shining, the tank system's water is pumped into boxes that contain algae-covered screens. The scrubbers filter out the animal wastes and radiant energy from the "sun" and convert them into a useable form. These scrubbers, or ones very similar, could be installed in other aquarium systems.

Moreover, they may have other uses besides that of maintaining the chemical balance in research tanks; Adey believes the scrubbers could be used for "ocean farming" of algae. The constant movement of water over the algae stimulates growth beyond its already rapid normal rate. By placing giant platforms around the tropical seas – oases in an otherwise minimally productive environment – scientists could grow large quantities of algae to be harvested and converted to methane fuel or cheap alcohol. The algae, unlike other potential biomass fuel sources, would require no fertilization, nor would it take cropland or food-crops in an increasingly food-short world. The Marine Systems Laboratory has applied for a patent on the algal scrubber, and meanwhile the staff scrapes algae off the scrubber every two days and weighs the accumulation. The amount harvested is, by dry weight, "much, much better than the best aquaculture," Adey says. Several years ago, giant kelp was hailed as the energy source of the future (see BioScience, May, 1976); Adey believes that algal turfs hold even greater promise.

Meanwhile, as the reef slowly changes, researchers continue to use its resources to investigate the marine environment. Current projects include one on the process of capturing solar energy in reefs and the possibility of adapting the principles for human use. The investigation of reef turfs is carried out in two ways: First, the researcher studies a miniature reef with an algal turf scrubber in which animals are largely omitted and the researcher serves as the grazer and harvester. Then the scientist examines the dry weight of the algae produced by the scrubbers as a function of the frequency of wave surge, day length, harvest rate, and other variables.

Another project is a study of the processes of reestablishing sea grasses in damaged lagoons. Once the grass is destroyed, whether by nature or by humans, the sterile sand is slow to recolonize and recovers only after many years. The lagoons serve as a nursery for many reef species, so it is important that they remain healthy. One possibility, the researchers say, is that certain fast-growing green algae act as colonial species and develop first, which provides the proper environment for more specialized sea grasses.

Adey and the staff now have a Maine rocky coast cold tank in the basement, complete with lobsters. That too, Adey hopes, will soon be moved upstairs. Other tanks will undoubtedly follow. Even with only the coral exhibit, Adey says, "There's no end to what we could do with this. We could fill the whole Whale Hall."

In the years since this tale was written, many newly developed technologies have made it possible to provide similar cutting edge coral reef experiences in educational settings beyond the Smithsonian Institute, including secondary school classrooms. For example, Brylske (2009) reports that for almost 30 years, teacher "Dave Makepeace has been building a niche at Tavernier's Coral Shores High School with an honors marine science curriculum." Rather than sitting in a classroom, Makepeace emphasizes "service learning." That means actually doing what you read about in textbooks. What's more, grades are based not on regurgitating facts but through submission of portfolios that include scientific reports, journals, photos, websites, and videos. And, Brylske continues "of course, being a certified diver is a requirement. Over the years, under Makepeace's guidance, Coral Shores students have studied and restored sea grasses with the Florida Fish and Wildlife Research Institute, created artificial reefs using Reef Balls®, monitored local coral reefs using the same scientific protocols as professionals, and have even explored the Dry Tortugas with none other than "Her Deepness," Dr. Sylvia Earle and the National Geography Society."

Brylske indicates that when Ken Nedimyer approached the school about participating in a Coral Restoration Foundation (CRF) project the only question was, "When do we start?" Since 2003 more than 130 Coral Shores HS students have assisted Nedimyer and the CRF, both during class and on their own time. "I have 30 more ready to dive in as soon as Ken is ready," Makepeace says. "This project is perfect for my students. They get an opportunity to apply and learn real science, help make a lasting contribution to the ocean, interact with extraordinary role models like Ken and, of course . . . their diving. It doesn't get much better than that."

Photo by Robert Jakus

In a *Dive Training* article, Brylske (2009) describes Ken Nedimyer, as the longtime proprietor of Sea Life Inc., a tropical fish-collecting business based in the Florida Keys. He furthermore explains that "Nedimyer is no reef pillager; in fact, just the opposite. A marine biologist by training, he has devoted as much of his life to marine conservation as to growing his business – if not more. He also sits on the Sanctuary Advisory Council of the Florida Keys National Marine Sanctuary, and heads a grass-roots conservation organization called the Coral Restoration Foundation (CRF). In the process, Nedimyer has become one of the world's leading experts in staghorn coral (*Acropora cervicornis*) propagation and restoration. And that's important because over the past three decades, more than 95 percent of all staghorn coral in Florida and the Caribbean has vanished. (Most of it has died off from a condition known as white band disease that started in the late 1970s.) In 2001, Ken's daughter, Kelly, was lamenting over what to do for her 4-H project. A few years earlier Ken began noticing young staghorn corals growing among his one-acre (0.4 hectare) live rock nursery offshore of Key Largo."

All stony corals are federally protected, but an exception in the live rock aquaculture laws allows the sale of any corals that settle and grow on cultured live rock. (At the time, this applied to staghorn coral until it, along with its cousin Elkhorn coral [Acropora palmata], made the federal endangered species list. Now, while Nedimyer still technically "owns" his corals, he can no longer sell these two species.) Brylske explains in this article that as Ken's daughter Kelly became convinced that rather than doing the usual 4-H project of raising goats, cows, or pigs—a real challenge in the Florida Keys—why not grow coral? The result: A first-place finish in the regional competition and second-place finish in the state contest. The body of Brylske's 2009 article is provided here to share the excitement of potential experiences students working on project five in *Beyond the Classroom* may accomplish.

> The next step in Kelly's 4-H career was a required community service project. For that Ken suggested inviting the marine biology class at the local high school to participate in raising the coral critters. Fortunately, the high school in question, Coral Shores, had on staff a veteran marine science teacher, Dave Makepeace, who was cut from a different cloth of educator. A marine scientist by training, Makepeace found teaching a much more rewarding career, and for more than 30 years has developed what is perhaps the most innovative high school marine science program in the nation. It turns out that scuba diving was already an integral part of the Coral Shore marine science curriculum, so the idea of raising corals was a perfect fit
>
> By 2003 what had started out as only five small fragments of staghorn coral in the Nedimyer nursery blossomed into nearly 50, and some had grown into formations the size of an extended hand. This gave Ken yet another idea, and it wasn't to make a boatload of cash selling corals. He wanted to use these hardy little survivors to help restore the reefs that, so long ago, had inspired his love of all things ocean. So, he contacted officials with the Florida Keys National Marine Sanctuary and asked to transplant them on Molasses Reef off Key Largo. Molasses Reef is what's termed a Special Preservation Area, or **SPA**, one of several set aside as a no-take zone to protect and preserve the marine environment. What's more, Molasses Reef is probably the most popular dive site in the Western Hemisphere, if not the world. But, unfortunately, a sizeable chunk of it was flattened and pulverized when the cargo ship 'Wellwood' went aground on the reef back in 1984.
>
> Since then, numerous efforts have been made to restore the area, but like most other reefs in Florida and the Caribbean, there's still almost no live staghorn coral. What better place, Ken thought, to give a helping hand. And so it began. "In the pilot project we took six clippings just to see what would happen and they all lived," Ken says. Soon folks ranging from scientists to the local tourism development office caught wind of what Nedimyer was up to, and it didn't take long to exhaust the capability of his singular effort. But money to continue was difficult to come by because, as Nedimyer himself says, "No one's going to give a grant to a fish collector."
>
> Taking the bull by the horns once again, he formed a nonprofit organization, the Coral Restoration Foundation, as a mechanism to accept and administer funding to support and continue the restoration. However, based on more out-of-the box thinking, Nedimyer's approach wasn't to take the normal route of finding only deep-pocket donors and financing organizations. Recognizing that divers are some of the most passionate people on the planet when it comes to protecting coral reefs, he had another idea. Why not marshal them to both help fund and do some of the physical work? Thus, in 2007 he launched a series of workshops to teach the divers about his work, and give them an opportunity to participate in an actual restoration project.
>
> Currently, the program involves two days, and is hosted by one of several Keys dive operators. (See "Want to Attend a Workshop?" on Page 35 for a list of dive centers that sponsor CRF workshops.) The workshop includes two lectures and two half-days of diving. Some workshops include an

additional Reef Environmental Education Foundation (REEF) day, with an evening lecture and fish survey dive. A Personal Perspective I had the opportunity to participate in one of the CRF workshops recently hosted at Amoray Dive Center in Key Largo and, even for someone with an extensive background in coral reef science, I found it a unique and fulfilling experience.

The lectures started out with Ken providing us with some often-humorous personal background on how the project got started. (With four daughters, all of whom are divers who at some time helped their dad, you can imagine the stories.) That was followed by a discussion of the problems facing coral reefs in Florida and around the world. Next was an overview of what's being done to address these problems, and finally a summary of what the CRF is doing to restore coral reefs in the Florida Keys. Of particular interest was Ken's perspective on how the Keys has fared over the years; and, take my word for it, only fish have spent more time on the local reefs than Ken Nedimyer. On balance, I can describe the first morning as both depressing and hopeful.

As enlightening as the classroom presentation was, the workshop highlight is clearly the open-water work. Our first dive was spent at Ken's nursery, which is now as much a home to staghorn coral as its intended purpose, raising live rock. Laid out before us was row after row of cinder blocks — reminiscent of a farm field — each holding a tray of 10 small-to medium-sized pieces of staghorn coral; Prior to our group, other volunteers clipped pieces, termed "frags" for fragments, of two or three centimeters in length, and mounted them to pedestals with special underwater epoxy. Having been unattended for several weeks, our job was cleaning and preparing the trays of coral for transplant on Molasses Reef the next day

As we learned in class, each coral started from what Ken describes as a "knuckle clipping" in the nursery. And this is no willy-nilly process. Consideration is given to make sure that there's **DNA** diversity at each transplant site. To accomplish this, **DNA** analysis was done to identify the genotype (the genetic constitution of an individual organism) of each family of corals growing at the nursery. This will help scientists identify which corals have the greater ability to survive under the conditions present. Each frag is tracked by genotype, the number of generations away from the parent colony and a fragment number, all critical information in determining the ultimate success of the project.

After about a year, when they have grown branches and stand about four inches tall, they're ready for transplant or for use as parent colonies for more clippings. Our class was composed of several divers from around the Southeast, plus a large group of high school students from Connecticut. Given the challenges of supervising a boatload of beginners, Ken used several assistants. On board to help out were Chris Calderwood, a Nedimyer colleague and marine science teacher from a local private school, and two of Dave Makepeace's marine science students, Thomas Tafoya, and Dannon Magrane. All have been working with Ken for some time. (On our second day of diving, proving that Ken's operation is truly a family affair, his wife, Denise, joined us as a fourth assistant.)

I entered the water a bit late and found the bottom already rife with busy divers going about the business of cleaning the coral trays and stands with wire brushes, chisels and putty knives. The trays of coral looked like serving dishes, each containing eight coral fragments. The cloud of debris around them was full of fish, many of which were the rarely seen juvenile stage angel and butterflyfishes that, in their adult form, inhabit nearby popular dive sites. Of course, there were also the ubiquitous triggerfishes and wrasses wrangling their way in for a free handout of the scrapings.

At that point the nursery looked as much like a tropical fish farm as a repository for coral and live rock. "Feeding frenzy" came to mind. I'm not sure who was busier; the divers cleaning or the fish eating. Day two began with another informative lecture on coral reef science, and the wide assortment of coral reef restoration techniques being used today around the world. Then, it was off for the second day of diving at a dive site called Snapper Ledge near Molasses Reef. Armed again with wire brushes, chisels and, this time, two-part epoxy, the objective was to prepare the sites where the new coral colonies would spend the rest of their lives.

Unlike most noiseless dives, this one was accompanied by the constant sound of banging, scraping and brushing. The cacophony was accompanied by other divers kneading the modeling claylike epoxy to affix the corals' bases. The epoxy team reminded me of a sculpture class in which dutiful students intently prepare their works of art for an approving instructor. Along with the worker-bee divers was a third group of data recorders who meticulously measured each branch of each coral to determine the total length so that, over time, growth can be accurately documented. This is an essential task — as critical as placing the corals — because the goal of the project is not only to transplant, but to document the corals' health and to what degree they're benefiting the local ecosystem.

A quick tour of the area, before heading back to the boat allowed me to spot dozens of previously transplanted colonies, all of which seemed to be healthy and prospering. Ken explained that from the initial six colonies placed as a proof-of-concept back in 2003, they have so far transplanted nearly 100 more on Molasses Reef and Snapper Ledge. From these immigrant parents about 500-600 new colonies have appeared. And all this is an area that hasn't seen significant live Staghorn coral in more than two decades. "It's amazing how fast they grow," Ken told me on our way back to the dock. "A two-centimeter [0.78-inch] cutting will be 30 or 40 centimeters [11.8 to 15.7 inches] in just a year." Ken also explained that he has recently turned his attention to the staghorn's cousin, Elkhorn, and has begun a preliminary transplant program with that endangered species as well.

Since my participation in the restoration workshop in June, there's been even more encouraging news. Continuing his commitment to involving youth, Ken has been working with a Tampa Bay-area organization called **SCUBA**nauts International. In August, under the supervision of Nedimyer and Florida Keys National Marine Sanctuary scientists, a group of 21 high school-aged **SCUBA**nauts documented the first known case of spawning of farm-raised coral in the entire Atlantic and Caribbean. "It's big news" Nedimyer told press representatives afterward. "This is what we have been trying to do. We are not just trying to restock the coral population. It's about letting them spawn and reproduce. We are just trying to give them a jump-start … I'm excited that it happened in just two years."

The SCUBAnauts group not only observed the spawning but also collected the staghorn gametes (eggs and sperm). Some of the larvae had already settled onto ceramic discs, which were taken back to the nursery for grow-out. Gametes were also collected from the water column by scientists for further research, and will be used for fertilization projects with the aim of eventually transplanting the offspring. Dr. David Palandro, a research scientist for the Florida Fish and Wildlife Conservation Commission and head of the **SCUBA**nauts program, described the experience as being like a great big circle of life. "Corals were transplanted here and we're collecting the gametes from those transplanted corals and we hope to take those gametes and transplant them someplace else."

But to Nedimyer there's something else that's equally important: the people who have done this. "The corals have been cut and moved by Coral Shores HS students," he said, "placed on restoration sites by [Florida Keys Community] college students and discovered spawning and reproducing by another group of students. Kids have done this whole thing." While local success is important in proving the transplantation concept, Nedimyer recognizes that Molasses Reef is just a tiny dot within the massive coral reef ecosystem of the Atlantic and Caribbean. So, his plans are to expand to other areas.

With modest resources, that means the next steps for **CRF** will probably take place elsewhere in the Keys. Already there are plans in the works to develop a partnership with the marine science department at the Florida Keys Community College to create and maintain a coral nursery in the Lower Keys. And, Ken has had a number of requests for programs throughout the Caribbean. "All actions have local impact," Nedimyer says, "and everybody can make a difference." As for me, it was great to see that there's a way to do something that truly makes a difference but involves more than just writing a check. And I have the epoxy stains on my wet suit to prove it.

http://www.lib.noaa.gov/retiredsites/docaqua/imageschool/liverock2.jpg

The husbandry of exotic marine species has seen dramatic changes over the past decades. In the 1950s, pioneers such as Lee Chin Eng still experimented with seaweeds and air stones which have now been upgraded to computerized aquaria equipped with strong lighting, protein skimmers, calcium reactors, and coolers. Nowadays, the husbandry of many coral and other invertebrate species has risen to a high level, allowing for high survival and growth. The next step however, reproduction, remains a major hurdle for most marine species. Sexual reproduction is a key step in the husbandry of any species as it generates genetic diversity. Closing the life cycle of corals is one of the main future challenges for marine aquarists.

Currently, large numbers of fish, coral, and other invertebrate species are exported annually for the American, European, and Japanese markets. This trade however is far from sustainable. According to a recent UNEP-WCMC report, only percent of exported corals have been raised by mariculture or aquaculture. This poses yet another challenge for coral reefs, aggravating current issues such as pollution, overfishing, habitat destruction, global warming, and ocean acidification.

The aquaculture of numerous marine species (vertebrate and invertebrate) will have to be upscaled significantly over the following years due to persisting demand for livestock. Quotas of wild-caught specimens, either corals or fish, will have to be reduced to prevent permanent ecological damage to coral reefs. This change toward a more sustainable approach is already ongoing, and aquarium hobbyists play an important role in this process. Species such as clownfish and sea horses are bred regularly by specialists who devote a great deal of time to this. Many corals are being fragmented and traded on Internet fora and events, and this has led to many commonly available species in the trade. Biologically speaking, fragmentation of corals is a form of vegetative or asexual reproduction.

Asexual reproduction

A marine aquarium which runs optimally quickly becomes a victim of its own success. Some corals grow out to impressive proportions, after which the colonies are subject to enthusiastic pruning. Fragmenting and selling ramets is a nice way to recuperate some of the investments made and also allows for expansion of a captive coral population.

"Fragmentation of corals is a form of vegetative or asexual reproduction."

In the wild, corals are also fragmented constantly; predators and storms continuously break off fragile colonies. Corals also release parts of their colonies, a process called autotomy. This often occurs during periods of stress and is observed frequently in aquaria. Autotomy can be regarded as a last resort for the colony to survive, in the form of a released part which often completely regenerates. Corals employ many different strategies of asexual reproduction, including intratentacular budding, extratentacular budding, "dripping", polyp bailout, and the formation of anthocauli displayed by members of the Fungiidae family.

Although regular manipulated coral fragmentation is a sign of success, there are disadvantages. Not only are these colonies prevented from reaching their natural sizes, fragmenting them may also lead to weakened colonies in the long run. Why is this? This has everything to do with genetics.

A coral ramet is an exact genetic copy of its parent colony and is thus considered a clone. Clones behave exactly the same under equal circumstances, which means they are equally strong but also equally weak as their parent colonies. Imagine a large Acropora colony sensitive to salinity fluctuations being fragmented into several small colonies. When these fragments are aquacultured, this latent weakness will present itself sooner or later. If these fragments are utilized for reef restoration and placed near an estuary, success may be limited. This happens when latent weaknesses become apparent too late. The same principle can be applied to corals sensitive to bacterial or parasitic infections, temperature fluctuations, or high irradiance levels.

These problems occur when corals display equal characteristics; this is called a genetically homogeneous population. It is the diversity of characteristics which bestows fitness upon a population of individuals, and this is called a genetically heterogeneous population. A population of coral colonies, with different resistant qualities, will therefore have a higher chance of survival in the long run. This is because for each disturbance, it is likely that several individuals will survive as they have an innate resistance to it. A genetically diverse population increases the chances of survival for any given species. There is only one natural way of generating diversity, and that is through sexual reproduction.

The field of coral husbandry has made significant advances over the last decades, with the advent of protein skimmers, calcium reactors, powerful pumps, high quality synthetic sea salts, and a thorough knowledge of water chemistry. The husbandry of any exotic plant or animal always seems to follow three distinct phases: survival (less than one year), growth, and finally reproduction. The first two phases have been successfully achieved for many marine species, save for non-photosynthetic corals such as *Dendronephthya sp.,* feather stars, tunicates, many bivalves, and sponges. However, when the final phase, reproduction, is considered, it becomes clear that coral aquaculture it still faced with interesting challenges. Of course, sexual reproduction of fish, corals, and other species is a time-consuming endeavor which is not open to all. Nevertheless, for the average hobbyist there still is room for much improvement. A fresh view on environmental reproductive cues, coral nutrition, and the usage of alternative filtration systems may yield new results. Zoos, public aquaria, universities, and research institutions have taken a leading role in the field of coral reproduction with projects such as SECORE (www.secore.org) and CORALZOO (Leewis, 2009). In Japan, much has been done in this field as well. For example, coral spawnings in Japanese public aquaria are more frequent, such as at the Okinawa Churaumi Aquarium. Coral larvae are also used around Okinawa to restore entire reef sections by pumping large numbers into plankton nets which are placed over demised parts of the reef. These larvae subsequently settle onto local substrates, after which they will start building new colonies.

Although impressive and pioneering, these efforts all depend on *in situ* field work or open aquarium systems. The development of a closed system aimed at large-scale sexual coral reproduction would be a major step in the field of marine aquaculture. New projects are in development at several institutions, and it is quite possible that captive sexual coral reproduction will become a standard practice in the near future.

Completing the Coral Propagation System

After the initial testing of the system for leaks in project three, it is now time to prepare it for fish and coral reef invertebrates. If any leaks were discovered in the original test make sure they have been eliminated. Again, PVC cement and hose clamps help keep the system watertight throughout the school year, but try to use the cement as little as possible. Remember that a day (now in the distant future), for disassembly and storage of materials will come, and it is good to reserve some thought to that time. When the time does come to disassemble the system, please be aware that students will not be able to separate pipes that are cemented together.

Determine temperature ranges that are necessary for system operation and ensure that properly sized water heaters are safely installed in the sump. Establish pH, DO, salinity, and calcium levels of system water (and protocols for maintenance of it). Provide a place for storage (and dechlorination, salinization, and pH adjustment) of water to be used to for water changes or to replace evaporated water. Ensure access to coral propagation system and species logbooks (enter data). Ensure the system lighting is operational. Install a protein skimmer to aerate and skim the water in the sump. Reserve a place in which additional water filtration equipment may later be installed.

A final but necessary step before working directly with fish and other organisms is to review information related to proper and ethical treatment of all classroom organisms. Among other things, students must be aware that bug sprays, sun block, perfumes, and many other substances on our skin kill aquatic organisms in an instant. Working with aquatic ecosystems requires a level of hygiene that many students are not familiar with. It is important that all students are guided to make every effort to work with aquatic ecosystems in ways that prevent exposure to substances that are toxic. In the effort to promote proactive aquatic ecosystem work, each student who is to work directly with the system must submit a signed form of agreement; the signature of a parent or guardian is required on this **Non-Human Vertebrate Endorsement Form** (see page 129). Students do not need to submit a separate form for each system. But each student who works with this system must submit this form, including a signature of a parent or guardian. Review protocols for healthy transfer of organisms to a new system (from the **Introductory Aquatic Ecosystem Inhabitant Form** - page 57).

Prepare fish and other invertebrates for introduction to the system by slowly acclimating them to the temperature and chemical properties of the water in it. Hopefully, differences between water conditions in new and old environments will not be too extreme. Organisms must be acclimated for transfer to the new system. In general, place them in a container that is half-full of water from their old system and then gently add a quarter of the remaining volume of water from the new system into the container. Wait for 15 minutes and then again add a quarter of the remaining volume of new system water to the bucket. After repeating this exposure to the new water three times, gently release the organisms to the system. Remember to keep data about the way that corals in the system grow (anthropometric data). Do not over feed the organisms in the system, especially when the nitrogen cycle is just beginning.

http://www.noaa.gov/features/04 resources

Coral Fragging Activities – Two weeks after "mother colonies" have been established in the coral propagation system (or one of the teacher's four model classroom systems), guide students through a discussion and review of the five coral fragging protocols provided below in order to prepare the class to engage in a coral fragging class. Remember to provide many towels in addition to other materials that are listed. Thoughtful discussion will help eliminate potential problems with working with super glue, epoxies, organic rocks, and mucus that will occur during the fragging procedure.

According to Osinga (2009), asexual propagation or "fragmentation" is a simple method for obtaining large numbers of coral colonies. New coral colonies are obtained by cutting a mother colony into smaller pieces and (if necessary) by attaching these fragments to a solid support. It makes use of the remarkable regeneration capabilities of corals; even a single, isolated polyp or a fragment that contains few polyps can grow out to a new adult colony. Hence, one adult coral colony can be cut into numerous small fragments that will all form healthy, new colonies. "This proliferation capacity seems to be nearly unlimited" (Bak, 1976), although some people have expressed their concern about repeated cloning of the same genetic individual.

Osinga (2009) further explains that many techniques have been developed to cut coral colonies into smaller fragments. Various tools are being used (e.g. knives, pliers, electrical cutters, scissors, saws - see Table 1), depending on the size and shape of the coral. It must be realized here that all fragmenting procedures impose stress on the corals. After cutting, fragments first have to invest energy into regeneration of wounded tissue before normal growth rates are restored. They may also have an increased susceptibility to infections during wound regeneration. This phase of regeneration is usually referred to as the lag-phase. A good fragmentation method is characterized by a high survival percentage of fragments (close to 100 percent), a short lag-phase, and good subsequent growth characteristics (a high growth rate and a natural growth form). Hence, these three parameters (survival, lag-phase length, and growth characteristics) should be monitored when optimizing a fragmentation method

According to Wray (2013), in terms of conservation, coral fragging—whether done by the amateur aquarist or the commercial coral farmer—promotes a sustainable hobby in which every wild collected coral can be "fragged" to produce many offspring. Conservation through propagation is very popular amongst conservation-minded aquarists. The fragmentation mode varies depending on coral architecture.

Fragging Protocols

The following protocol describes how to fragment **branching stony coral species** such as *Stylophora pistillata* and *Seriatopora caliendrum* (Schutter and Shafir, 2009).

Required materials:

- Wire cutter
- Colonies
- Gel based cyano-acrylate (superglue, see Annex for manufacturer)
- Two-component epoxy (see Annex for manufacturer)
- Substrate such as small (live) rocks or PVC or student developed aragocrete plugs

Procedure

1. Take a coral colony, and carefully cut off the desired branch size by using a wire cutter (see picture).
2. Fix the branch onto a substrate to allow for easy transportation by using cyanoacrylate or two-component epoxy above water.
3. After gluing, immediately place the branch in system water at an area with low flow rate.
4. If possible disconnect the flow pumps for 30 minutes.
5. Repeat the process for every branch.
6. Do not forget to reactivate any flow pumps after 30 minutes, and make sure the newly fragmented coral branches remain in areas with low flow rates for several hours.

Notes:

1. Ideal substrates are small PVC pipes/sticks and small (live) rock.
2. Cyano acrylate is commercially available in both water and gel-based forms. Make sure to use gel-based glue which does not wash away easily in marine water.
3. Two-component epoxy should be mixed manually and left to settle for several hours according to the manufacturer's protocol. Use dry hands when manipulating the epoxy, and prevent it from covering the coral tissue as much as possible.
4. Cyano acrylate will react with marine water, and will quickly solidify. After several minutes, the branch will already be firmly glued to its substrate. Two-component epoxy will take several hours to solidify but will already be quite solid after 30 minutes.

The size of the fragments may vary with the colony size, and with the aim of the cut. There is no pre-determined size for fragments. The procedure is different for different coral species. This mainly depends on the strength and thickness of the branches.

The following protocol describes how to frag relatively **delicate species of branching corals** like Acropora (Santos & Baylina, 2009).

Materials

- Pliers
- Two component epoxy "glue"
- Pieces of rock or plastic support (Bushes)
- Latex gloves
- Buckets / 50 L container

Procedure

1. Cuts are done inside the tank.
2. Place the pliers in the area that you want to break, rotate the pliers, and let the fragment break. Sometimes (like for fine branching colonies) you can also use a specific kind of pliers that cuts like scissors.
3. Using "Reef Construct," glue the fragments to pieces of rock.
4. Fragments are then transferred and acclimatized to the water of the propagation tank. The water quality of the exhibition tanks and the propagation tank is kept similar in order to reduce the acclimation process. Nevertheless, fragments are placed in a bucket or a 50 l container with half of water from the tank where they were, the other half is filled with water from the destination tank using a small hose. Half an hour later, the fragments are transferred. Depending on the size of the fragments, they may be glued to pieces of rock or put on plastic supports (bushes).

Coral fragments that are smaller than three cm and have fine branches can be inserted directly in the plastic bushes. But this only works for species with tissue in the base that grows over the support (which doesn't happen with Pocillopora damicornis). Therefore, it is preferred to attach all fragments to pieces of rock, using "Reef Construct." This can be done with fragments as small as two cm.

The following protocol explains steps for fragging relatively **robust species of branching corals** such as Euphyllia, divisa (Baylina & Santos, 2009).

Materials

- Pliers
- Hammer
- Bucket / 50L container
- Small hose
- Latex gloves

Procedure

1. With this species, work inside the tank.
2. Determine which polyps you need, and/or where you have to break the coral: search for the weakest point in the skeleton, where chances of breaking it are highest.
3. Separate the polyps you want with a quick blow with the hammer in the zone that you want to break.
4. The pieces you have broken off are then picked up by hand using latex gloves, although if they are grabbed by the skeleton there is no need for gloves.
5. Transfer the fragments to the water of the propagation tank.
6. Let acclimatize.
7. Place the fragments of this species in holes in the live rock – do not glue.

The following protocol (Baylina & Santos, 2009) shows how to proceed if **a sizable piece has to be cut off from a plate like species** like Montipora or other rather sturdy coral species, when a fragment might have to be cut in order to prevent the death of the colony.

Materials

- Pliers
- Sharp knife
- Two component epoxy "glue"
- Pieces of rock
- Buckets / 50 L container
- Fine hose
- Latex gloves

Procedure

1. Place the pliers in the area that you want to break
2. Rotate the pliers and let the fragment break.
3. Sometimes you can also use a very sharp knife and take "slices" of the colony.
4. Transfer the fragments to the water of the propagation tank.
5. Let acclimatize.
6. Depending on the size of the fragments, they can be glued to pieces of rock or just put loose on the rock or sand: pieces smaller than 15 cm are glued; pieces bigger than 15 cm are left lying loose.

In yet another fragging technique, it is by use of an electric saw that corals are cut in two or more pieces. This technique allows precise fragmentation of corals and thus can also be applied on plate corals like Montipora sp. and Echinopora sp. Massive corals are more difficult to fragment due to their dense skeletons and relative slow growth. When using a sawing protocol on massive corals they may be fragmented without too much harm. Also, there is more control when fragmenting than with breaking or tearing techniques.

At Burgers' Zoo, in Arnhem, Nederland, an electric power saw is used on the following coral species to fragment them: Acanthastrea sp., Coeloseris mayeri, Favia sp., Favites sp., Fungia sp. Lobophyllia sp., Montastrea sp., Platygyra daedalea, Porites sp., Tubastrea sp., and Turbinaria sp. This technique can also be used for fragile plate corals (like Montipora sp., Echinopora sp.) if more control over the line of fragmentation is required.

It is preferred to cut through as little living tissue as possible to decrease the wound size. However, no adverse effect has been noticed from cutting through large colonies. Under good husbandry circumstances, corals recover very well from this procedure. Usually the wound heals within a few days, after which the fragments continue growing straight away. When fragmenting Fungia sp. using this technique, it takes years before the oval or round shape of the coral is resumed.

Below, the sawing fragmentation technique is described for **massive or encrusting corals**, like Favia sp., Favites sp., Lobophyllia sp., Fungia sp., and many other species (Janse, 2009).

Materials

- Dremel or electrical cutter or saw
- Table to work on
- Safety glasses
- Safety gloves
- Dust mask
- Ear protection
- Two component epoxy or cyano acrylate glue
- Tweezers
- Small pieces of rock or other substrates to fix the fragments on
- Bucket (2x)
- A vice or clap to stabilize the colony while cutting. Also a tray can be used where the coral is pushed into a corner. The equipment very much depends on the size and shape of the coral.

Procedure

1. Place the coral on a table.
2. Fix the coral on the table.
3. Use safety equipment.
4. Saw the coral.
5. Rinse fragments in bucket with tank water (to get rid of saw dust and mucus).
6. Fix the fragments on rock with two component epoxy or cyanoacrylate.
7. Place the coral with substrate in a second bucket with tank water to leach excess material.
8. Wait five minutes, then place the coral back in the tank. Fungia sp. regular shape doesn't recover until several years after fragmenting.

According to Wray (2013), corals are generally fragged out of the water, which should not pose a problem so long as the coral is returned to it within 15 minutes and the air temperature is in the mid-70s F (a humid environment is best). If you are fragging a coral with large polyps, gently wave your hand over the coral before removing it from the water so that the polyps retract. Major tissue damage can occur if you lift some corals with large polyps out of the water while the polyps are fully extended. It is best to use gloves to protect yourself and the coral while fragging.

Wray (2013) suggests the use of epoxy on most LPS corals because epoxy can handle weightier frags better and can be formed to contact only dead calcareous portions of the frags skeleton. Since the epoxy curing process kills any tissue it comes into contact with, Wray indicates that Crazy Glue is the better adhesive choice for SPS corals. He also indicates that it is important to never frag more than about 25 to 30 percent of a mother colony at one time. Wray also adds that he has found waiting three to four days after fragging from the broodstock colony before mounting the new frag increased the survival rate, color, and growth. He reminds us to dip both the mother colony and the new frag in a coral dip after fragging. This dip will help ward off bacteria, fungus, and protozoans. Some aquarists also use a vitamin C additive to aid in healing.

*** It is important that students performing the activity on this page have sufficient skill and background understanding of the coral propagation system and general coral fragging skills and knowledge presented here to effectively lead students in other groups so that they can also understand certain coral propagation techniques and perform them (per the "Sharing of Knowledge and Accountability" activity on page 141, 165, and 186).**

Coordinated Sharing of Coral Propagation System Development and Operation Information with Classmates

For classes that followed the division of labor approach to assembling their classroom aquatic ecosystems, it should now be apparent that, working independently of one another, three student groups (or more) have each completed the design and assembly of one classroom aquatic ecosystem. The project five concluding activity described here is designed to be synchronized with project four and six concluding activities (see pages 141 and 186). As with the chapter two concluding activity on page 121, it is again assumed that students from each respective project team possess sufficient skill and background understanding of their ecosystem and activities presented in their respective projects to effectively guide students from other ecosystem teams to also understand their ecosystem and to capably perform the respective techniques featured in their project

Once all systems are operating and students in each "home group" are familiar with their system related information, students are prepared to engage in a coordinated cycle of information sharing and accountability. This activity begins by having students in each of the three groups (aquaponics, coral propagation, and broodstock larval rearing) count off by twos. The following chart is designed to illustrate this verbal description. It is important to provide each student group with examples of paperwork and forms that were used by the students in each group as they completed their project activities.

According to the chart, the "ones," who assembled the aquaponics system will work to guide students ("number ones") who assembled the coral propagation system through the seed germination and respiration related activities. According to the chart, as the X1 students lead the Y1's, the Y2 students will lead the Z1's through calcium carbonate egg plug production and coral fragging related activities, and the Z2 students will lead the X2's through setting up black worm culture systems and decapsulating brine shrimp cysts. As this activity concludes, students shall have completed the first cycle of sharing specific aquatic ecosystem-related knowledge and skills. In cycle two, students in the X1 group lead students in the Z1 group. Students in the Y2 group lead the X2 group, and students in the Z2 group lead students in the Y1 group. As the chart below indicates, at the end of cycle two, the ones and twos in each group exchange leader/learner roles and pass through cycles three and four.

Leaders Cycles 1 & 2	Learners Cycle 1	Learners Cycle 2	Leaders Cycle 3 & 4	Learners Cycle 3	Learners Cycle 4
X1 seed germination photosynthesis and respiration	**Y1**	**Z1**	**X2** seed germination photosynthesis and respiration	**Y2**	**Z2**
Y2 calcium carbonate egg plug production and coral fragging	**Z1**	**X2**	**Y1** calcium carbonate egg plug production and coral fragging	**Z2**	**X1**
Z2 setting up black worm culture systems and decapsulating brine shrimp cysts	**X2**	**Y1**	**Z1** setting up black worm culture systems and decapsulating brine shrimp cysts	**X1**	**Y2**

This coordinated sharing activity is designed so that at its conclusion, students from each team will be familiar with all topics and issues that are relevant to the design, assembly, and operation of each of the three student-designed classroom aquatic ecosystems. Students may be asked to use details of each classroom system to develop a comparison and contrast poster.

Write a list of **science-based** activities you have participated in during this project and briefly describe evidence that verifies your interaction with and knowledge of scientific concepts for each item on your project five list.

Write a list of **technologically-oriented** activities you have participated in during this project and briefly describe evidence that verifies your interaction with and knowledge of technologically-oriented concepts for each item on your project five list.

Write a list of **engineering-oriented** activities you have participated in during this project and briefly describe evidence that verifies your interaction with and knowledge of engineering-oriented concepts for each item on your project five list.

Write a list of **mathematical** activities you have participated in during this project and briefly describe evidence that verifies your interaction with and knowledge of mathematical concepts for each item on your project five list.

Write a brief list of **Non-STEM-related** activities you have participated in during this project and briefly describe evidence that verifies your interaction with and knowledge of Non-STEM related concepts for each item on your project five list.

Rate your level of involvement in STEM-related activities in this project (1=lowest, 10=highest).

1—2—3—4—5—6—7—8—9—10

Beyond the Classroom Aquarium **Daily Journal Guidelines**

Entries students place in their daily journals should be:

1. Dated accurately and logged chronologically.
2. Kept in a notebook and written neatly in ink or typed.
3. Written or typed in complete sentences using correct capitalization, spelling, and punctuation.

In their journals, students should keep a daily record of:

1. Who they worked with each day by including the names of school staff, fellow students, community members, and volunteers.
2. What they did each day, such as lecture topic, activities or project design, assembly, or maintenance.
3. What they learned each day by asking questions, following instructions, and/or observing activities.
4. How *Beyond the Classroom Aquarium* experiences relate to past, present, or future life experiences.

For bonus points, students may complete these journal options, including illustrations, diagrams, or creative expression relating to the journal entry. For project one, journal writing bonus topics may include answers to the following questions or descriptions of a project activity or assignment. Reflection on these topics may also help students prepare for exams, quizzes, and other assessments.

- What do you want to understand better in order to contribute more to your team's productivity as it independently designs, assembles, and then manages the coral propagation system?
- Create a new drawing and a story about an organism or system that is interesting to you.
- Create additional posters and signs so that people who visit the coral propagation system will be able to read about new and important system components and organisms.
- Create a journal entry that tells what information you learned during project five. Also, write down what you would like to know more about and if you know someone else who might help you learn more if you asked them.

- Select a book or magazine to read that is related to a topic in project five. Write a report on this book and include your report in your *Beyond the Classroom Aquarium* portfolio.
- Explain the purpose of zooxanthellae in coral, and describe how this knowledge influences the way the classroom coral propagation system is designed.
- Make a list of agencies (governmental or not) that are concerned about coral reefs. Describe measures they are taking and progress they are making. Explain which measures you believe need to be adopted.
- Explain three important ideas and/or skills you gained by participating in the coral propagation system fieldtrip/workshop.

Journal entries will be evaluated on a point system, with four points being the maximum number a student can earn. A half-point will be awarded for each of the first eight journal requirements above. Bonus points will be awarded when journal options are completed. Journal points will be totaled and divided by the number of journal dates.

Additional Project Five Assessment Activities and Questions

Find sources of information that explain factors that affect coral growth in closed systems, including: lighting, feeding strategies, seawater circulation, and water quality. Develop a presentation of how students have considered these factors in designing the classroom coral propagation system.

Review information related to destruction of native prairieland ecosystems across the United States. Develop a report on this topic. Include descriptions of prairie restoration efforts. What do prairie and coral reef preservation efforts have in common?

Create updatable water quality data graphs from the aquatic ecosystem data log books. Report your interpretations of the data and over time (and updates) explain patterns and trends in each system.

Research organizations involved in preservation and restoration of coral reefs. Describe two that seem to be doing the best job, and explain why you think they are more successful than others.

After reading "The Reef's Tale" (Walton, 1980), explain the factors that made Adey's efforts to keep a living coral reef in a manmade environment successful. Describe contributions that research on the system and the creatures in it have helped make possible. Report on this for the class to critique.

After reading "Coral Restoration: Giving Mother Nature a Kick Start" (Brylske, 2007), describe your impressions of the value of the service learning activities that the honors students at Coral Shores High School are engaged in. Report on similar service learning that is being performed by students in areas near you.

Review *Beyond the Classroom Aquarium* information about Ken Nedemeyer, and develop a presentation that updates this information and provides additional related information

Compare and contrast techniques for fragging different species of coral. Explain reasons for different techniques and create a chart/diagram to illustrate the techniques. Create a report on this topic and prepare this information for a class critique. Include photos of your own fragging experiences when possible.

Work with a group of students in your team, and design a web page layout that you could use to feature the classroom coral propagation system. Explain the value of the information contained in your web design, and describe who might find your web site valuable. Prepare this layout for classroom critique.

Develop a report that explains the procedure for testing saltwater for calcium concentrations. Explain the role calcium plays in coral polyp metabolism and coral reef growth. Describe factors that reduce concentrations of calcium in saltwater. Develop a proposal for a calcium reactor design that could be attached to the coral propagation system. Present this report for the class to critique. Build the class approved design and collect data to demonstrate its effectiveness.

Review activities and information that interested you in this project and determine if this information may fit on the "Research-ability" Potential Topics Form on page 59. If so, add it to the list.

Beyond the Classroom Aquarium STEM Self or Peer Assessment Rubric

This self-assessment is valid only to the extent to which you can document and display evidence that supports it. For each category in the left hand column, write a STEM-Oriented example in the box beneath 3 if you can display evidence that your STEM-Oriented example represents work that exceeds expectations. If your evidence shows that your STEM-Oriented statement meets but does not exceed expectations, write the statement in the box beneath 2. Similarly, for STEM-Oriented examples that demonstrate some progress toward expectations.

Name: **Project Number:** **Date Project Work Began:** **Date of Project Completion:** **List Math Topics and/or Objectives Currently Under Study:**		**3** **Work Exceeds Expectations**	**2** **Work Meets Expectations**	**1** **Work Is Progressing Toward Expectations**
Write examples of questions and defining problems related to your STEM work.				
Describe how you have developed and used models in your STEM work.				
Write examples of STEM investigations you planned and carried out.				
Describe STEM related data that you analyzed or interpreted.				
Give examples of how you used mathematics and computational thinking related to the five standards listed to the right.	Problem Solving (Math Process Standard One)			
	Reasoning & Proof (Math Process Standard Two)			
	Communication (Math Process Standard Three)			
	Connections (Math Process Standard Four)			
	Representation (Math Process Standard Five)			
Describe STEM-Oriented explanations and solutions you designed				
Explain evidence you used while engaging in a STEM project-based argument.				
Give examples of STEM information you obtained, evaluated, and/or communicated				
TOTAL POINTS				

The prospect of propagation of broodstock pairs and subsequent survival of hundreds of aquatic larvae challenges students to study many factors that influence the delicate and dynamic balance of light, temperature, and chemistry upon which broodstock productivity and larval survival depend. Toward this end, the primary student task for project six involves completion of the assembly of the student-designed broodstock and larval rearing system(s) that began at the end of project one. In project six, students also design and assemble an aquatic black worm propagation system or a brine shrimp hatchery and growout system, introduce broodstock pairs or reproductively active adult aquatic organisms to one of the three broodstock tanks in the system, and apply knowledge of a given species to the task of optimizing their procreative activity. Along with this, students also study the potential for profitable production of aquatic organisms for local markets (business-related calculations) and consider how aquaculture may be used to raise money for aquatic ecosystem projects. Ongoing data collection occurs with the broodstock and larval rearing system and many student teams find it crucial in their upcoming research project. Both digital and less-techy "old school" data collection methods are included.

Photo by Robert Jakus

The broodstock larval rearing system with which students work in project six permits them to look much more closely into the life cycle of fish and other system inhabitants. This system is a multi-tank system with a common source of recirculating water. It allows for a number of species to coexist and makes it possible for students to establish separate tank conditions (for each broodstock pair) that are optimal for reproduction to occur. Creating a multi-tank system increases the probability that students will experience reproductive success with at least one of the species in the system.

Initially, organisms held in the broodstock and larval rearing system will be those for whom techniques and protocols for successful reproduction are common. Such organisms, primarily fish, include many species of freshwater fish (such as tilapia and other tropical cichlid species, Corydoras catfish, zebras, characins, Siamese fighting fish, and livebearing fish) or a limited number of saltwater species (such as clownfish, peppermint shrimp, or neon gobies). As a result of this initial emphasis on fish, a deeper study of fish is also part of the information that is also provided in project six's text.

Photo by Robert Jakus

For more advanced aquaculture students, there are many recently developed techniques for successful culture of aquatic organisms that could be attempted in one or more of the three tanks in the broodstock larval rearing system. These aquatic organisms may include, but are not limited to various species of molluscs, crustaceans, echinoderms, and fish.

How to Culture Lumbriculus, variegatus - Project Six Opening Activity – (Freshwater)

The information below is from a web page maintained by Iowa State University-Culturing Lumbriculus variegatus (California blackworms) by Charlie Drewes [http://www.eeob.iastate.edu/faculty/DrewesC/htdocs].

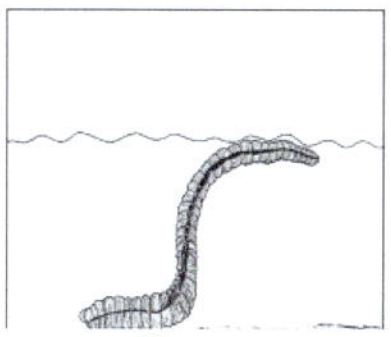

Drewes

"Lumbriculus is easily maintained and continuously cultured in small aquaria with minimal time, expense, and space. At room temperature, populations double in about 3 to 4 weeks, or less. A culture can be started using a deep pan or small aquarium, filled with about 2 to 3 inches of spring water. (Note: Well-aged tap water may be an acceptable alternative to spring water in many but not all cities.) Using a plastic disposable pipette, transfer a few dozen, undamaged, healthy worms into the aquarium. Never attempt to handle or transfer worms with a forceps or hook. They are easily injured. Next, add enough strips of brown paper towel to just cover the bottom of the container. The towel serves as a fibrous substrate of decomposing material, both for the worms and for numerous microscopic organisms that may inhabit the culture, such as bacteria, protozoans, rotifers, and ostracods. Sinking fish food pellets (Spirulina or Spirulina chips) may be used as the primary food source for this simple aquatic ecosystem. Start by adding one or two small pellet fragments. After a few days, additional food may be added, but only if the others have been consumed. Do not overfeed, since decomposition of uneaten food may contaminate the aquarium and cause a mass die-off of worms. Worms will not be harmed however by irregular feeding or weeks of starvation as long as culture water is not foul. Continuous, gentle aeration is recommended, and this becomes increasingly important as biological decomposition of the paper occurs and as the worm population increases. Water lost to evaporation is replaced by adding spring (or distilled) water.

"As the paper towel disintegrates and waste residues accumulate, the culture water should be regularly replaced (about every two weeks). Do this by slowing decanting the culture water down a drain, being careful not to pour out the paper and worms that remain at the bottom. After rinsing the paper and worms again with spring water and decanting, the aquarium should be refilled with spring water to the original level and new pieces of towel added. "Occasional 'harvesting' of surplus worms is advised, as live food for fish, or for starting duplicate cultures. Maintenance of at least one duplicate culture is strongly suggested. Under these laboratory conditions, worms will reproduce continuously by asexual reproduction (fragmentation) and cultures may be sustained for years."

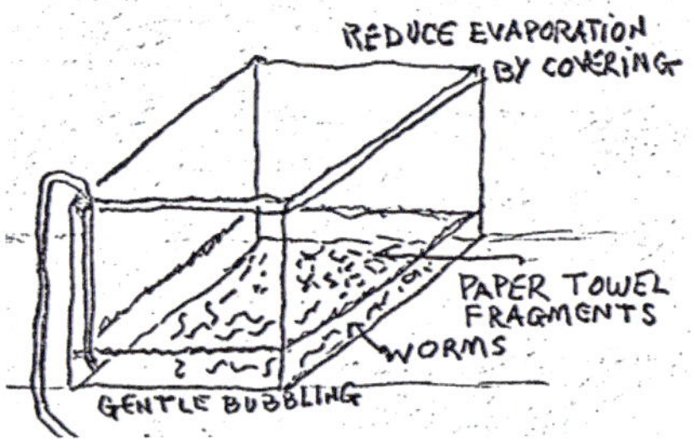

Based on Drewes

In response to a question about the reputation of black worms as disease transmitters, specifically that black worms might transmit the microscopic young of Camallanus, an insidious parasitic nematode, Dr. Drewes replied: ". . . The answer to your question about Lumbriculus transmitting Camallanus infection is a definitive "NO." I have checked with our resident parasitologist who has more than 30 years of research and teaching experience in parasitology and life cycles of parasites. Camallanus is transferred by copepods of several different species, not by oligochaetes. Perhaps, it is possible that your tanks already had a few copepods in them that "bloomed," which these small invertebrates can do. Or, perhaps the water and residues that worms were in when you purchased them were somehow contaminated with copepods. I have been doing research on oligochaete worms (both terrestrial and aquatic) for about thirty years and do not know of any parasites that are transmitted by Lumbriculus (black worms). This, of course, does not mean that they cannot transmit parasites of some type. I do know that whirling disease, in fish, can be transmitted by tubificid worms and that organisms such as small leeches, and tubificid worms, occasionally show up in bulk Lumbriculus that are sold.

"Here is what I would recommend to help minimize problems: When Lumbriculus are purchased, do not just throw some into your fish tanks. Rather, immerse and rinse them in spring water (dechlorinated tap water) in a shallow white-colored pan. Use an eyedropper to remove and discard decaying worm carcasses, damaged worms, or alien life forms. Withdraw the intact lively Lumbriculus worms and use them to feed your fish. Discard (flush) damaged worms or worms whose identity or health is doubtful. I know this is tedious, but with a little practice, it can be done quickly and efficiently and it may save time, expense, and headaches. To establish long-term lab cultures, subdivide worms into several large-volume containers to create replicate cultures."

Drewes says that, "one good way to clearly prove that a worm is Lumbriculus and not a tubificid (tubifex worm) is to lightly tweak its tail end while it is freely crawling underwater. If it quickly attempts to get away by swimming a short distance using helical twisting movements, then it is definitely Lumbriculus. If it is unable to swim, then it is either a messed up Lumbriculus or a tubificid worm - discard it. Tubificids cannot swim."

". . . I would always check with pet shop owners to find out when their shipments of Lumbriculus come in, so that you can get the freshest ones possible. They usually store them in the refrigerator and if they have been in there a long time, they may be full of dead worm carcasses and really stinky. I certainly would worry about bacteria, fungi, and toxins that could build up in a mass of half-dead worms!"

Photo by Robert Jakus

In this activity, students will be guided to investigate the artemia life cycle and to gather information (including images) they may use to help them design systems for hatching and rearing artemia in the classroom. It is important to be alert to the possibility that some students may have allergic sensitivities to crustaceans - such as brine shrimp. It is best for these students to observe the activity but to not come in contact with the shrimp. For this reason among others, it is necessary to keep work areas clean. Information students gather may be organized as a report that explains:

- Reasons for raising artemia
- Important environmental factors that impact artemia hatching, growth and development (in project nine, this information may form the basis for many student research projects)
- Design drawings that students will use to construct and operate an artemia hatchery and rearing system in the classroom.

After students and groups present and critique their artemia reports in class and after students have assembled their systems, they are then guided to create posters that may be used to provide viewers with information about artemia and the plan they followed during the assembly of their classroom systems.

Information (language/vocabulary and concepts) in student developed artemia reports should include (but not be limited to):

- A good narrative description of the Artemia life cycle (from cyst to old age) - including descriptions of the changes in physical appearance of artemia as they mature; include an explanation of the phototrophic nature of new hatched nauplii.
- A description of artemia size/growth (anthropometric data) - including a labeling of the Artemia's larval stages with information about its lecithotrophic gut development
- Artemia diet (variation in nutrient content of new hatched nauplii during maturation) including a description of the various foods that may be provided (pros and cons of each)
- Artemia temperature (optimal temperature and a temperature range)
- Artemia reproduction (ovoviviparous and parthenogenesis)
- Explanation of the influence of salinity on growth and reproductive productivity of adults (free swimming nauplii or nauplii cysts) - optimal salinity and a salinity range
- Explanations of optimal light and water aeration parameters for use in hatchery and growout systems (include an explanation of why aeration may harm adult artemia).
- Ideal density of cysts in an artemia hatchery (why this may vary with intended purpose of artemia)
- Time and quantity (expectations for harvesting new hatched artemia)
- Protocols to follow in harvesting nauplii from the culture container (including a brief introduction to the use principle of siphoning (with air line)
- Special instructions about how to transfer nauplii from the hatchery to a larval culture tank containing larval fish/invertebrates (including how culture water should be discarded).
- Development and listing of a multi-step protocol for batch culture of adult Artemia (including the design for a container that may be used to hold artemia as they mature to

adulthood). Include a description of ways to ensure that enough food is provided each day and to ensure that water quality in the artemia hatchery is adequate (it may be worthwhile to describe the procedure for using a density measuring stick to measure water transparency and thereby indirectly estimate food level).

- Explanation of important conditions that must be monitored and adjusted in order to maintain maturing and adult artemia in a growout system (listing important artemia system maintenance protocols).
- Report a multi-step artemia decapsulation protocol students may safely follow (explaining what decapsulation is and ways it may be of value in a fish hatchery setting).

Ask students to create an illustration depicting the sequence that a dehydrated artemia cyst passes through, over a 24 hour period, as it hydrates in warm and agitated water (passing through an umbrella stage before hatching into an Instar I nauplius (the illustration should include swimming appendages, antennae, and an eye spot).

Dehydrated Cyst	**Hydrated Cyst**	**Beginning Umbrella Stage**	**Late Umbrella Stage**	**Nauplius Instar I**

*** It is important that students performing the activities on the previous pages in project six have sufficient skill and background understanding of the Broodstock and Larval Rearing System, setting up black worm culture systems and decapsulating brine shrimp cysts techniques to effectively lead students in other groups so that they can also understand the techniques and perform them (per the "Sharing of Knowledge and Accountability" activity on page 141, 165, and 186).**

Introduction to Aquaculture Related Topics

Aquaculture may be defined as the "propagation and rearing of aquatic organisms for commercial, recreational, or public purposes. This definition covers all authorized production of finfish, shellfish, plants, algae, and other aquatic organisms for 1) food and other commercial products; 2) wild stock replenishment for commercial and recreational fisheries; 3) rebuilding populations of threatened or endangered species under species recovery and conservation plans; and 4) restoration and conservation of aquatic habitat" (Department Of Commerce, 2011).

The U.S. aquaculture industry (currently valued at about $1 billion per year) is dominated by the production of freshwater fish for human consumption. The marine aquaculture segment (about 20 percent of current production) is mainly comprised of shellfish, but also includes farming of finfish and algae in coastal waters and on land. Another major component of U.S. aquaculture is hatchery production to replenish stocks of important commercial, recreational, and endangered species and to restore habitat (e.g., oyster reefs). Emerging technologies include land-based recirculating systems, algal rearing technologies for production of biofuels and other non-food products, systems that integrate different types of aquaculture or combine aquaculture with other uses, and systems in exposed open-ocean waters.

Aquaculture in the United States can make major contributions to the local, regional, and national economies by providing employment and diverse business opportunities from coastal communities to the agricultural heartland. By implementing this policy and working with partners, including the Department of Agriculture, U.S. Food and Drug Administration, Department of the Interior, and the Joint Subcommittee on Aquaculture, DOC can also help to make the United States a world leader in developing, demonstrating, and employing innovative and sustainable aquaculture technologies and in encouraging worldwide adoption of sustainable aquaculture practices and systems.

Background - Fisheries and seafood are important components of American life. In 2007, Americans consumed a total of nearly five billion pounds of seafood, which equates to approximately 16 pounds per person per year. The most recent Dietary Guidelines for Americans (2010) June 2011 recommend that Americans more than double their current seafood consumption from 3.5 ounces per week to eight ounces per week and that women who are pregnant or breast feeding consume eight to 12 ounces of seafood per week from a variety of seafood products. The United States is the third largest consumer of seafood in the world, and demand exceeds domestic supply. Currently, the United States imports 84 percent of its seafood, and about half of those imports are from aquaculture (United Nations Food and Agriculture Organization, 2009). The current trade deficit in seafood is approximately $9 billion. Growth of domestic aquaculture would support fishing and agricultural communities and new aquaculture-based industries in the United States.

Policy Implementation - To implement this policy, the DOC and its bureaus will work in partnership with other federal agencies, Congress, state, local, and tribal governments, industry, academia, non-governmental organizations, and other stakeholders at the national, regional, and local levels to:

1. Deliver U.S. government services to aquaculture and related industries in a comprehensive and coordinated manner.
2. Provide technical assistance to state and local governments to help plan and develop marine resource infrastructure needs including in coastal communities.
3. Accelerate the implementation of sustainable aquaculture production methods by developing pilot, demonstration, and technology transfer projects with seafood and related industries, nongovernmental organizations, state and local governments, federal agencies, and other partners.
4. Enhance the capabilities of federal research laboratories and participating research partners to provide the necessary ecological, technological, economic, and social data and analysis to effectively and sustainably develop, support, manage, and regulate private and public sector aquaculture.
5. Develop an efficient, coordinated, transparent, and timely science-based permitting process and regulatory structure to ensure that marine aquaculture facilities are properly designed and sited and incorporate appropriate technologies and practices to minimize adverse impacts.

Benefits and Challenges - As interest in commercial aquaculture production and wild species restoration in the marine environment has increased, so too have debate about potential economic, environmental, and social effects of aquaculture and the need for better public understanding with respect to these issues. Benefits of sustainable aquaculture may include species and habitat restoration and conservation; nutrient removal; provision of safe, local seafood that contributes to food security and human health and nutrition; increased production of low trophic-level seafood; and synergies with fishing (e.g., using fish processing trimmings in aquaculture feeds). Sustainable aquaculture can also contribute economic and social benefits by creating jobs in local communities and helping to maintain the cultural identity of working waterfronts. Environmental challenges posed by aquaculture, depending upon the type, scope, and location of aquaculture activity may include nutrient and chemical wastes, water use demands, aquatic animal diseases and invasive species, potential competitive and genetic effects on wild species, effects on endangered or protected species, effects on protected and sensitive marine areas, effects on habitat for other species, and the use of forage fish for aquaculture feeds. Economic and social challenges may include market competition affecting the viability of domestic aquaculture and/or the prices U.S. fishermen receive for their wild seafood products, competition with other uses of the marine environment, degraded habitats and ecosystem services, and impacts to diverse cultural traditions and values.

Growing consumer demand for safe, local, and sustainably produced seafood, increasing energy costs, increasing seafood demand in countries that currently export seafood to the United States, and growing interest in maintaining working waterfronts are emerging drivers that support sustainable domestic aquaculture production. U.S. aquaculture production—both small-scale and large-scale—has evolved and improved over time through regulations at the federal and state levels, scientific advancements, consumer demand, technological innovation, industry best management practices, and protocols for responsible stock replenishment and hatchery practices. This policy will allow NOAA to further advance these developments through the priority actions described below.

Science and Research

- Expand NOAA's research portfolio to (1) provide the necessary ecological, technological, economic, and social data and analysis to effectively and sustainably develop, support, manage, and regulate private and public sector marine aquaculture and species restoration, including technologies deemed necessary under recovery and conservation plans for depleted, threatened, and endangered species and habitat; (2) monitor, assess, and address the environmental and socioeconomic effects of marine aquaculture, including cumulative impacts; and (3) complement the scientific work of our federal, state, and academic partners.
- Evaluate alternative protein and lipid sources to be used in lieu of wild fish and fish oil in aquaculture feeds and develop cost-effective alternative feeds that maintain the human health benefits of seafood and reduce reliance on the use of wild forage fish in the diets of farmed fish.
- Develop and evaluate the cost-effectiveness of methodologies to prevent, minimize, and mitigate potential adverse ecosystem and socioeconomic impacts of aquaculture.
- Monitor and assess the effects of ocean acidification and climate change on marine aquaculture and develop adaptation strategies

Regulation

- Actively engage federal agencies, Fishery Management Councils, federal advisory councils or committees, coastal states, tribes, other stakeholders, and Congress to clarify NOAA's regulatory authority related to aquaculture in federal waters in the context of other federal, state, and tribal authorities and to establish a coordinated, comprehensive, science-based, transparent, and efficient regulatory program, taking into account relevant international standards, as appropriate, for aquaculture in federal waters consistent with the President's Executive Order on Improving Regulation and Regulatory Review.
- Work with federal, state, local, tribal, and regional agencies and organizations to clarify regulatory requirements and to establish coordinated, comprehensive, science-based, transparent, and efficient processes for permit reviews, permit consultations, and other regulatory and management actions for marine aquaculture in state waters – taking into account existing authorities, international standards, and regional, state, and local goals, policies, and objectives.
- Engage in coastal and marine spatial planning with other agencies and jurisdictions, including the Regional Planning Bodies being created under the National Ocean Council, to ensure siting of marine aquaculture that reduces conflicts among competing uses, minimize adverse impacts on the environment, and identify activities for potential colocation with aquaculture operations.

Innovation, Partnerships, and Outreach

- Collaborate with federal partners, coastal communities, states, tribes, the aquaculture industry, non-governmental organizations, and other stakeholders to transition innovative aquaculture technologies from laboratory studies to commercial and restoration projects and document and assess their environmental, ecosystem, and socioeconomic impacts. Focus on projects that will create jobs in coastal communities, produce healthful local seafood, revitalize working waterfronts, support traditional fishing communities, avoid impacts to protected areas, and restore depleted species and habitat.
- Work with extension and outreach services to interpret technical and scientific data and provide informational products to transfer that knowledge to other stakeholders and the public.
- Support restoration and commercial shellfish aquaculture initiatives to restore shellfish populations that provide locally produced food and jobs, help improve water quality, and restore and conserve coastal habitat.
- Develop synergies among NOAA's fisheries management, enforcement, financial assistance, aquaculture, seafood inspection, Coastal Zone Management, National Marine Sanctuaries, and National Sea Grant programs to rebuild wild fish stocks and support alternative or supplemental economic options for fishermen.
- Engage within the Joint Subcommittee on Aquaculture and National Ocean Council to promote coordination among federal agencies on marine aquaculture regulatory and science issues and pursue opportunities for collaboration, such as integrating aquaculture with other ocean uses and using aquaculture facilities as a platform for more comprehensive environmental monitoring.

International Cooperation

- Work with other federal agencies to establish a coordinated, consistent and comprehensive international strategy on sustainable marine aquaculture that supports and is consistent with U.S. policies and priorities regarding food security, international trade, healthy oceans, and economic well-being.
- Work with other nations, as appropriate, to adopt sustainable aquaculture and seafood safety approaches using the best practices.
- Exchange scientific insights with other nations and promote joint participation in cooperative research that is of potential multinational value, including addressing impacts of aquaculture that breach international

Introductory Information about Fish

This section of text (p. 169 – 178) is largely an extraction of information produced by Jason Buchheim (1998d) from his Odyssey Expeditions - Marine Biology Learning Center Publications on fish and fish ecology.

"The origin of fishes dates back over 480 million years. Fish evolved in freshwater; the chondrichthyes moved to the sea early in evolution, while the bony fishes went through most of their evolution in fresh water and spread to the seas at a much later period. Fish are the dominant free-swimming animals of the seas. The structure of a fish body is designed for ease of movement. This ability to move about easily, without relying on water currents to carry them about, has enabled fishes to exploit most parts of the world's oceans. This is reflected in an extraordinary variety of sizes and shapes. Fish are found in both fresh and saltwater worldwide and are a very important food source for many nations.

FISH: Any of a large group of cold-blooded, finned aquatic vertebrates. Fish are generally scaled and respire by passing water over gills. Modern fish are divided into three classes:

I. **AGNATHA**, primitive jawless fish. Lampreys and Hagfish
II. **CHONDRICHTHYES**, the jawed fish with cartilaginous skeletons. Sharks, Rays, Rat-Fishes
III. **OSTEICHTHYES**, fish with bony skeletons. Lungfish, Trout, Bass, Salmon, Perch, Parrot Fish

Fish come in all shapes and sizes. Some are free swimming while others rest on the bottom of the sea. Some are herbivores and others are carnivores. Some lay eggs while others give live birth and parental care to their young. Buchheim (1999d) provides the information in this table.

FISH - class Agnatha Fish of the class agnatha ("no jaw") are the most "primitive" of the fishes; they lack a jaw and a bony skeleton. The hagfish and the lamprey are the only living representatives of this once large class. As they lack true bones, these fish are very flexible, the hagfish can actually tie itself in a knot to rid itself of a noxious slime it can produce to deter predators. They have a smooth, scale less skin and are soft to the touch. In place of the jaws is an oral sucker in the center of which is the mouth cavity. Many of the agnathas are highly predatory, attaching to other fish by their sucker like mouths, and rasping through the skin into the viscera of their hosts. The juvenile lamprey feeds by sucking up mud containing micro-organisms and organic debris - as did the primitive agnatha. Agnathas are found in both fresh and salt water and some are anadromous [living in both fresh and salt water at different times in its life cycle]. The hagfish has no eyes, while the lamprey has well-developed eyes.	**Characteristics- Agnatha** Primitive No jaws Cartilaginous skeleton Scaleless skin Oral sucker in place of jaws Predators and filter feeders anticoagulating saliva fresh and salt water some anadromous
FISH - class Chondrichthyes Members of the class Chondrichthyes ("cartilage-fish") include the sharks, skates, rays, and ratfish. These fish have a cartilaginous skeleton, but their ancestors were bony animals. These were the first fish to exhibit paired fins. Chondrichthyes lack swim bladders, have spiral valve intestines, exhibit internal fertilization, and possess 5-7 gill arches (most have 5). They have cartilaginous upper and loosely attached lower jaws with a significant array of teeth. Their skin is covered with teeth like denticles which give it the texture and abrasive quality of sandpaper.	**Characteristics- Chondrichthyes** Cartilaginous skeleton Skin covered with denticles, not scales Five to seven gill slits per side No swim bladder Internal fertilization Spiral valve intestines Five to seven gill arches Cartilaginous jaws, loosely attached lower jaws

Sharks bring out some of man's greatest primordial fears. To be attacked by a shark is many people's greatest fear of the water, especially after watching the movie *Jaws*. The motif of being eaten alive by a brainless killing machine has been a major Hollywood success, much to the shark's disfavor. While there have been some documented and highly publicized cases of people being attacked and even eaten by sharks, most sharks do not deserve the title "man-hunters".

In fact, most sharks are entirely incapable of this feat. The largest fish of all, the Whale Shark that, can reach lengths of up to 59 feet and weights of 88,000 pounds, is a very calm and approachable plankton feeder. There are many species of sharks which can inflict severe bodily injury and require the utmost of respect. The most feared of all, the Great White Shark, has been responsible for most of the fatal shark attacks off the California and Australian coastlines. While the Great White gets all the notoriety, pound for pound, the Bull Shark is probably the most ferocious. The Great White generally attacks a person because it has confused it with its favorite food—seals and sea lions—but the Bull Shark will attack a person just because they are there. Even with these dangerous animals roaming the ocean, your chances of getting attacked by a shark are very remote.

Worldwide, there are only about 300 documented shark attacks a year. The chances that you will be hit by a drunk driver while driving to the beach are much higher than you encountering a dangerous shark when you get there. There are some activities that will greatly increase your chance of a shark attack, such as carrying speared fish with you while diving or collecting abalone in turbid waters. Statistics of 1,652 shark attacks show that males are much more likely to be attacked than females (10 to one ratio); this is probably because males are much more active in the water, surfing and going to deeper depths where sharks are more common.

The presence of large numbers of fish, or fish behaving in an unusual manner, has been reported preceding many attacks. In 40 percent of the reported shark attacks, people were pole-fishing or spear-fishing in the area of an attack. A comparison of the number of people swimming to those fishing and spear-fishing seems to show that these two pastimes have by far the highest risk of inducing an attack. While swimming, the chance of drowning is more than 1,000 times greater than that of dying from a shark attack.

Most shark attacks occur in shallow water, where most bathers are, and in 94 percent of the cases the attack was by an individual shark acting alone. About 10 percent of reported shark attacks are on divers; since the number of divers in the water at one time must be much smaller than 10 percent of beach bathers, the odds of being attacked must be significantly greater for divers.

Close passes were seldom made before the attack, and in the majority of the cases there was only one strike. Few attacks involved more than one bite. This indicates that in many cases the attacking shark mistook the victim for a more usual kind of food and did not attack any further when the error was discovered. It is fortunate that sharks, in most cases, do not consider humans to be suitable food. This information also refutes the long-standing notion that fresh human blood is a powerful attractant that excites sharks into a feeding frenzy. If this were so, the presence of blood would certainly have induced that attacking shark to strike the victim repeatedly. Most wounds occur on the appendages: the hands, arms, legs, and feet. Lacerations of varying severity are the most common types of injury. About 25 percent of attacks kill the victim. The most usual cause of death is shock, combined with a severe loss of blood.

Swimmers and divers can reduce the chance of being attacked by following a few simple rules: Never swim in areas where sharks are known to be common. Never enter the water where people are fishing, either from the beach or from inshore boats. If there are a number of people in the water, do not separate yourself from them. There is safety in numbers. Avoid swimming near deep channels or where shallow water suddenly becomes deeper. Do not swim alone or at dusk or after dark when sharks are feeding actively and are likely to be closer to the shore. Do not enter the water, or if in the water, leave immediately, if large numbers of fish are seen or if fish seem to be acting strangely. Be alert for unusual movements in the water. Do not wear a watch or other jewelry that shines and reflects light. Do not enter the water with an open wound, and women should not swim during their menstrual periods.

FISH – Class Chondrichthyes, (Sharks)
Sharks are animals that are superbly adapted to their environment. Almost all are carnivores or scavengers, although the species that live close to the sea floor feed mostly on invertebrates. Most possess a keen sense of smell, a large brain, good eyesight, and highly specialized mouth and teeth. Their bodies are usually heavier than water, and they do not have an air-filled swim bladder for buoyancy like most bony fishes. All sharks have an asymmetric tail fin with the upper lobe being larger than the lower one. This feature, together with flattened pectoral fins, and an oil-filled liver compensates for the lack of a swim bladder. There are 344 known species of sharks living in all parts

of the ocean, shallow to deep and tropical to polar. A few even venture into freshwater and have been found in rivers and lakes. Contrary to popular belief, most sharks are harmless to humans. Sharks are classified into eight orders:

1. Sawsharks (Pristophoriformes), one family, eight sp. They live on the bottom in warm temperate or tropical seas. They are easily recognized because of tubular, blade-like snouts. They bear live young.
2. Dogfish Sharks (Squaliformes), three families, 119 sp. Bottom dwelling deep water sharks distributed worldwide. Bear live young and eat bony fishes, crustaceans, squid, and other sharks. Dogfish sharks are harmless to humans.
3. Angel Sharks (Squatiniformes), one family, 19 sp. Flattened, bottom dwelling sharks. Found on continental shelves and upper slopes of cold temperate and tropical seas. Have very sharp, awl-like teeth that are used to impale small fish and crustaceans.
4. Bullhead Sharks (Heterodontiformes), one family, nine sp. They live on rocky reefs where there are plenty of cracks and crevices. Found in Pacific and Indian Ocean. Eat invertebrates.
5. Gilled Sharks (Hexanchiformes), two families, six sp. Deep-water, bottom-dwelling sharks distributed worldwide. Gilled sharks have six or seven gill slits. Bear live young and eat bony fish, crustaceans, and other sharks.
6. Mackerel Sharks (Lamniformes), seven families, 15 sp. Small, highly diverse order. Found in tropical to cold temperate or even Arctic waters. Range includes oceanic and coastal regions. Most very large, eat bony fish, other sharks, squid, and marine mammals. This group includes the Mako and Great White and the plankton eating Megamouth and Basking Sharks.

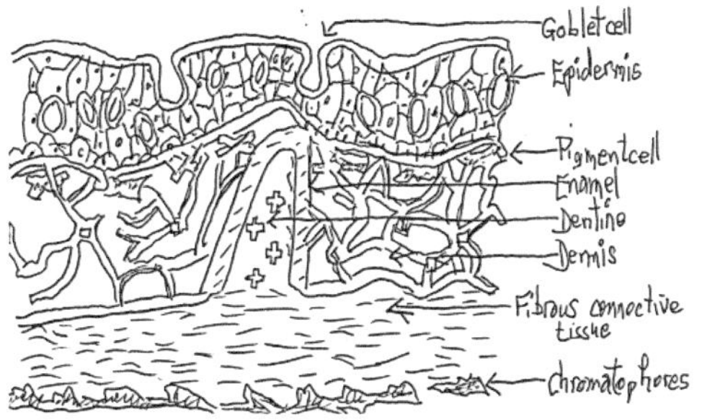

7. Carpet Sharks (Otectolobiformes) seven families, 31 sp. Warm tropical to temperate waters. All members except whale shark live on bottom. Bodies are flattened. Most eat small fish and invertebrates. Whale shark is plankton feeder. Some bear live young and others lay eggs.
8. Ground Sharks (Carcharhiniformes) 8 families, > 270 sp. Largest order of sharks. This order of shark has a worldwide distribution in temperate and tropical waters. Most live near coast, although some found in deeper waters. Eat bony fishes, other sharks, squid, and small invertebrates. This order includes the dangerous Tiger shark.

Shark Anatomy - Sharks have numerous structural and physiological features that make them unique among the fishes. They have a simple cartilaginous skeleton with no ribs, and a cartilaginous jaw, backbone, and cranium. Thick skin supports the flimsy skeleton. The skin is elastic and aids in movement; when the tail is arched, it pulls on the skin which pulls back like a rubber band. The jaws are not connected to the skull and become unhinged, protruding forward from the skull allowing for a wider gape when feeding. The teeth are ossified with minerals known as apatite. They form a conveyer belt with as many as eight teeth in a row. When a shark loses a tooth, another one just pops up. Sharks go through up to 2,400 teeth a year.

Sharks have placoid scales which are fixed, slightly ossified and layered. They are smooth to the touch in one direction and extremely course in another. Just rubbing a shark the wrong way can inflict serious wounds. All sharks, rays, and skates are carnivores. They have normal sensory modalities, a small brain (most of which is dedicated to the olfactory lobes, giving them an acute sense of smell) and well-developed eyes with color vision and adaptation to low light levels.

Some sharks lay eggs (all skates and ratfish do), but most are ovoviviparous (all rays are). The young develop with their yolk sacks within the mother but without a placenta or umbilical cord. Some sharks (the Great White) are oviphagous; the young eat the other developing young and embryos inside their mother and only the fiercest is born! A few sharks (hammerheads and reef sharks) are viviparous; like mammals, the young are nourished with a placenta within the mother. The gestation period is around 22 months and two to 80 pups are born per litter. Because most sharks are ovoviviparous or viviparous, they do not produce mass numbers of young like other fish do. They are slow to develop, and for this reason shark population numbers have been decreasing rapidly due to the recent popularity of shark fin soup. Fishermen are taking many more sharks than the maximum sustainable yield will allow. Some sharks will soon be endangered species.

Rays - Rays in general are physiologically exactly like sharks except the ray's pectoral fins are fused to their heads. Their gills are ventrally located. They swim with their ventral fins, like wings. Their eyes are dorsally (top) located and have spericules behind them. The spericules are used to breathe in with. Rays are modified as bottom feeders, feeding on invertebrates found in the sand. Sometimes you can watch a ray making quite a ruckus on the sand bottom in search of the invertebrates. Manta rays are planktivores and cruise the open water filter feeding out small animals. Mantas are the largest of the rays. Electric rays swim with their caudal fin and use their modified pectoral fins to electrically shock and stun their prey. Sawfish look like sharks but have true fused pectoral fins and gills on the ventral surface. Stingrays have a toxin filled spine at the base of their tail. Stingrays are not mean creatures roaming the waters to hurt swimmers as many people believe. Stingrays are actually very approachable and can be hand fed and petted, just don't step on them!

THE BONY FISH, Class - OSTIEICTHYES - The bony fish comprise the largest section of the vertebrates, with over 20,000 species worldwide. They are called bony fish because their skeletons are calcified, making them much harder than the cartilage bones of the chondrichthyes. The bony fishes have great maneuverability and speed, highly specialized mouths equipped with protrusible jaws, and a swim bladder to control buoyancy. The bony fish have evolved to be of almost every imaginable shape and size and exploit most marine and freshwater habitats on earth. Many of them have complex, recently evolved physiologies, organs, and behaviors for dealing with their environment in a sophisticated manner.

REPRODUCTION - Fish have three modes of reproduction depending on the method they care for their eggs.

Modes of Reproduction
Ovopartity - Lay undeveloped eggs, external fertilization (90 percent of bony fish), internal fertilization (some sharks and rays)
Ovoviviparity - Internal development: without direct maternal nourishment, advanced at birth (most sharks + rays), Larval birth (some scorpeaniforms-rockfish).
Viviparity - Internal development: direct nourishment from mother, fully advanced at birth (some sharks, surf perches).

In fish, oviparity is most common. The eggs are inexpensive to produce and, as eggs are in the water, they do not dry out (oxygen/nutrients are not scarce). The adult can produce many offspring which they broadcast into the plankton column. When the offspring settle out of the plankton column, they may be in totally new environments, allowing for a great area in which the young may survive. This mode also comes with its disadvantages. When born, the fish must first go through a larval stage for growth before they transform into the adults. In this larval stage, they must fend for themselves in obtaining food and avoiding predation. They may not find a suitable environment when they settle out of the plankton column. The survival of individual eggs is very low, so millions of eggs must be produced in order for the parent to successfully produce offspring. The other modes have their advantages. The eggs are much less prone to predation when carried within the mother, and the young are born fully advanced and ready to deal with the environment as miniature adults. These advantages come with a price-tag also; the adult must supply nutrients to its offspring and can only produce a few eggs at a time. The young are limited to the environment that their parents were in, and if this environment is deteriorating, they are stuck with it.

Parental care: In fishes, parental care is very rare as most fish are broadcast spawners, but there are a few instances of parental care. Male gobies guard the eggs in a nest until they are born. The male yellowhead jawfish actually guards the eggs by holding them in his mouth!

Hermaphroditic Fish

- Hermaphroditism: Some fish individuals are both males and females, either simultaneously or sequentially. There is no genetic or physical reason why hermaphroditism should not be present. About 21 families of fish are hermaphrodites.
- Simultaneous hermaphrodite: There are some instances where being a member of both sexes could have its advantages. Imagine all the dates that you could have! In the deep sea, the low light levels and limited food supply make for a very low population density; meaning that potential mates are few and far between. Members of the fish family Salmoniformes (e.g. salmon) and Serranidae (hamlets) are simultaneous hermaphrodites; they can spawn with any individual encountered.
- Sequential hermaphrodite: Very strange life histories develop in species whose individuals may change sex at some time in their life. They may change from being males to females (protandry) or females to males (protogyny).

An example of protandry is found in the anemonefish. The fishes live with anemones in a symbiotic relationship: the anemone provides shelter and protection from predation, and the fish supply the anemone with food. Groups of fishes will live with one anemone and will not switch anemones. Only the two largest fish spawn. The largest (a female) mates with the second largest (the male). With the female being the largest, she can produce the most eggs. When the female dies the male becomes (the largest egg producing) female. The rest of the fish are immature males.

A classic example of protogyny is found in the wrasses and parrotfishes. The males in these species form harems, with one large male sequestering and defending a group of smaller females. The male enjoys spectacular reproductive success as it has many females to mate with. The females also enjoy a limited reproductive success, producing as many eggs as they can, all fertilized by the one male. The male has the advantage over the females; it has many females producing eggs for him to fertilize, whereas the females only have themselves.

As only the largest male (the "super-male") gets to mate, a smaller male would experience zero reproductive success. There is no advantage to being a small male, and this is where the hermaphrodism comes in. If all the smaller fish were females, they could all enjoy a limited reproductive success while they are growing. If the male dies, the one that has grown to be the largest female will change sexes and become the male, in turn enjoying a much greater reproductive success than if she did not switch. So there are no small males and everything is all said and done, but wait! Evolution has a keen ability in finding weaknesses in any system, and it has done so with the parrotfish. In nature, we do find smaller male parrotfish, why should this be so? It has to do with the kind of thing that if a parrotfish was a human, could get the parrotfish into a great deal of trouble. The "super-male" has to run around all of the time keeping track of and protecting all of his females as well capturing and eating food himself, so he does not necessarily have time to pay close attention to the details. When parrotfish mate, they form a spawning aggregation where the super-male will release his sperm into the water and the many females release their eggs. The sperm and egg find each other in the water column and fertilization takes place, and this is the weakness of the system. Along comes the smaller male, who has evolved to look just like a female. Most of the time, a smaller male makes itself completely inconspicuous by behaving just like a female. But during the spawning aggregations, he will be releasing sperm instead of eggs. The super-male will probably not even know that he has been conned. Everything gets really mixed up as males are changing into females changing into males.

FISH - Schooling Behavior - Everyone has heard of a school of fish, an aggregation of fish hanging out together.... They are obviously not learning reading, writing, and arithmetic. Schools of fish may be either polarized (with all the fish facing the same direction) or non-polarized (all going every which way). There are some factors that can make it advantageous to hang out with other fish.

Antipredator: by hanging out with other fish, each individual fish may gain an advantage in not being eaten by other fish.

A. **Confusion effect.** A large school of fish may be able to confuse a potential predator into thinking that the school is actually a much larger organism.
B. **Dilution effect.** If a fish hangs out with a lot of other fish and a predator does come around, the predator must usually select one prey item. With so many choices, the chances are that it will not be you. This is known as the "selfish herd".
C. **Predator detection.** A bunch of fish has many times the eyes and other senses than a solitary fish, so a school of fish may be better at detecting predators. But a school may also attract predators due to its large size.

Spawning Aggregation - Many fish species form schools only when it comes time to mate. They will form a huge school and release their eggs and sperm in mass quantities. Releasing a massive onslaught of fertilized eggs in the water may be advantages over a solitary egg, because a massive onslaught may be enough to overwhelm the egg predators. The predators will eat as many as they can, but some eggs will inevitably survive.

Enhanced Foraging: A school of fish may have better abilities to acquire food. With many more eyes to detect food, many more meals may be found, but there would also be many more mouths to feed. By working as a team, the school may be able to take larger food items than any one individual could manage to capture.

Migration: The migration abilities of fish in schools may possibly be enhanced due to better navigation, etc.

Hydrodynamic efficiency: Due to the complex hydrodynamic properties of water (properties the fish probably discovered only by accident), a fish may gain a swimming advantage by being in a school. The slipstream from the fish ahead of it may make it easier to pass through the water. Good for all the fish except for the ones in front.

How fish swim - The density of water makes it very difficult to move in, but fish can move very smoothly and quickly. A swimming fish is relying on its skeleton for framework, its muscles for power, and its fins for thrust and direction. The skeleton of a fish is the most complex of all vertebrates. The skull acts as a fulcrum, the relatively stable part of the fish. The vertebral column acts as levers that operate for the movement of the fish.

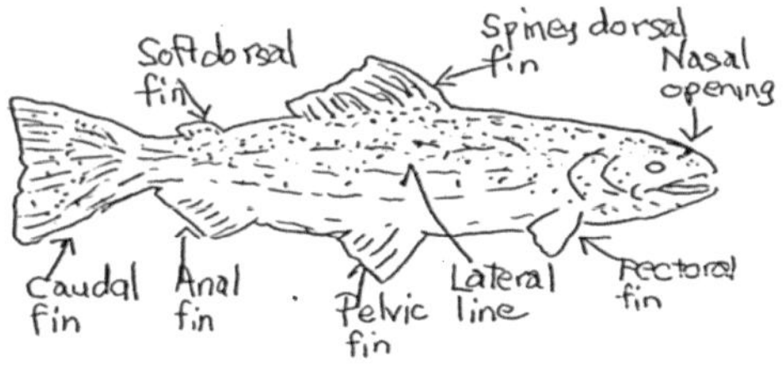

The muscles provide the power for swimming and constitute up to 80 percent of the fish itself. The muscles are arranged in multiple directions (myomeres) that allow the fish to move in any direction. A sinusoidal wave passes down from the head to the tail. The fins provide a platform to exert the thrust from the muscles onto the water.

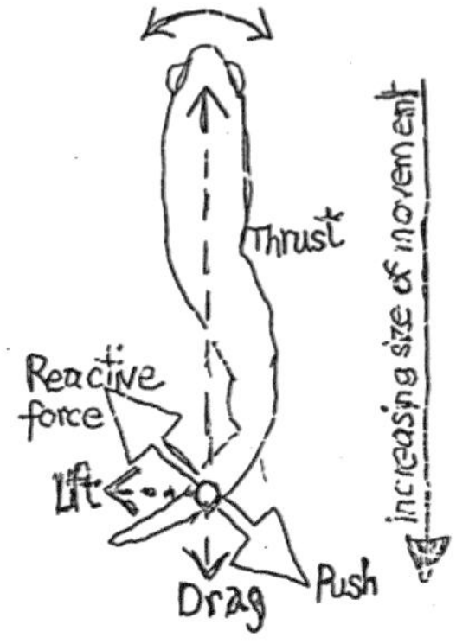

Diagram of forces when a fish swims

Thrust- force in animal's direction
Lift- force opposite in right angles to the thrust
Drag- force opposite the direction of movement
All lift forces cancel out over one complete tail stroke

Drag

Drag is minimized by the streamlined shape of the fish and a special slime fishes excrete from their skin that minimizes frictional drag and maintains laminar (smooth) flow of water past the fish.

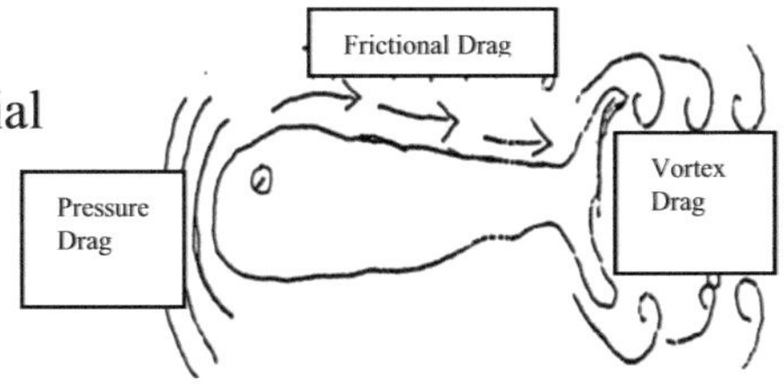

When Thrust > Drag, we have swimming!

Two swimming types in fishes - Defined by their method of living and reflected in their physiology.

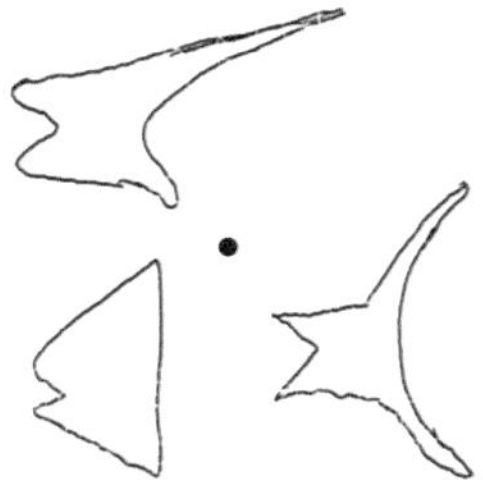

Cruisers: These are the fish that swim almost continuously in search for food, such as the tuna. Red muscle is richly vascularized (blood-carrying capacity), rich in myoglobin (oxygen holder and transferor into the muscles active sites) * able to sustain continuous aerobic movement.

Burst Swimmers: These fish usually stay relatively in the same place such as most reef fish.

Fins - fins give a fish control over its movements by directing thrust, supplying lift and even acting as brakes. A fish must control its pitch, yaw, and roll.

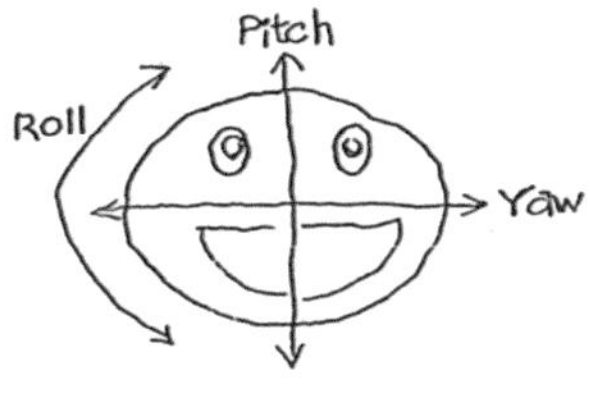

- Caudal fin—provides thrust, and controls the fishes direction
- Pectorals—act mostly as rudders and hydroplanes to control yaw and pitch. Also act as very important brakes by causing drag.
- Pelvic fins—mostly controls pitch
- Dorsal/anal—controls roll

VARIATIONS IN BODY FORM - **Fish shape has a great bearing on ability to move through the water.**

- A tuna fish, which has a fusiform similar to a torpedo, can cruise through the water at very high speeds.
- The attenuated shape of the eel allows it to wiggle into small crevices where it hunts prey.
- The depressed shape of the angler fish is advantageous for its "sit and wait" strategy of hunting.
- The compressed shape found on many reef fishes such as the butter fish gives the fish great agility for movement around the reef and can support sudden bursts of acceleration.

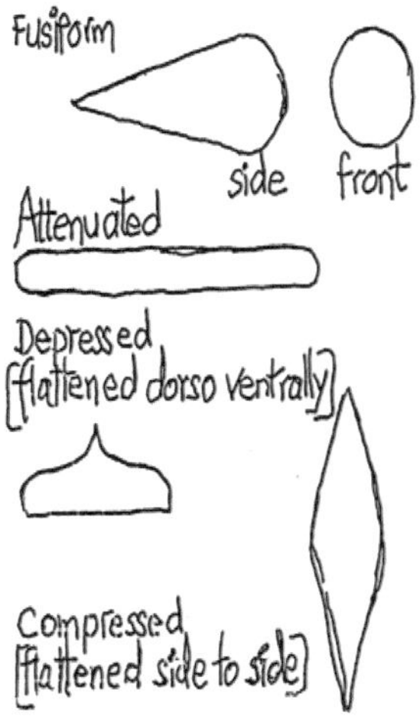

Fish Thermal Strategies - In general, fishes are cold blooded. They derive their body heat from their environment and conform to its temperature. As water has a high heat capacity, it is able to easily suck any excess heat out of a fish and into the environment.

- Ectothermic: fish derive their heat from the environment
- Poikilothermic: fish conform to the heat in the environment

Some large, fast-swimming fish are not ectothermic. The tunas and mackerel sharks can actually have core body temperatures ten to twenty degrees Celsius higher that the surrounding water. They are endothermic and derive their body heat from their metabolism, but they are still poikilothermic. Their body temperature may be higher than the surrounding water, but they still conform to the temperature of the water, just 10 to 20 degrees above it.

They maintain a higher body temperature through the use of a specialized counter-current heat exchanger called a rete mirabile. These are dense capillary beds within the swimming muscle that run next to the veins leaving the muscles. Blood passes through the veins and arteries in a counter current (opposite) direction. The heat produced from the muscle contraction flows from the exiting veins into the incoming arteries and is recycled.

Having an elevated body temperature increases the speed of the fish; the higher the body temperature, the greater the muscular power. 30°C is the optimum temperature for muscular speed. With increased speed, the tuna can capture the slower, cold blooded fish it preys upon. Tuna have been clocked at record speed of 50 to 70 mph!

Swim Bladders - Bony fish have swim bladders to help them maintain buoyancy in the water. The swim bladder of modern fish is an ellipsoidal, gas-filled sac that arises as an extension of the gut (Ramel, 2013). In many species of fish, the swim bladder remains connected to the gut, and the fish can use this connection to control the amount of gas in the swim bladder; this is called the **physostomou**s condition. Fish with gas bladders of this sort mostly live in shallow waters and swallow air at the surface of the water. This air is then passed into the gut and then forced into the swim bladder. In some fish, the gut connected physostomous bladder is constricted near the middle to make it look like two sacs; however, gas freely passes from one section to the other. In other species, the connection with the gut is closed and no gas can be moved from the gut to the swim bladder, this is called the **physoclistous** condition. These fish are able to control the amount gas in their swim bladders by means of one or more areas where the membrane is very thin and richly supplied with capillaries, gas exchange is through the capillaries and the membrane. The capillaries are both arterial and venous, and they are arranged in a counter-current system to facilitate the creation of high gas pressures within the swim bladder. These areas are known as rete mirabile, and some physostomous species also possess one or more rete mirabile.

If you have ever caught a fish and wondered why its eyes are bulging out of its head, it is because the air in the swim bladder has expanded and is pushing against the back of the eye. Oxygen is the largest percentage of gas in the bladder; nitrogen and carbon dioxide also fill in passively.

A physoclistous swim bladder is completely closed up, allowing no gas to be moved from the gut to the swim bladder. The gas enters the bladder through a gas gland, fed by venous blood in the front of the swim bladder. Gas leaves the bladder through an oval body filled with venous blood in the back of the swim bladder.

Inside the physoclistous gas bladder are giant secretory cells that secrete lactate in capillary clusters called rete mirabile. Increased lactate levels from the giant secretory cells lower the surrounding pH, causing blood hemoglobin to release oxygen. Released oxygen diffuses back into the incoming capillary, increasing the partial pressure of oxygen in the incoming capillary. This continues until the partial pressure of the oxygen in the capillary is higher than that of the swim bladder (which has a high concentration of oxygen). This complex system is necessary because the concentration of oxygen is higher in the swim bladder than it is in the blood, so simple diffusion would tend to pull the oxygen out of the bladder instead of pushing it in. If the fish wants more buoyancy, it must tell its secretory cells to release more lactate. Since oxygen diffuses easily with oxygen-poor venous blood, the gas can be forced out.

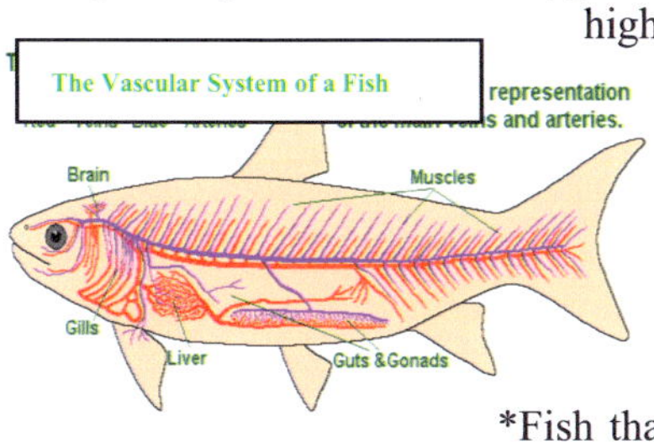

(Ramel, 2013)

*Fish that migrate vertically tend to have high oxygen levels in their bladders because it fills in faster and leaves faster.

*Fish that maintain a stable depth tend to have more nitrogen because it is inert, enters slowly, and exits slowly.

How Fish Breathe - How in the heck can a fish, which is underwater, breath (if there is no air)? When we go under water, we have to bring air with us to survive. Whales and dolphins have lungs that store air from the surface. Fish don't have lungs, and they rarely ever venture into the air, so how do they survive? We all know it has something to do with gills, but what exactly?

The water surrounding a fish contains a small percentage of dissolved oxygen. In the surface waters there can be about five ml. of oxygen per liter of water. This is much less than the 210 ml. of oxygen per liter of air that we breath, so the fish must use a special system for concentrating the oxygen in the water to meet their physiological needs. Here it comes again, a counter current exchange system, similar to the one we found in the fish's swim bladder and in the tuna's muscles.

The circulation of blood in fish is simple. The heart only has two chambers, in contrast to our heart which has four. This is because the fish heart only pumps blood in one direction. The blood enters the heart through a vein and exits through a vein on its way to the gills. In the gills, the blood picks up oxygen from the surrounding water and leaves the gills in arteries which go to the body. The oxygen is used in the body and goes back to the heart. The gills are a simple closed-circle circulatory system composed of a gill arch (which gives the gill rigid support), gill filaments (always paired), and secondary lamellae (where gas exchange takes place).

Fish can breathe the water by swimming forward and letting some of the passing water flow in through the mouth, across the gills, and then out. They can control the amount of blood flowing through their gills, increasing the amount of oxygen they take from the water if they swim more actively. When moving slowly, or resting in the water, fish can breathe the water by synchronously expanding and contracting the buccal cavity (the mouth and throat) and the opercular cavity, which is the space behind the operculum where the gills are. In doing this they can pump a continuous stream of water across their gills.

Sharks do not have an operculum or an opercular cavity but have instead a branchial cavity and a series of branchial flaps, one for each set of gill but which work in a similar way. Sharks also have a pair of spiracles in front of their gills (one on each side); they can pump water in through these spiracles and over their gills. In those species of sharks and rays which often, or even habitually, rest on the sea floor, these spiracles are on the top of the animal's head, allowing it to breathe without sucking in sand which would damage the delicate gills.

Most fish can also absorb some oxygen through their skin, and while they are in their larval form this is often the main source of respiration. The smaller a fish, the higher is its ratio of skin area to body mass. However, as it grows it increases its mass much faster than its surface area and soon the skin is not able to take in all the oxygen the body needs and so other forms of respiration/gaseous exchange take precedence. But in some fish, scientists have learned that the skin continues to take up oxygen throughout the animal's adult life, although the oxygen taken up is only used by the skin and not passed on to the muscles. Two examples of this are the Crucian Carp (*Carassius carassius*) and the Yellow Perch (*Perca fluviatilis*). In adult fish, only the Ameirus melas (check genus spelling) the Scaleless Black Bullhead uses its skin to take up oxygen for the body, in this species about five percent of the total gaseous exchange occurs through the skin.

Blood Flow, Counter Current Exchanger

- The blood flows through the gill filaments and secondary lamellae in the opposite direction from the water passing the gills. This is very important for getting all of the available oxygen out of the water and into the blood.
- If the blood flowed in the same direction as the water passing it, then the blood would only be able to get 17 percent of the available oxygen from the water. The blood and water would reach equilibrium in oxygen content, and diffusion would no longer take place.
- By having the blood flow in the opposite direction, the gradient is always such that the water has more available oxygen than the blood, and oxygen diffusion continues to take place after the blood has acquired more than 50 percent of the water's oxygen content. The countercurrent exchange system gives fish an efficiency of 80 to 90 percent in acquiring oxygen from water.

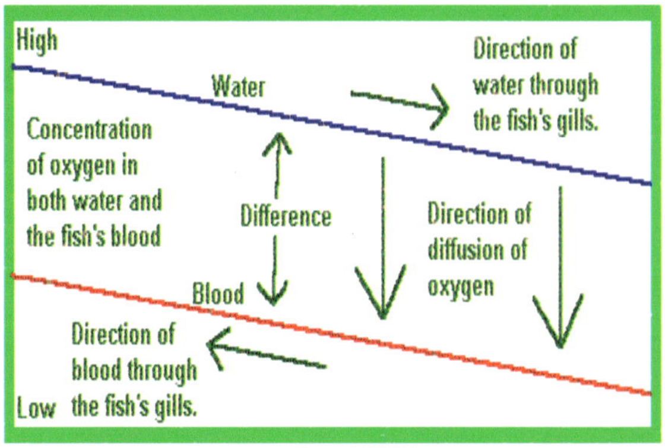

Ramel (2013)

- When fish are taken out of the water, they suffocate. This is not because they cannot breathe the oxygen available in the air, but because their gill arches collapse, and there is not enough surface area for diffusion to take place. There are actually some fish that can survive out of the water, such as the walking catfish which have modified lamellae allowing them to breathe air.

- It is possible for a fish to suffocate in the water. This could happen when the oxygen in the water has been used up by another biotic source such as bacteria decomposing a red tide.

To ventilate their gills, fish must pass new water over their gills continuously to keep a supply of oxygenated water available for diffusion. Fish use two different methods for keeping a continuous supply of new water available, one is very simple and the other complex.

- Ram Ventilation: Swim through the water and open your mouth. Very simple, but the fish must swim continuously in order to breathe, not so simple.
- Double Pump System: An elaborate system of passing water over the gills while the fish stays stationary. There is no moment throughout the cycle when the water stops flowing.

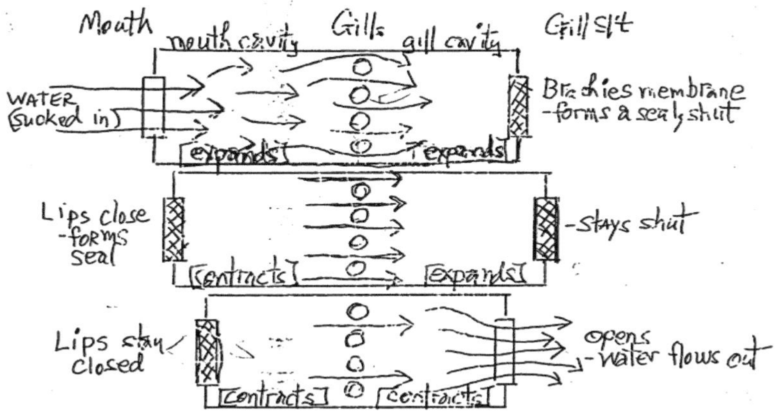

How Do Fish Sense - Successful survival in any environment depends upon an organism's ability to acquire information from its environment through its senses. Fish have many of the same senses that we have. They can see, smell, touch, feel, and taste. And they have developed some senses that we don't have, such as electroreception. Fish can sense light, chemicals, vibrations, and electricity.

Light: Photoreception (vision). Fish have a very keen sense of vision which helps them to find food, shelter, mates, and avoid predators. Fish vision is on par with our own vision; many can see in color, and some can see in extremely dim light.

Fish eyes are different from our own. Their lenses are perfectly spherical which enables them to see underwater because it has a higher refractive index to help them focus. They focus by moving the lens in and out instead of stretching it like we do. They cannot dilate or contract their pupils because the lens bulges through the iris. As the depth at which fish are found increases, the resident fish's eye sizes increase in order to gather the dimmer light. This process continues until the end of the photic zone, where eye size drops off as there is no light to see with. Nocturnal fish tend to have larger eyes then diurnal fish. Just look at a squirrelfish, and you will see this to be so. Some fish have a special eye structure known as the Tapetum lucidum which amplifies the incoming light. It is a layer of guanine crystals which glow at night. Photons which pass the retina get bounced back to be detected again. If the photons are still not absorbed, they are reflected back out of the eye. On a night dive, you may see these reflections as you shine your light around!

Chemicals: Chemoreception (smell and taste). Chemoreception is very well developed in fishes, especially the sharks and eels which rely upon this to detect their prey. Fish have two nostrils on each side of their head, and there is no connection between the nostrils and the throat. The olfactory rosette is the organ that detects the chemicals. The size of the rosette is proportional to the fish's ability to smell. Some fish (such as sharks, rays, eels, and salmon) can detect chemical levels as low as one part per billion.

Fish also have the ability to taste. They have taste buds on their lips, tongue, and all over their mouths. Some fish, such as the goatfish or catfish, have barbels which are whiskers that have taste structures. Goatfish can be seen digging through the sand with their barbels looking for invertebrate worms to eat and can taste them before they even reach their mouths.

Vibrations: Mechanoreception (hearing and touch). Though not visible, fish do have ears located within their bodies as well as a lateral line system that actually lets them feel their surroundings (Odyssey, 1999). Fish do not have external ears, but sound vibrations readily transmit from the water through the fish's body to its internal ears. The ears are divided into two sections, an upper section (pars superior) and a lower section (utriculus). The pars superior is divided into three semicircular canals and give the fish its sense of balance. It is fluid-filled with sensory hairs. The sensory hairs detect the rotational acceleration of the fluid. The canals are arranged so that one gives yaw, another pitch, and the other roll. The utriculus gives the fish its ability to hear. It has two large otoliths which vibrate with the sound and stimulate surrounding hair cells.

Fish possess another sense of mechanoreception that is kind of like a cross between hearing and touch. The organ responsible for this is the neuromast, a cluster of hair cells which have their hairs linked in a glob of jelly known as **cupola**. All fish possess free neuromasts which come in contact directly with the water. Most fish have a series of neuromasts not in direct contact with the water. These are arranged linearly and form the fish's lateral lines. A free neuromast gives the fish directional input.

A lateral line receives signals stimulated in a sequence and gives the fish much more information (feeling the other fish around it for polarized schooling and short-range prey detection: "the sense of distant touch").

Electricity: Electroreception. Sharks and rays possess special organs for detecting electrical potential (voltage). A set of pits comprise the electroreceptive system called the ampullae of Lorenzini. These are canals in the skin filled with a gelatin-like material that also contain sensory cells. Movements or disturbances near the shark change the voltage drop along the canals, which allows the shark to sense other organisms nearby. These sensors are so sensitive that if there were not any other distortions a shark could detect the heartbeat of a fish 500 miles away! They can detect muscular contractions of struggling prey and even the earth's magnetic field (which sharks use for navigation)."

After the initial testing of the system for leaks, in Project Three, it is now time to prepare it for broodstock fish invertebrates. If any leaks were discovered in the original test, make sure they have been eliminated. Again, PVC, cement, and hose clamps help keep the system watertight throughout the school year, but if possible, try to use the cement as little as possible. Remember that a day (now in the distant future) for disassembly and storage of materials will come, and it is good to reserve some thought to that time. When the time does come to disassemble the system, please be aware that students will not be able to separate pipes that are cemented together.

Photo by Robert Jakus

Determine temperature ranges that are necessary for system operation and ensure that properly sized water heaters are safely installed in the sump. Establish pH, DO, salinity, and hardness levels of system water (and protocols for maintenance of it). Provide a place for storage (and dechlorination, salinization, and pH adjustment) of water to be used to for water changes or to replace evaporated water. Ensure access to system and species logbooks for each broodstock pair in the system (enter data). Ensure the system lighting is operational. For a saltwater broodstock larval rearing system, install a protein skimmer to aerate and skim the water in the sump. Reserve a place in which additional biological water filtration equipment may later be installed.

A final but necessary step before working directly with fish and other organisms is to review information related to proper and ethical treatment of all classroom organisms. Among other things, students must be aware that bug sprays, sun block, perfumes, and many other substances on our skin kill aquatic organisms in an instant. Working with aquatic ecosystems requires a level of hygiene that many students are not familiar with. It is important that all students are guided to make every effort to work with aquatic ecosystems in ways that prevent exposure to substances that are toxic. In the effort to promote proactive aquatic ecosystem work, each student who is to work directly with the system must submit a signed form of agreement; the signature of a parent or guardian is required on this **Non-Human Vertebrate Endorsement Form** (see page 129). Students do not need to submit a separate form for each system. But each student who works with this system must submit this form, including a signature of a parent or guardian. Review protocols for healthy transfer of organisms to a new system (from the **Introductory Aquatic Ecosystem Inhabitant Form** - page 57).

Prepare fish and other invertebrates for introduction to the system by slowly acclimating them to the temperature and chemical properties of the water in it. Hopefully, differences between water conditions in new and old environments will not be too extreme. Organisms must be acclimated for transfer to the new system. In general, the procedure for this is to place them in a container that is half-full of water from their old system, and then gently add a quarter of the remaining volume of water from the new system into the container, wait for fifteen minutes, and then again add a quarter of the remaining volume of new system water to the bucket. After repeating this exposure to the new water three times, gently release the organisms to the system. Remember to keep data about the way that corals in the system grow (anthropometric data). Do not over feed the organisms in the system, especially when the nitrogen cycle is just beginning.

Coordinated Sharing of Broodstock and Larval Rearing System Development and Operation Information with Classmates

For classes that followed the division of labor approach to assembling their classroom aquatic ecosystems, it should now be apparent that, working independently of one another, three student groups (or more) have each completed the design and assembly of one classroom aquatic ecosystem. The project six concluding activity described here is designed to be synchronized with project four and five concluding activities (see pages 141 and 165). As with the chapter two concluding activity on page 121, it is again assumed that students from each respective project team possess sufficient skill and background understanding of their ecosystem and activities presented in their respective projects to effectively guide students from other ecosystem teams to also understand their ecosystem and to capably perform the respective techniques featured in their project.

Once all systems are operating and students in each "home group" are familiar with their system related information, students are prepared to engage in a coordinated cycle of information sharing and accountability. This activity begins by having students in each of the three groups (aquaponics, coral propagation, and broodstock larval rearing) count off by twos. The following chart is designed to illustrate this verbal description. It is important to provide each student group with examples of paperwork and forms that were used by the students in each group as they completed their project activities.

According to the chart, the students who assembled the aquaponics system will work to guide students who assembled the coral propagation system through the seed germination and respiration related activities. According to the chart, as the X1 students lead the Y1's, the Y2 students will lead the Z1's through calcium carbonate egg plug production and coral fragging related activities and the Z2 students will lead the X2's through setting up black worm culture systems and decapsulating brine shrimp cysts. As this activity concludes, students shall have completed the first cycle of sharing specific aquatic ecosystem-related knowledge and skills. In cycle two, students in the X1 group lead students in the Z1 group. Students in the Y2 group lead the X2 group, and students in the Z2 group lead students in the Y1 group. As the chart below indicates, at the end of cycle two, the ones and twos in each group exchange leader/learner roles and pass through cycles three and four.

Leaders Cycles 1 & 2	Learners Cycle 1	Learners Cycle 2	Leaders Cycle 3 & 4	Learners Cycle 3	Learners Cycle 4
X1 System Overview seed germination photosynthesis and respiration	**Y1**	**Z1**	**X2** System Overview seed germination photosynthesis and respiration	**Y2**	**Z2**
Y2 System Overview calcium carbonate egg plug production and coral fragging	**Z1**	**X2**	**Y1** System Overview calcium carbonate egg plug production and coral fragging	**Z2**	**X1**
Z2 System Overview setting up black worm culture systems and decapsulating brine shrimp cysts	**X2**	**Y1**	**Z1** System Overview setting up black worm culture systems and decapsulating brine shrimp cysts	**X1**	**Y2**

This coordinated sharing activity is designed so that at its conclusion, students from each team will be familiar with all topics and issues that are relevant to the design, assembly, and operation of each of the three student-designed classroom aquatic ecosystems. Students may be asked to use details of each classroom system to develop a comparison and contrast poster.

Write a list of **science-based** activities you have participated in during this project and briefly describe evidence that verifies your interaction with and knowledge of scientific concepts for each item on your project six list.

Write a list of **technologically-oriented** activities you have participated in during this project and briefly describe evidence that verifies your interaction with and knowledge of technologically-oriented concepts for each item on your project six list.

Write a list of **engineering-oriented** activities you have participated in during this project and briefly describe evidence that verifies your interaction with and knowledge of engineering-oriented concepts for each item on your project six list.

Write a list of **mathematical** activities you have participated in during this project and briefly describe evidence that verifies your interaction with and knowledge of mathematical concepts for each item on your project six list.

Write a brief list of **Non-STEM-related** activities you have participated in during this project and briefly describe evidence that verifies your interaction with and knowledge of Non-STEM related concepts for each item on your project six list.

Rate your level of involvement in STEM-related activities in this project (1=lowest, 10=highest).

1—2—3—4—5—6—7—8—9—10

Beyond the Classroom Aquarium **Daily Journal Guidelines**

Entries students place in their daily journals should be:

1. Dated accurately and logged chronologically.
2. Kept in a notebook and written neatly in ink or typed.
3. Written or typed in complete sentences using correct capitalization, spelling, and punctuation.

In their journals, students should keep a daily record of:

1. Who they worked with each day by including the names of school staff, fellow students, community members, and volunteers.
2. What they did each day, such as lecture topic, activities or project design, assembly, or maintenance.
3. What they learned each day by asking questions, following instructions, and/or observing activities.
4. How *Beyond the Classroom Aquarium* experiences relate to past, present, or future life experiences.

For bonus points, students may complete these journal options, including illustrations, diagrams, or creative expression relating to the journal entry. For project one, journal writing bonus topics may include answers to the following questions or descriptions of a project activity or assignment. Reflection on these topics may also help students prepare for exams, quizzes, and other assessments.

- What do you want to understand better in order to contribute more to your team's productivity as it independently designs, assembles, and then manages an aquatic ecosystem?
- Create a new drawing and a story about an organism or system that is interesting to you.
- Create a journal entry that tells what information you learned during Project Six. Also write down what you would like to know more about and if you know someone else who might help you learn more if you asked them.
- Create additional posters and signs so that people who visit the broodstock and larval rearing systems will be able to read about new and important system components and organisms.

- Select a book or magazine to read that is related to a topic in Project Six. Write a report on this book and include your report in your *Beyond the Classroom Aquarium* portfolio.
- Explain how and why governmental agencies are concerned about aquaculture. Explain progress they are making and what measures need to be adopted in the future to improve the effectiveness of their policies and actions.
- Explain three important ideas and/or skills you gained by participating in the broodstock and larval rearing system fieldtrip/workshop.
- Explain the value of nutritious and live food for broodstock organisms.

Journal entries will be evaluated on a point system, with four points being the maximum number a student can earn. A half-point will be awarded for each of the first eight journal requirements above. Bonus points will be awarded when journal options are completed. Journal points will be totaled and divided by the number of journal dates. A possible journal grade based on points is included in the project one assessment section.

Additional Project Six Assessment Activities and Questions

Review information that is to be entered on the pair history data sheet (provided below) for broodstock organisms. Explain how the information on the sheet may be a useful tool in monitoring the productivity of each broodstock organisms in the system.

___________ - PAIR HISTORY **TANK LOCATION- ___________**
SPECIES

Pair History Data Sheet								
Spawn Number	Date	Days Since Last Spawn	Nest Size	Hatch Time	Percent Missing	Nest Site	Nest Collected	Comments
1								
2								
3								
4								
5								
6								
7								

Create a display that illustrates the steps to follow for producing a black worm culture. Include photos of the steps you followed as you produced your own culture. Explain how to monitor the rate at which the culture grows. Create a chart to use for monitoring culture growth. Update each culture's chart weekly.

Explain the life cycle of Artemia, *Salina*. Create a poster with drawings of the life stages (labeled). Note the phylum this organism belongs to.

Measure out a small mass of artemia salina cysts. Create a poster that illustrates the procedure for decapsulating artemia cysts as provided in *Beyond the Classroom Aquarium*. Include photos of the procedure you followed and the resulting hatch. Using the same mass of artemia cysts, use the procedure described in this book to produce a batch culture of artemia. Include this procedure with photos of your steps and the resulting hatch. Compare and contrast results of the two techniques. Which do you prefer?

State the three classes of fish and give examples of fish belonging in each class. Explain what criteria are used to classify a fish into one of these three categories.

After reviewing information provided about sharks, list and describe five important shark related ideas. If possible to obtain a viable bamboo shark egg case, design a system to contain the shark after it hatches. Have the class critique the design and bamboo shark research you performed before the system is built.
Explain the mode of reproduction of fish in the clownfish species.
Explain the dynamics involved when fish swim. Place this information on a poster with labeled drawings. Include descriptions of the types of swimming that different fish use.

What is a swim bladder? Where is it located? How do fish use their swim bladders?

Explain how fish get oxygen from the water. Include a large, labeled diagram on a poster.
Describe the senses that fish have and explain how they function.

Question – What is a fish?
List six defining features of a fish.

Research and explain protocols for nest collection and or larval collection that relate to broodstock contained in the classroom broodstock larval rearing system. Prepare a report for classroom critique. Include an explanation of the techniques you believe are most effective.

Describe protocols to follow to raise zebra danios in the classroom. Develop and present a report on reasons ways that zebra danios are being used for research.

Work with a group of students in your team to research and develop an Aquatic Ecosystem Daily Data Log Form that is applicable to all systems in the classroom. Prepare this information for a class critique. After the critique, follow up with each group to ensure that daily data from each system is logged into the format that the class agreed to use.

Once a week, work with a small group of students (one or two from each team) and review daily reflections and system logs for the classroom aquatic ecosystems. Create a chart that reports on the nature of the data that has been entered in each. Prepare a report for class critique that describes the data. How much data in the reflection logs is quantitative? How much qualitative? How consistently have students been entering data in the logs? Describe important information about species and systems that has been included and omitted? Recommend procedures to ensure that data is recorded into the sheets every day so that missing data does not become an issue.

Use data collected during the first three weeks following each systems first setup to create posters that illustrate the nitrogen cycle (relative concentrations of ammonia, nitrite, and nitrate) in each system.

Update water quality data from each classroom system data log book using an excel spreadsheet. Use Excel descriptive statistics applications to graphically depict the water quality of each system. Interpret the data and explain patterns and trends. Be prepared for a class critique of your data discussion.

Work with a group of students in your team and design a web page layout that you could use to feature a classroom broodstock and larval rearing system. Explain the value of the information contained in your web design and describe who might find your web site valuable. Prepare this layout for classroom critique. Would it be possible to include daily water quality data on the web site?

Review activities and information that interested you in this project and determine if this information fits on the "Research-ability" Potential Topics Form on page 59. If so, add it to the list.

Beyond the Classroom Aquarium STEM Self or Peer Assessment Rubric

This self-assessment is valid only to the extent to which you can document and display evidence that supports it. For each category in the left hand column, write a STEM-Oriented example in the box beneath 3 if you can display evidence that your STEM-Oriented example represents work that exceeds expectations. If your evidence shows that your STEM-Oriented statement meets but does not exceed expectations, write the statement in the box beneath 2. Similarly, for STEM-Oriented examples that demonstrate some progress toward expectations.

<table>
<tr><th colspan="2">Name:
Project Number:
Date Project Work Began:
Date of Project Completion:

List Math Topics and/or Objectives Currently Under Study:</th><th>3
Work Exceeds Expectations</th><th>2
Work Meets Expectations</th><th>1
Work Is Progressing Toward Expectations</th></tr>
<tr><td colspan="2">Write examples of questions and defining problems related to your STEM work.</td><td></td><td></td><td></td></tr>
<tr><td colspan="2">Describe how you have developed and used models in your STEM work.</td><td></td><td></td><td></td></tr>
<tr><td colspan="2">Write examples of STEM investigations you planned and carried out.</td><td></td><td></td><td></td></tr>
<tr><td colspan="2">Describe STEM related data that you analyzed or interpreted.</td><td></td><td></td><td></td></tr>
<tr><td rowspan="5">Give examples of how you used mathematics and computational thinking related to the five standards listed to the right.</td><td>Problem Solving (Math Process Standard One)</td><td></td><td></td><td></td></tr>
<tr><td>Reasoning & Proof (Math Process Standard Two)</td><td></td><td></td><td></td></tr>
<tr><td>Communication (Math Process Standard Three)</td><td></td><td></td><td></td></tr>
<tr><td>Connections (Math Process Standard Four)</td><td></td><td></td><td></td></tr>
<tr><td>Representation (Math Process Standard Five)</td><td></td><td></td><td></td></tr>
<tr><td colspan="2">Describe STEM-Oriented explanations and solutions you designed</td><td></td><td></td><td></td></tr>
<tr><td colspan="2">Explain evidence you used while engaging in a STEM project-based argument.</td><td></td><td></td><td></td></tr>
<tr><td colspan="2">Give examples of STEM information you obtained, evaluated, and/or communicated</td><td></td><td></td><td></td></tr>
<tr><td colspan="2">TOTAL POINTS</td><td></td><td></td><td></td></tr>
</table>

CHAPTER IV: Students Design, Assemble & Maintain Auxiliary Aquatic Ecosys-STEM Components (Projects 7-8)

After completing the first six projects, many students have an immense sense of accomplishment. The classroom aquaponics, coral propagation, and broodstock larval rearing systems are cycling. Organisms in the systems are healthfully acclimating. And the student daily workload increases. As students practice daily observation, maintenance, and performance assessments of their classroom systems (and the organisms in them), they may begin to recognize an obvious need for supplementary systems and lab equipment. Students may begin to wonder what happens when eggs hatch and larvae survive once there are hundreds of hungry larval mouths to feed. Students may wonder what morphological features are present from day to day as eggs and larvae develop. They may be curious about the kinds of feeds that promote larval health and growth. Information presented throughout this chapter provides many ideas from among which students may find a topic upon which to base the design of their chapter V aquatic ecosystem research project.

Chapter IV is designed to begin with field trip four. During this fourth field trip, students are introduced to natural planktonic environments, microscopic identification of aquatic plankton, plankton systems in hatcheries and/or aquatic labs, and computer automation systems in aquatic labs. In order to prepare for this field trip students may study about planktonic organism that reside in local bodies of water and then depending upon the site which is to be visited student teams may design, critique and construct plankton collection nets and containment vessels that may be used to transport organisms back to the classroom for further study. When possible it is nice to also be able to include a visit to a nearby aquatic lab where students may meet people who study plankton and get to observe them applying their work. An afternoon segment of field trip four should be used to introduce students to a local computer automation educator/expert who can introduce students to microcomputer mounting, wiring, proper attachment of input/output sensors and probes, and ladder logic circuit code. Following the field trip, students (or student teams) are guided to work with tools, equipment, and materials that help them contribute further to successful growth and development of their aquatic organisms as they are guided to develop auxiliary algae and rotifer plankton culture systems in the classroom and to construct their own working compound microscopes.

Project seven focuses students upon the skills and knowledge they must develop in order to provide their larval aquatic organisms with proper (live) food. Upon returning to class from the field trip and critiquing the field trip experience and prior to developing an algae rotifer culture system, students are engaged in lab activities that lead to the creation of quantities of Guillard's f/2 nutrient solution. As students develop this nutrient solution they are guided in the proper use of hot plates with magnetic stirrers and a digital analytical balance. The importance of water quality and sterile lab procedures in plankton culture is emphasized. After learning more about algae and rotifers and designing, critiquing and assembling algae rotifer systems, students are next challenged to explore and design protocols for developing stock single species algal cultures.

During this period, time will also be available to guide students through activities related to microcomputer setup, ladder logic programming of electrical circuits, and the design and assembly of a computer automated rearing and collection system for planktonic copepods. The design, critique, set up and operation of a computer automated plankton culture and collection system is the final activity in project seven. If set up as a workbench, the ecosystem PLC may be considered the "base" of an expandable aquatic ecosystem automation/robotic laboratory. Please note that development of all necessary plankton rearing systems and microscope construction activities in *Beyond the Classroom Aquarium* can be performed without reliance upon computer automation of any kind for teachers who choose to bypass this set of activities for any reason. For these classes, once students have set up operational algae/rotifer culture systems in their classroom, they may just skip past the microcontroller-related activities and begin work on project eight.

Although some educators will find the use of a microcontroller in class to be a new experience, information is provided so that no one needs to be overwhelmed by it. As described later, in detail, to operate the computer-automated plankton system, students will use the PLC and three electronic float switches to synchronize the operation of an air pump, a LED light, a solenoid valve, and two water pumps. Some students may find that this aquatic ecosystem-related PLC experience leads them to consider pursuing a career in industrial, infrastructure, and facility-based human machine interfacing (HMI).

An optional field trip (field trip five) to a space observatory or planetarium may be taken at this point in the curricular sequence. Ideally, students will not only get an idea about telescopes, microscopes, lenses, and magnification but they will also get an opportunity to interact with a knowledgeable member of the staff who can introduce the microscope construction activity to the students and assist them in designing pieces correctly for assembly and further work back in the classroom.

In project eight, students are guided through the process of learning about various topics in optics in order to create their own compound microscopes. The plans for design, construction, and use of the student-built compound microscopes are from a book by Alan Curry (1982). Basic optics activities were obtained from Melvin Berger (1987). As the live food systems begin to produce food, students should be able to use their microscopes and organisms from their systems to assess the productivity of their systems and to modify system characteristics in order to maximize productivity. In this project students will be guided to use their microscopes and a hemocytometer to document egg and larval development, count algal cell densities, and sample rotifer cultures.

Drawn by R. Jakus

From my days as a youthful but dedicated reader of *Tropical Fish Hobbyist* at the age of ten, to the present, my interest in aquaculture-related projects continues. After overcoming my reluctance to enter the world of salt-water aquariums, I discovered that many people who were having success with aquaculture of aquatic organisms were culturing unicellular algae and rotifers. Principles of high quality nutrition, the use of purified water, a long list of chemicals, and an analytical balance are used in this project to introduce students to important chemical lab protocols and concepts as they are guided to create Guillard's f/2 nutrient solution.

Photo by Robert Jakus

In order to culture eggs, or aquatic larvae, that have been collected and placed in larval rearing containers developed in previous projects students must provide newly hatched larvae with food that is abundant and healthful. In general, the larval feeding protocol requires that nutrient-packed rotifers be available at first feeding. As larvae grow, they benefit from larger foods such as Artemia naupli, some powdered foods, and eventually, Artemia. Sufficient growth allows aquaculturalists to wean larvae from exclusively live food diets to diets that consist principally of commercially prepared flake foods.

To provide food for the larvae of a single target species, many hatcheries culture three different live foods (microalgae, rotifers, and Artemia). The primary student task for this project involves study of nutrition, food chains, and the process of laboratory production of nutrient solutions for use in a 2-shelf algae/rotifer system. Students will study the design of a 2-shelf algae/rotifer system and construct it. Students will be introduced to the use of mechanical and digital analytical balances and the chemicals necessary to produce Guillard's nutrient solution. Students will become familiar with various algae species and topics (including harmful algal blooms) and engaged in the algae/rotifer culturing process. Students are also introduced to other species of larval foods (such as copepods) and work in teams to design a system for their automated culture and harvest. Emphasis will include sterile laboratory procedures to prevent culture contamination.

Laboratory Production of Guillard's f/2 Formula – Project Seven Opening Activity

Preparing f/2 Media - Instructions given here were developed by Rippingale and Payne (2001). Their protocol differs from those described by others. It is included here because I believe it is presented in such a way that students will more comfortably follow the steps. By following this procedure, students will produce 50 aliquots of concentrated f/2 medium, each of 20 mL, which are to be frozen. Each 20-mL aliquot will contain f/2 nutrients for two liters of algae culture. The quantities should be adjusted if larger cultures or more than 50 aliquots are needed. For example, to make 50 aliquots with each aliquot containing nutrient for 20 liters of f/2 media, use 10 times the amounts given below.

The chemicals are available from most suppliers. Be prepared to shop around to get the best price. Chemicals should be stored dry in a sealed container with a dessicant. Some of these chemicals tend to absorb water from the atmosphere (especially Ferric Chloride) and they cannot be weighed accurately after this occurs.

Trace Elements - The first step is to prepare a standard stock solution of the concentrated trace elements. Only a portion of this is used for each batch of f/2, the remainder can be frozen for later use.

Trace element	Weigh out	Volume of water
$CuSO_4.5H_2O$	100 mg	100 mL
$ZnSO_4.7H_2O$	220 mg	100 mL
$CoCl_2.6H_2O$	100 mg	100 mL
$MnCl_2.4H_2O$	1800 mg	100 mL
$Na_2MoO_4.H_2O$	60 mg	100 mL

Rippingale and Payne (2001)

If the amount cannot be weighed exactly, weigh the amount close to this and calculate the volume of water required to end up with the same concentration. For example, if 112 mg of $CuSO_45H_2O$ is weighed, add it to 112 mL of water and then only use 100 mL of that solution.

There are now five solutions, each of a trace element

Take exactly 50 mL of each solution and pool them in a 500-mL cylinder of volumetric flask and make the volume up to 500 mL with water. The pooled solution of trace elements is more than needed to prepare 50 X 20 ice-blocks; only 10 mL is required for two L of culture.

Keep 10 mL of this trace element solution. Freeze the fifty 10-mL aliquots for later use (or continue the production process before freezing.)

Vitamins - Prepare standard stock solutions of concentrated vitamins. As for the trace elements, most of this can be frozen for later use.

Vitamin	Weigh out	Volume of Water
Thiamine HCl	500 mg	50 mL
Biotin	50 mg	1000 mL
B_{12}	50 mg	1000 mL

Rippingale and Payne (2001)

As above, but 50 mL of Thiamine is needed. Do not weigh out less than 500 mg.
Take 50 mL of each of the three vitamin solutions and pool them in 500 mL of water.
Keep 10 ml of this vitamin solution. Fill the 50 half-full aliquots with 10 mL of vitamin solution. Freeze the 50 aliquots for later use.

Major Chemicals - Weigh exactly each of the following. Keep them in separate labeled vials.

$NaNO_3$	8.3 g
$NaH_2PO_4.H_2O$	0.5 g
$FeCl_3.6H_2O$	0.315 g *caution hygroscopic
EDTA	0.436 g

Rippingale and Payne (2001)

Preparation of 20 mL aliquots

To about 500 mL of distilled or deionized water;

- Add 0.436 g of the EDTA. Stir to dissolve. **EDTA must be added first**!
- Add 0.315 g of the iron chloride. Dissolve.
- Add 0.5 g of the phosphate salt.
- Add 8.3 g of the nitrate salt.
- Add 10 mL of the pooled trace metal solution.
- Add 10 mL of the pooled vitamin solution.

Pour 20-mL aliquots of this solution into snap top vials, seal, and freeze. These will keep almost indefinitely. The culture of diatoms required the addition of a silica solution to the solution and is not described here.

The information that follows provides a thorough review of unicellular algae culture. It is important to establish a routine of daily steps to successfully operate an algae/rotifer culture system that is capable of continually producing high quality live food (rotifers) for newborn fish larvae. Before going over this information in detail, students may find the following informal demonstration activity of interest. Students from class may be guided to perform the following steps to create the first classroom algae culture in a remote corner of the classroom.

As these steps are taken, try to ensure that all materials and equipment are clean throughout the process.

- Prepare a 1600 mL saltwater solution that contains about 15 ppt of salt in a two-liter beaker (the amount of salt does not need to be exact) but each group should record the salinity level of water they use).
- Place one 20-mL aliquot of nutrient solution in the beaker of saltwater and mix the solution.
- Fill an empty, clean, plastic, two-liter soda bottle with 1620 mL of the nutrient treated saltwater.
- Pour 80 mL of algae starter culture (obtained from the teacher's algae culture) into the two-liter bottle.
- Attach a good length of airline hose to an air pump (for each two-liter algae culture).
- Drill a hole through the two-liter bottle's cap so that the airline hose fits through the hole snugly.
- Insert the hose (already through the hole in the bottle cap) into the bottle and tighten the cap.
- Send the hose to near the bottom of the jar.
- Turn on the air pump and adjust the bubbles so they are gently aerating the water (the algae will be removing carbon dioxide from the water and the bubbles will be replacing the carbon dioxide the plants remove).
- Position a common fluorescent light in a position that permits each algae culture to be illuminated.
- For documentation, take photos of the cultures progress each day as you begin to review the following information.

Algae and Rotifer Biology and Production – Auxiliary System Designs

Photo by Robert Jakus

The work of Hoff and Snell (1987) is one example from an expanding field of reference literature educators may use to help them guide students to succeed with their efforts to design and assemble systems in which to culture aquatic plankton. In their book's introduction, they claim their "goal is to clearly describe the technical aspects of culturing phytoplankton and zooplankton so that anyone who is careful and follows these instructions can grow all the microalgae, rotifers, Artemia and daphnids they desire. The information is presented at a level that should be comprehensible to those with minimal experience in aquaculture. Every attempt has been made to keep the descriptions simple but with sufficient detail so that most people can succeed on their first try." They claim to have provided technical data and illustrations to support the rationale behind their instructions. They also state their "overall intention is not only to provide technology for growing live food, but also to promote a more thorough understanding of the aquatic environment and the ecological interrelationships of the food web."

In this activity, students will be guided to investigate the life cycle of various microalgae and rotifer Brachionus species and to gather information (including images) they may use to help them design systems for culturing microalgae and rotifers in the classroom. Information students gather may be organized as a report that explains:

- Reasons for raising microalgae and rotifers (objective in producing live food).
- Important environmental factors that impact algae and rotifer culture—growth and development (in project nine, this information may form the basis for many student research projects).
- Create design drawings that students will subsequently use to construct and operate (appropriately sized) microalgae and rotifer culture and rearing system in classrooms by methods that do not require stringently controlled temperatures, salinities, or light levels and can be accomplished in non-sterile environments.

After students and groups present and critique their microalgae and rotifer culture system reports in class and after students have assembled their systems, they are then guided to create posters that may be used to provide viewers with information about the culture system plans they followed during the assembly of their classroom systems. Students may be encouraged to organize information about microalgae and rotifer species in such a way that it may be placed in tables.

Information (vocabulary and concepts) in student developed reports may include (but not be limited to):

- Good narrative descriptions of microalgae and rotifer life cycles. Some algae species of interest may include: Chaetoceros muelleri; Phaeodactylum tricornutum; Dicrateria sp.; Isochrysis galbana; Skeletonema costatum; Nannochloropsis oculata; Pavlova virdis; Tetraselmis subcordiformis; Tetraselmis tetrahele and; Chlorella ellipsoidea.
- A description of microalgae and rotifer sizes/growth (anthropometric data) including images displaying the labeling of the organism's anatomy.
- Microalgae and rotifer diets (standards and variability for nutritional value of microalgae and rotifers) including a description of the various foods that may be provided (pros and cons of each).
- Microalgae and rotifer temperature (optimal temperature and a temperature ranges).
- Microalgae and rotifer reproduction strategies.
- Explanation of the influence of salinity on growth and reproductive productivity of Microalgae and rotifers—optimal salinity and a salinity ranges.
- Explanations of optimal light (intensities) and water aeration (and consequent water circulation) parameters for use in hatchery and growout systems (air-stone use?).
- Fatty acid composition of different microalgae species cultured under identical conditions.
- Explanation for why some species are more commonly used (and are preferred; a table may be created that includes a display of the most common aquaculture uses for each species of microalgae).
- Time and quantity (expectations for harvesting Microalgae and rotifers).
- Multi-step protocols to follow in harvesting microalgae and rotifers from the culture container.
- Special instructions about how to transfer cultured microalgae and rotifers from the hatchery to a larval culture tank containing larval fish/invertebrates (including how culture water should be discarded).
- Development and listing of a multi-step protocol for batch culture of microalgae and rotifers (including the design for container that may be used to hold microalgae and rotifers as they mature to adulthood).
- Explanation of important conditions that must be monitored and adjusted in order to maintain maturing and adult microalgae and rotifers in a growout system (listing important system maintenance protocols).

- Description of why cultured microalgae grow best in media with quite different primary and trace nutrient composition than natural waters.
- An overview of nutrient enrichment medias that have been developed for microalgae.
- Create a table that shows optimal salinity ranges for various species of algae and how they overlap with optimal salinity ranges for species of rotifers.
- Techniques that may be used to increase the duration of the exponential growth phase of a microalgae culture.
- Reasons why newly inoculated cultures may take a few days to adapt to the characteristics of local water and grow slowly at first.
- Describe methods for finding the source of biologic, chemical, and/or physical contamination of a microalgae culture and common ways such problems can be solved by making careful observations, considering everything as a possible source of contamination.
- Applications of microscopic technologies to the maintenance of microalgae cultures may be developed and described.
- Standard multi-step protocols used to pre-treat culture water may be developed and described (including: heat sterilization, mechanical sterilization, gross filtration, cartridges, ultraviolet sterilization, diatomaceous-earth pleated cartridges, microwave sterilization and, chlorination).
- Description of equipment and materials that students may use to assemble the classroom algae and rotifer systems (preferably materials that are easy to obtain, disposable, and recyclable.
- The importance of cleaning culture equipment (describing techniques for students to follow).
- Maintaining liquid stock cultures - describe multi-step protocol to follow to preserve a source of cells for stocking future cultures.
- Develop and describe a generalized practical protocol for using rotifers to rear larvae of several marine fishes.
- Create an illustration of a small scale, continuous culture system, using two-liter cola bottles for algae and five-gallon aquaria for rotifers. Describe how the system operates.
- Create and describe a ten day protocol to follow for continuous production of rotifers. Explain why it is necessary to produce rotifers continuously.
- Explain how collection and use of detritus from old rotifer cultures may be used to re-establish rotifer cultures after an active culture has died out.
- Develop protocols for sampling plankton cultures and quantifying the density of living organisms in the culture, including protocols for graphing the density of each culture and comparing it to a sigmoid curve graph (on the next page) to help identify which of four commonly recognized growth phases a culture is in (lag phase, exponential growth phase, stationary phase, and the death phase), including a detailed explanation of each phase and common knowledge about possible difference in the nutritional value of organisms that are collected from cultures in any of these growth phases.

Photo by Robert Jakus

Generalized population growth curve (over time "T") as demonstrated by microalgae culture.

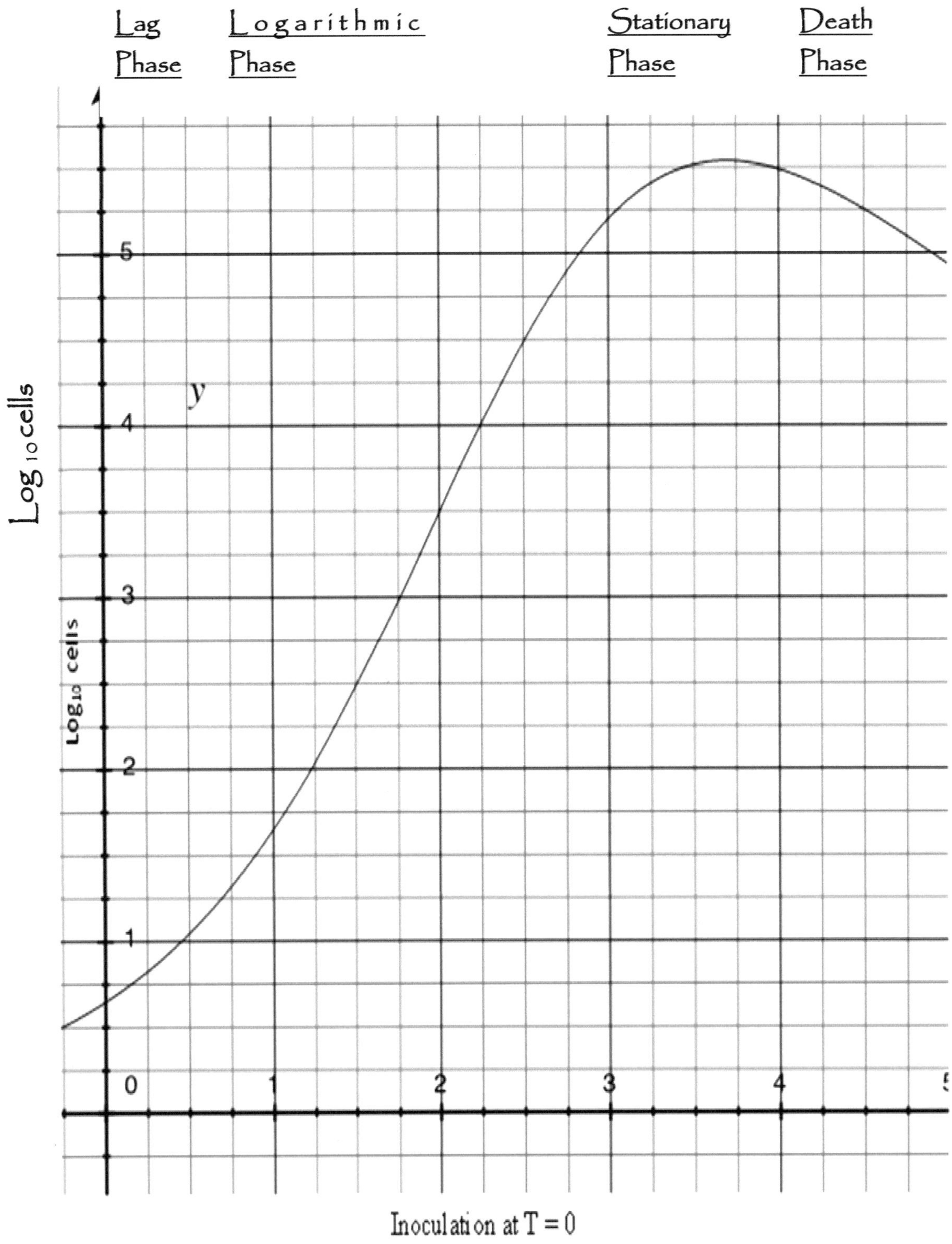

In addition to accessing and reviewing the work of Rippingale and Payne (2001) upon which the design and construction of an automated plankton culture and collection system is based, educators approaching computer automation and programmable logic controllers for the first time will benefit greatly by reviewing the papers referenced in the bibliography (Fowler, et. al. (April, 1994); Losordo, et. al. (1988); Scott, and Rust, (1996); Dickinson, and Johnson, (2006); and Kheiralla, A.F; (2007)). It is also often possible to solicit the assistance of an electrical engineer who is a member of the local community.

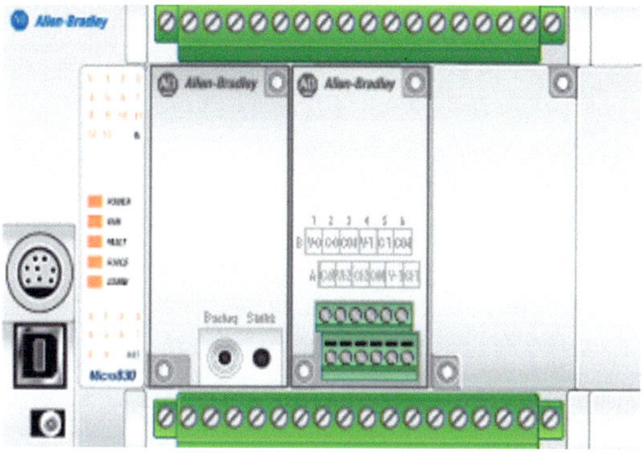

Micro830 – PLC;
Courtesy of Rockwell Automation, Inc.

Standard principles of automation are applicable to this project regardless of the specific manufacturer of the PLC used for this activity. Information provided here is based on the use of a Rockwell Micro 830 PLC. Some educators may prefer to structure this activity by pre-wiring the PLC and positioning it securely in a location that will minimize cable distances between the PLC and input / output devices on the soon-to-be assembled plankton culture and collection system. In this way, students may design and assemble the nonelectrical components of the automated system and just plug the systems devices into the proper PLC circuits. Other educators may choose to involve their students in the process of wiring and setting up the PLC. Many manufacturers have developed a range of valuable introductory lessons on PLC function and setup for public viewing (on YouTube).

A number of educators have suggested using the highly adaptable and relatively inexpensive entry level Arduino microcontroller for this project and others that are part of project 13.

To begin, students working in a design team are given a description of an automated plankton culture and collection system that they are to design, including a detailed idea of the physical properties that their system design must include. Teams are given time to design a system on paper. They may also use mocked-up small scale models to illustrate their design ideas. After these designs are complete, students present them for the class to critique. Each team then uses feedback from their critique to improve their design. The class then holds a final critique session in order to decide on the one design they will build. Depending on the accessibility of microcontrollers, each team can build the agreed upon system or perhaps the class will build only one.

As the students are involved in the physical assembly of the containers, lights, pumps, hoses, bulkheads, and other necessary materials and equipment that comprise a plankton culture and collection system, they must then be carefully guided to understand how to use ladder logic coding on a computer and to transfer it from the computer to the PLC to tell the system when to turn pumps, lights, and solenoids on or off.

For the Micro 830, the training materials students follow guides them through the process of developing ladder logic coding skills related to the control of traffic lights at a four way intersection. This ladder logic portion of the activity is performed in the computer lab. Each computer will have a version of "connected components" software with which students who read carefully and follow directions well will develop good coding skills. The coding instruction is contained in a multipage document that students will access through a Rockwell web site at http://literature.rockwellautomation.com/idc/groups/literature/documents/qr/2080-qr001_-en-p.pdf, pages 13 through 79. Students should be able to transfer their successful traffic light coding experience (with a bit of trial, error, and experimentation) to develop and send code to the PLC that correctly operates plankton culture and collection system devices. Students may find the following checklist useful.

Components of Physical Design	Components of PLC Programming
Three properly-sized containers with proper elevations relative to one another	Follow guidelines developed by Allen Bradley leading to the successful operation of a four way traffic signal
Proper screening for culture and collection containers	Create a new program "Plankton." Include code to turn on a light, turn off an air pump, and open a solenoid valve based on the push of a button
Lid for culture system	Include code to close a solenoid valve, turn on a water pump based on solenoid condition, and water level in collection tank,
Positioning light above screen in culture system	Include code to turn off one water pump based on water level in the filtration tank, and turn on another water pump in the filtration tank
Placement of siphon hose in screen area of culture container	Include code to turn off a water pump, turn on an air pump, and turn off a light based on water level in the culture tank
Attachment of solenoid valve to siphon hose and structure frame	Follow guidelines to connect microcomputer to the PLC and upload functional program to the PLC
Location of other end of siphon hose in the collection container (in properly sized bag or screen)	Follow guidelines to run the program through the PLC with all devices connected
Proper placement of float switches in each of the three water containers	Troubleshoot the programming to make it work more smoothly
Submersible pump placement in collection container – pump and hose to filtration container firmly attached to system frame	Place copepods in the culture container and remember to check their food supply every day
Submersible pump placement in filtration container – pump and hose to culture tank firmly attached to system frame	After one week, collect copepods from the system; provide fresh food if needed
Protein skimmer and other filtration devices in filtration container	Collect fresh copepods daily

The Allen Bradley manufactured Micro830-PLC is capable of synchronizing the operation of 10 output devices, 14 digital input devices, and with additional upgrades students can be guided to use this PLC to synchronize input from a number of analog sensors and monitor/accessed system performance data via wireless remote networks and cellular telephones.

Photo by Robert Jakus

There are numerous small grants available for educational purposes that might agree to fund the acquisition of additional sensors, probes, and wireless networking devices.

Photo by Robert Jakus

Micro 830 images shown here may be helpful to the classroom instructor for the purpose of illustrating how the PLC may be mounted to a plywood board using rails. I have used the wiring configuration illustration to the left as depicted in the diagram below to operate the automated plankton system described in this activity. I believe the diagram provides useful ideas about the placement of wires to and from various input/output barrier strips and associated

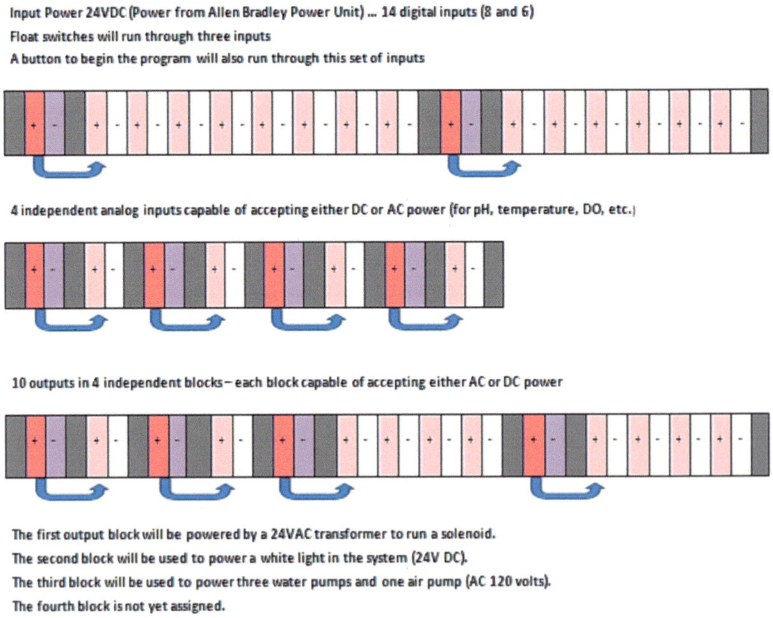

Diagram by Robert Jakus

devices (and the use of a surge protector).

As a result of reported nutritional and physical limitations of rotifers as a first food, Knuckey and Semmens, et.al (2005) have reported on development of technologies that may be applied to the culture of copepods, minute crustaceans that are nutritionally superior to rotifers. Although copepods cannot be grown to the high densities that rotifers can, students who develop the computer automated copepod culture and collection system described below will most likely be able to provide enough copepod larvae for any larval rearing project that they wish to study and develop.

Computerized Automation System for Culturing and Harvesting Copepods (Description of a Model)

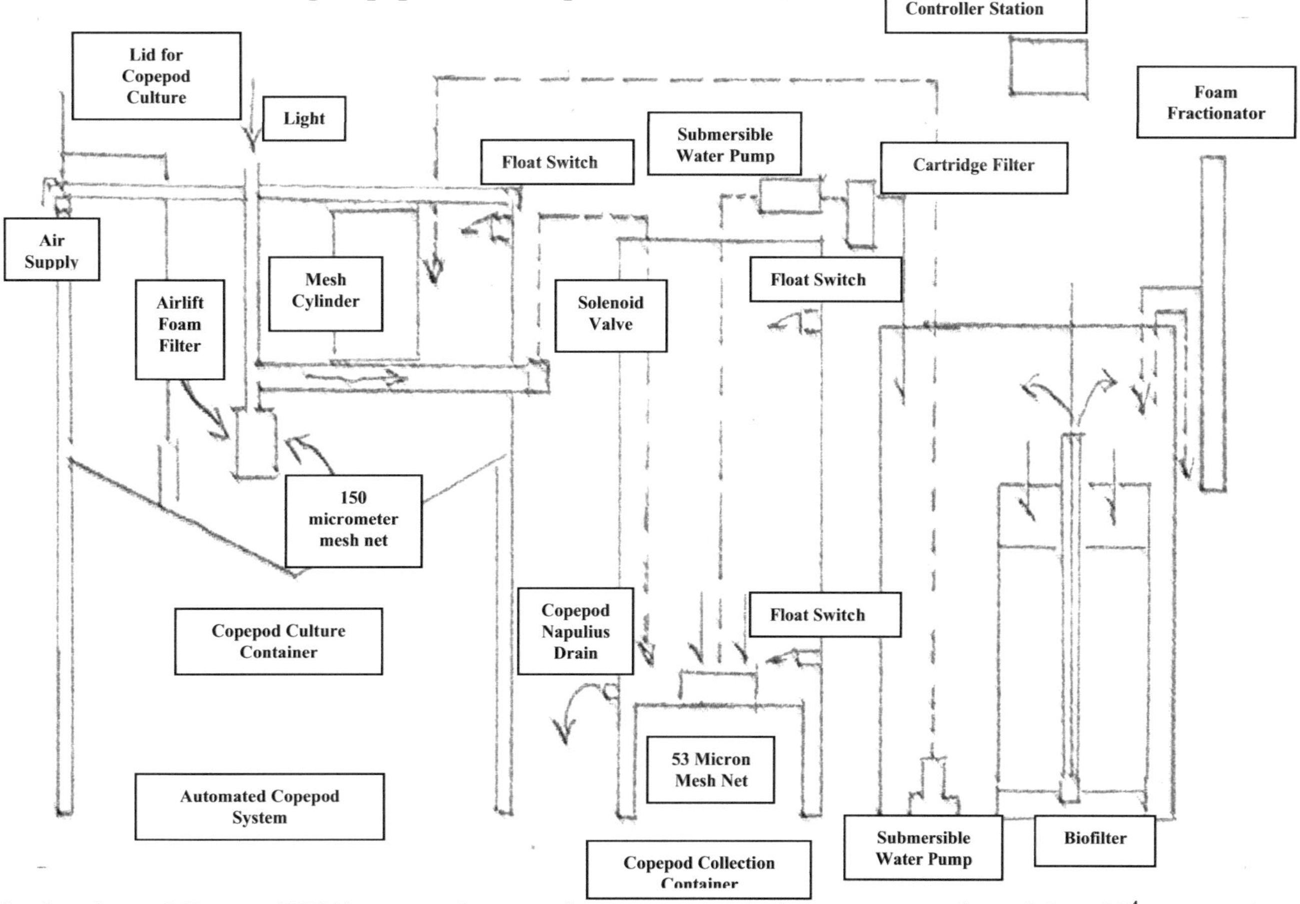

Rippingale and Payne (2001) report that a culture system as shown can produce 5.0 x 10^4 copepod nauplii per day. The image represents a summary diagram of a culture system based on a 500 L vessel, recirculated water, and a programmable logic control unit. Procedures to follow to collect copepod naupli and to recondition water in this system are described below. The PLC controls circuits that operate a solenoid water valve (through which a siphon is maintained), two water pumps, an air pump, and a light. Their function is sequenced as shown in the figure and described in the eight steps below. Note also that Nauplii that collect in 10 L of water at the bottom of the nauplius container are removed by manually operating a tap. A one mL sample of this is collected for counting and estimating total numbers of copepods in the culture.

Nauplius collection and water recycling procedures are described here (timings will vary depending on hose diameters, container volumes and other equipment used in any given system).

1. The harvest sequence is initiated by pressing a "start" button connected to a PLC input. Pressing the start button allows the PLC to deactivate power supplies to the air pump and one water pump. Power to a light above the plankton culture container is activated.
2. Air supply to the culture vessels is turned off (for 10 minute) to stop water movement and promote sediment settling.

3. An LED light is also turned on for 10 minutes to illuminate 150 um area of net mesh, submerged in the culture vessel.
4. After the light has been on for 8 minutes, the PLC opens a solenoid valve so that water may pass through it via a siphon (containing naupli within the screened off area of the plankton culture vessel [made of 150 um mesh]) The siphoned water (containing plankton) flows into a harvest vessel.
5. A float switch (FS#1) connected to another PLC input is activated thereby signaling the PLC to close the solenoid valve, turn the light off, and turn pump #1 on.
6. FS#2 (connected to another PLC input) is activated in order to turn pump #1 off and pump #2 on (drawing water through an area enclosed by a 53um mesh).
7. FS#3 (connected to another PLC input) is deactivated in order to turn pump #2 off and the air supply on.
8. A concentrated solution of naupli is drained from the harvest vessel (by a human).

If two 500 L culture vessels are kept in parallel, a cohort of naupli collected from newly mature adults in the first vessel, can be used to begin a culture in the second. An inventory of items needed to design and assemble a functional PLC copepod culture system is provided here.

The following list of procedures (1-7) for establishing a 500L culture of copepods culture was developed by Rippingale and Payne (2001).

Number	Item
1	Allen Bradley Micro 830 PLC
1	computer
1	Surge Protector
1	12 V AC Transformer
2	fuses to protect PLC AC outlets
1	24 V DC Transformer
2	120 V Submersible Maxi-Jet Pumps
2	24 V Pneumatic solenoid valve (water)
1	120 V air pump
1	12 V DC LED light (from boat marina)
3	Nylon Float Switches
2	Electronic button to initiate sequence
Cut to fit	Airline hose and water hoses
	53 and 150 micron netting
	Sockets to run 120 V pumps to PLC
	Red/Black Cable (26 x 0.3 mm)

1. Fill the culture vessel to the 300 L mark with 5 um of filtered seawater diluted to 27 ppt with carbon filtered tap water. Add 25 mL hypochlorite (12.5 percent) and leave it overnight without aeration.
2. Deactivate hypochlorite with 25 mL Sodium Thiosulfate (100g/L). Turn airlift circulation on and leave for 1 hour. Test for additional chlorine.
3. Stock the culture vessel with 500,000 naupliii (1 copepod/mL when full). Ensure that aeration is gentle and culture vessel is covered to exclude light. Add algae (T-Iso) as food at the rate of 30,000 cells/mL
4. Increase feed rate by approximately 10,000 cells per mL every 2 or 3 days until the feed rate is 100,000 cells/ml. The feed rate may be further increased if cultures remain healthy. Do not exceed 200,000 cells per ml.
5. Monitor copepod health daily by observing water clarity; and by viewing copepod activity and gut contents with a stereoscope. Perform 80 percent water change if the culture is weak.
6. Almost all naupli should have metamorphosed into copepodids by days 5 or 6. When this has occurred, increase airlift circulation, add plastic mesh substrate and add the foam filter. Ensure that the filter is placed centrally, near the bottom of the vessel, and operates with very low aeration.
7. Commence daily nauplius collection when adults carrying egg sacs are abundant in a sample taken from the culture vessel. Continue collection until nauplius numbers decrease.

Volume of Copepod Culture	Minimum Volume of Algal Culture Required
0.1 – 0.5 L	2 L (2 x 1 L)
30 L	8 L (2 x 4 L)
60 L	16 L (4 x 4 L)
500 L	80 L (5 x 20 L)
1000 L	250 L (5 x 50 L)

Also please note that the quantity of algae needed to maintain copepod cultures depends on the level of copepod production that is required. As a general rule, the total volume of algal cultures must be at least 25 percent of the volume of the copepod culture if high copepod production is required. This table indicates the volume of algal culture required to supply copepod cultures of different volumes.

Write a list of **science-based** activities you have participated in during this project and briefly describe evidence that verifies your interaction with and knowledge of scientific concepts for each item on your project seven list.

Write a list of **technologically-oriented** activities you have participated in during this project and briefly describe evidence that verifies your interaction with and knowledge of technologically-oriented concepts for each item on your project seven list.

Write a list of **engineering-oriented** activities you have participated in during this project and briefly describe evidence that verifies your interaction with and knowledge of engineering-oriented concepts for each item on your project seven list.

Write a list of **mathematical** activities you have participated in during this project and briefly describe evidence that verifies your interaction with and knowledge of mathematical concepts for each item on your project seven list.

Write a brief list of **Non-STEM-related** activities you have participated in during this project and briefly describe evidence that verifies your interaction with and knowledge of Non-STEM related concepts for each item on your project seven list.

Rate your level of involvement in STEM-related activities in this project (1=lowest, 10=highest).

1—2—3—4—5—6—7—8—9—10

Beyond the Classroom Aquarium **Daily Journal Guidelines**

Entries students place in their daily journals should be:

1. Dated accurately and logged chronologically.
2. Kept in a notebook and written neatly in ink or typed.
3. Written or typed in complete sentences using correct capitalization, spelling, and punctuation.

In their journals, students should keep a daily record of:

1. Who they worked with each day by including the names of school staff, fellow students, community members, and volunteers.
2. What they did each day, such as lecture topic, activities or project design, assembly, or maintenance.
3. What they learned each day by asking questions, following instructions, and/or observing activities.
4. How *Beyond the Classroom Aquarium* experiences relate to past, present, or future life experiences.

For bonus points, students may complete these journal options, including illustrations, diagrams, or creative expression relating to the journal entry. For project one, journal writing bonus topics may include answers to the following questions or descriptions of a project activity or assignment. Reflection on these topics may also help students prepare for exams, quizzes, and other assessments.

- What more would you like to know about the organisms and systems that you observed?
- What do you want to understand better in order to contribute more to your team's productivity as it independently designs, assembles, and then manages an aquatic ecosystem?
- Create a new drawing and story about an organism or system that is most interesting to you.
- Create a journal entry that tells what information you learned during project seven. Also, write down what you would like to know more about and if you know someone else who might help you learn more if you asked them.

- Select a book or magazine to read that is related to a topic in project seven. Write a report on this book and include your report in your *Beyond the Classroom Aquarium* portfolio.
- Along with topics already under study, individualized applications mathematical topics may be used to include mixture problems.
- Describe your experience of the process of developing Guillard's f/2 nutrient solution.
- Report on the consistency and success you are having with your efforts update classroom aquatic ecosystem water quality data in data log books and in the classroom water quality spreadsheet.
- Explain the factors that influence the nutritional value of microalgae and how these factors can be optimized.
- Describe successes, failures, and complications you experienced as you developed protocols to ensure the production of sterile single species algae cultures.

Journal entries will be evaluated on a point system, with four points being the maximum number a student can earn. A half-point will be awarded for each of the first eight journal requirements above. Bonus points will be awarded when journal options are completed. Journal points will be totaled and divided by the number of journal dates. A possible journal grade based on points is included in the project one assessment section.

Additional Project Seven Assessment Activities and Questions

Explain the most important considerations for the successful growth of algae cultures, rotifer cultures, copepod cultures, and other cultures. In each case, describe what might easily go wrong.

Explain why rotifers are a much more commonly used first food for many aquatic larvae.

Describe the life cycle and productivity (fecundity) of the rotifer. Explain factors that you should try to optimize as you culture rotifers in the aquatic ecosystem classroom/lab.

Describe the components of the computer automated plankton culture system, the function each component performs, and how the components interface with one another.

Explain what, to you, were the largest obstacles in learning and developing ladder logic code.

Use Excel to graphically represent the water quality of each system. Interpret the data and explain patterns and trends. Create a chart/poster that summarizes your interpretations.

Monitor daily population estimates of culture growth (algae & rotifer) and use this data to develop a population growth curve. Enlarge the graph to fit on a poster and label the four stages of the population growth curve as they relate to the culture samples you are reporting on.

Explain the range of ways you have been able to refine and strengthen your knowledge of math, science, technology, and engineering by designing and constructing various plankton culture systems.

Explore the history of computer automation in aquatic ecosystems. Develop a report that describes and compares systems from the past with systems of the present and possible new directions that computer automation may make possible in the future (to start, you may consider the work of Scott & Rust [1996] whose work helped inspire me to write this book).

Review activities and information that interested you in this project, and determine if this information may fit on the "Research-ability" Potential Topics Form on page 59. If so, add it to the list.

Beyond the Classroom Aquarium STEM Self or Peer Assessment Rubric

This self-assessment is valid only to the extent to which you can document and display evidence that supports it. For each category in the left hand column, write a STEM-Oriented example in the box beneath 3 if you can display evidence that your STEM-Oriented example represents work that exceeds expectations. If your evidence shows that your STEM-Oriented statement meets but does not exceed expectations, write the statement in the box beneath 2. Similarly, for STEM-Oriented examples that demonstrate some progress toward expectations.

Name: **Project Number:** **Date Project Work Began:** **Date of Project Completion:** **List Math Topics and/or Objectives Currently Under Study:**		**3** **Work Exceeds Expectations**	**2** **Work Meets Expectations**	**1** **Work Is Progressing Toward Expectations**
Write examples of questions and defining problems related to your STEM work.				
Describe how you have developed and used models in your STEM work.				
Write examples of STEM investigations you planned and carried out.				
Describe STEM related data that you analyzed or interpreted.				
Give examples of how you used mathematics and computational thinking related to the five standards listed to the right.	Problem Solving (Math Process Standard One)			
	Reasoning & Proof (Math Process Standard Two)			
	Communication (Math Process Standard Three)			
	Connections (Math Process Standard Four)			
	Representation (Math Process Standard Five)			
Describe STEM-Oriented explanations and solutions you designed				
Explain evidence you used while engaging in a STEM project-based argument.				
Give examples of STEM information you obtained, evaluated, and/or communicated				
TOTAL POINTS				

While working for a summer camp, I was assigned to teach a class called microscopy. As I sought supplementary information to share with my campers, I discovered two useful books; one includes an explanation of how to construct a microscope. The second guides young people through activities related to optics. Some information from those books is presented in project eight of *Beyond the Classroom Aquarium*. Microscopes are awesome tools. The construction and assembly experience guarantees student access to the microscopic world. Some teachers may be able to make it possible for students to own their own personally constructed microscope. Beyond having students construct their own microscopes in class, I recommend that each class acquire a pair of modern digital microscopes (compound and dissecting) to promote use of microscopic photographs.

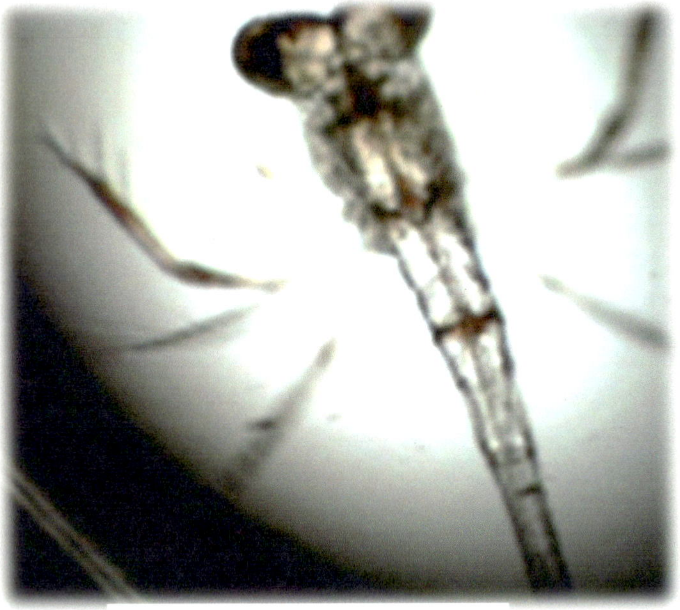

Photo by Robert Jakus

By using microscopes in the aquatic ecosystem classroom, students will develop a much clearer idea about the many creatures that play a major role in system aquatic ecosystem stability although they cannot be seen by the naked eye. In project eight students will learn to use lenses and magnification to observe morphology of miniscule plants and animals. When possible, students will be provided access to a digital microscope and use of all of its advanced features will be introduced. Microscopy skills will be refined. The study of population growth and sampling will be introduced using invertebrates from mini-tanks and records being kept about them. Students will gain new perspectives on measurement units, practice in measurement, an introduction to topics of magnification in optics, ideas to incorporate microscope use in the chapter V research project, and opportunities to study a range of simple invertebrate cultures, specifically: brine shrimp, daphnia, black worms, snails, limpets, hermit crabs, sea urchins, brittle stars, grass shrimp, copepods, and also plants, and algae. Students may construct stage micrometers. Activities in this project reinforce the formal study of phyla and morphology of plants and simple animals. Students are introduced to the process of microscopic culture analysis, data collection on cultures, and use of ongoing data collection in reports of these cultures.

Applications of individualized mathematical topics may include the Newton Scale, focal length, photo-micrograph calculations, calculation of the refractive index of optically dense substances, angles of incidence and refraction, and relations between a) lens magnification, b) focal length, c) focal length, d) numerical aperture, e) working distance, and f) chromatic aberration.

In order to appreciate the design of the microscope that students will construct it is necessary to understand basic properties of lenses. In order to introduce students to important lens information, I found the book: *Lights, Lenses and Lasers* (Berger, 1987) quite useful.

Magnifying Glass - You have probably used a magnifying glass at home or in school. As you know, magnifying glasses are double convex lenses that make objects appear larger. People use magnifying glasses to see tiny details that are not clearly visible to the naked eye.

Magnifying glasses differ in the magnifying power, or magnification, according to the exact shape of the lens. The magnifying power is the number of times larger the image appears to be than the actual size of the object. To find the magnifying power of a glass, place the magnifying glass on a flat ruler so that the center of the glass is on the measuring edge of the ruler. Count the number of eighth-inch markers. Write down the number.

Next, without moving your eye, raise the glass until you get the largest, most distinct image. Count the number of eighth-inch markers you can now see through the glass. Write down this number and then divide the original number by this magnified number. The resulting figure is the magnifying power. Try this with different magnifying glasses; the higher the magnifying power, the greater the magnification.

People usually look through magnifying glasses to see something better. But you can also use a magnifying glass to project an image. Stand with your back to a window. With one hand, hold up a blank sheet of white paper. With the other, hold a magnifying glass between the paper and the window. Move the glass back and forth until you get a clear image on the paper. The image should be sharp and distinct, but you will notice that is upside down.

With the help of a friend, measure the distance from the magnifying glass to the paper. This is the focal length of the glass. Here is how you can use this figure to show how a simple magnifying glass can be used to produce a variety of images.

In a darkened room, open a newspaper on a table. Use some clay to hold your magnifying glass erect in the middle of the paper. Draw a straight line that extends both in front and behind the glass. On both segments of the line measure out and mark both the focal length (f) and double the focal length (2f).

Place yourself by one end of the line looking toward the glass, and put a candle about two inches beyond 2f in front of the glass. Hold a piece of typing paper behind the glass and between f and 2f on that side. Move it back and forth until you get a sharp image.

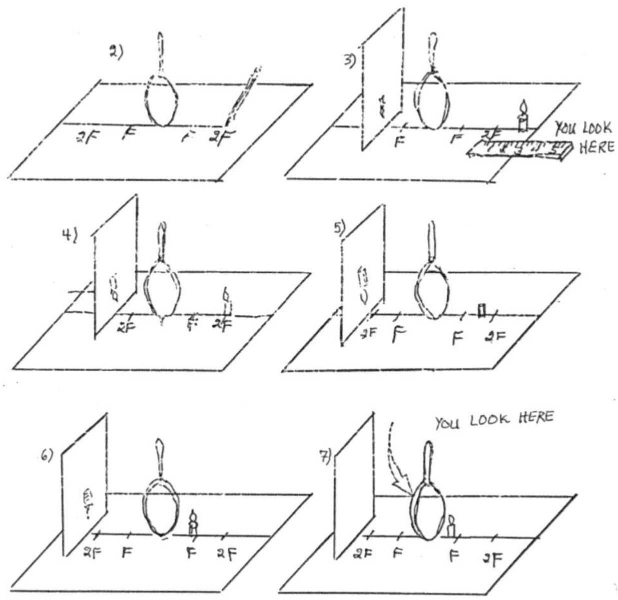

You'll see that the image of the candle is smaller than the original candle and inverted, or upside down. The lens systems in cameras work in this way. If you can think of replacing the paper with film, you can see how the film captures the image.

Next move the candle to the 2f position, but keep the paper in the same place. The image is still inverted, but this time it is the same size as the original candle. Some telescopes include a lens that is used in this way.

What happens when you place the candle between f and 2f? The image is clearest when the paper is behind the 2f part. It is still inverted, but it is larger than the original. The lens systems of slide and movie projections cast a large image on the screen this way. Other lenses in the projector turn the image right side up.

If you place a candle exactly at the focal point, you will not be able to get any image on the paper. An object at the focal point of a lens does not form an image since the rays are all parallel and therefore do not come to a point.

Finally, position the candle between the glass and f. Again, you will see no image on the paper. But, if you look through the looking glass in the other direction, at the candle, you will see an image. It is larger than the original, not inverted, and appears between f and 2f in front of the glass. This is actually what you see when you use a magnifying glass in the usual way.

Lenses are curved pieces of transparent material, such as glass or clear plastic that bends rays of light that pass through them.

Lenses have many important uses in modern life. The lens in your eye lets you see the world around you. Eyeglasses, contact lenses, and magnifying glasses help you see better. The lenses in microscopes and telescopes make it possible to see the very small and the very distant. Slide and movie projector lenses make it possible for many people to see the same pictures at the same time. And camera lenses allow you to take clear, distinct photographs of objects near and far.

All lenses work in the same way. They all bend, or refract, beams of light. Light is refracted because it travels through different transparent materials at various speeds. As a light beam passes from one material to another, it actually changes direction.

Light travels through air at 186,000 miles per second but through water at 140,000 miles per second. When you look at a long pencil placed into a half glass of water, you see the top part through the air. But you see the lower part through the air and through the water. Because the light bends as it goes from one medium to another, the pencil looks like it has two distinct parts.

You can use the principle of light refraction to fool a friend. First ask whether he or she thinks a penny can float. The answer will probably be no. Then drop a penny into a small bowl. Tell your friend to raise the bowl until the penny just disappears from his or her sight. Now slowly pour water into the bowl. Soon your friend will see the penny "float" into view.

Of course the penny does not actually float. But the light, passing from water to air, is bent so that it makes the penny appear to be floating on the water.

The Shape of Lenses - Lenses come in two basic shapes: concave and convex. Lenses that are thin in the middle and thick at the edges are concave. Lenses that are thick in the middle and thin at the edges are convex. (An easy way to remember the difference is that concave lenses look "caved-in" at the center.)

Because of the difference in shape, the two types of lenses have opposite effects on light. A concave lens spreads out the rays of light that pass through. A convex lens brings together rays of light.

Since a concave lens is thinner in the middle than around the edge, it causes the beams of light to spread out, or diverge. Diverging rays of light make objects appear smaller than they really are.

When light beams enter a convex lens, they are refracted on entering and refracted again on leaving. Those passing near the edges of the lenses are bent sharply; those near the center are bent slightly. As a result, all the light rays come together, or converge. The point at which the rays converge is called the focus. The distance from the lens to the focus is the focal length.

Six Lens Variations	Double	Plano	Inner Edge – Outer Edge
Concave	Double-Concave	Plano-Concave	Convexo – Concave
Convex	Double-Convex	Plano-Convex	Concavo – Convex

The Lens in Your Eye - The human organ of sight is, of course, the eye. Your eye is shaped like a ball. It is filled with a transparent jelly-like substance known as vitreous humor. Covering the front part of the eye is the transparent cornea. Directly behind the cornea is a colored disk called the iris with a dark, round opening called the pupil. This controls the amount of light that can enter they eye.

Just behind the pupil is a double convex lens through which light enters your eye. Made of clear tissue, the lens acts like a magnifying glass. It brings to a focus the rays of light that pass through the cornea and the pupil. If the lens is properly shaped, it forms a sharp image on the inner lining of the eyeball called the retina. The retina sends tiny electrical impulses along the optic nerve to the brain. The images formed on the retina are always upside down. But the brain interprets the images so that we see everything right side up.

The lens in your eye has a difficult job to perform. The lens has to bring the words from the pages of a book to a focus, from a distance of about 18 inches to your eye. But the same lens has to be able to focus on road signs that might be as far away as 1,800 feet or more. How does the lens manage to focus on both near and far objects, as well as on everything in between?

To see objects at different distances, the lens in your eyes actually changes its shape. This ability to change is called accommodation. Tiny muscles in the eye, called ciliary muscles, can pull on the lens in one way or another to make the lens thicker of thinner. To see things that are close by, the muscles make the lens thicker or thinner. To see things that are close by, the muscles make the lens thicker. For more distant objects, the lens is made thinner, or flatter.

Some people cannot see things sharply. There are those who can see things close up perfectly, but cannot form clear images of distant objects. They are said to be nearsighted. Nearsightedness may be caused by an eyeball that is too long, or it may be caused by a lens that is too convex. In either case, rays of light from distant objects come to a focus in front of the retina. Those who are farsighted can see faraway objects well but cannot form a sharp image of things close up, The condition may be caused by an eyeball that is too short or by a lens that is not convex enough. In either case, rays of light from close objects come to a focus behind the retina.

Astigmatism is a condition in which a person sees clearly in some parts of his field of vision but with blurring in other parts. Astigmatism is usually caused by an uneven curve of the lens.

Ophthalmologists are highly trained medical doctors who specialize in the eye. These doctors can determine the kinds of lenses that can help normalize focal distances for those who are nearsighted, farsighted, or astigmatic. Clarity of focus is also of concern for people who design microscopes.

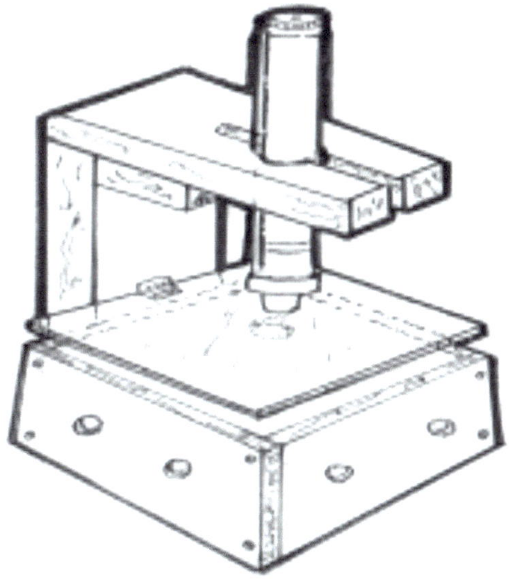

Image by Mike Graff

Microscopes and Telescopes – Berger (1987, p. 42) explains "while the microscope is used to examine the very small, the telescope helps us see the very distant. Like the microscope, the telescope uses two lenses. But instead of both being convex, the one at the far end of the tube (the objective lens) is convex, and the one near the eye (the eyepiece lens) is concave."

In 1609, Galileo developed the first telescope. According to Berger "the crude instrument had two lenses attached at the opposite ends of a long cardboard tube. Galileo's most powerful telescope magnified things only 33 times. Nevertheless, with it he could see the rings of Saturn, the satellites of Jupiter, and the mountains of the moon." (p.42)

The simple telescope used by Galileo and other astronomers is called the refracting telescope. The following is a description of how lenses in refracting microscopes operate. "A large convex lens at the far end of the tube gathers light from a star, for example. The lens bends the light rays and brings them to a focus at the eyepiece. The concave eyepiece lens then magnifies the image and makes it appear very close. But the image appears upside down. Where necessary to see the image right side up, a special lens is placed between the two lenses to invert the image." (Berger, 1987 42 – 43)

Design and Construction of a Homemade Microscope

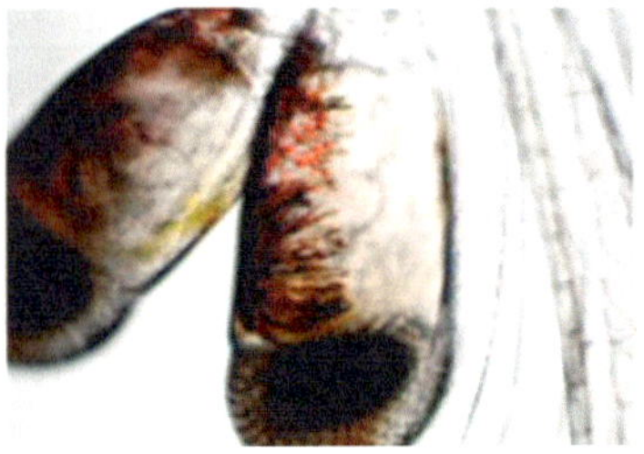

Photo by Robert Jakus

In his introductory microscope discussion, Curry (1982, p.8) states: "From its simple beginnings in the seventeenth century the microscope has evolved into a complex and sophisticated instrument. Its use has become a discipline in its own right, with its own terminology and methods; and as an item of laboratory equipment the microscope is virtually indispensible.

Early microscopes bear little resemblance to those available today. They consisted of a lens arranged to view objects impaled on a pin. Their modern descendants are the hand lens and the dissecting microscope, which are suitable for observing whole objects, but which do not achieve high magnification.

The beginning of the eighteenth century saw an innovation that led to a great increase in the popularity of the microscope – the use of light transmitted through the specimen, for viewing items mounted upon a suitable slide. This is the principle upon which the modern compound microscope works, and most of the procedures involved in microscopy are to do with preparing the specimen in a form suitable for examination.

The twentieth century has seen the development of a fundamentally different microscope, which uses electrons rather than light for producing an image. The electron microscope can resolve features that are much smaller than those that are resolvable in the light microscope, because the wavelength of an electron bean is much shorter than that of light.

Although the microscope has contributed to all branches of science, its impact has probably been greatest in biology and medicine. In the nineteenth century, the light microscope promoted the science of histology (the study of tissues) just as today the electron microscope has promoted the study of cell ultrastructure. Methods of sectioning and staining tissues were developed in the nineteenth century, and new insights into the anatomy of living organisms resulted. Today these techniques have a new importance, in the diagnosis of disease by the removal for examination of a small sample of tissue (a biopsy) at operation.

However, the use of the microscope is by no means restricted to the examination of tissues. Many whole organisms and inorganic materials such as rock sections can be investigated with the microscope . . . ."

Curry (1982, p. 9) states: "Units of measurement used in this table are SI and are now accepted as standard. Inches are given in brackets where appropriate. In microscopy, the basic units of length in the SI system are the millimeter, micrometer, and nanometer and their relationships to one another are shown below."

One meter (m)	1000 millimeters (mm)
One millimeter	1000 micrometers (um)
Limit of resolution of the human eye	100 micrometers (um) approximately
Diameter of a bacterium	1 micrometer (um) approximately
One micrometer (um)	1000 nanometers (nm)
Red light	740 - 620 nanometers (nm) wavelength
Orange light	620 – 585 nanometers (nm) wavelength
Yellow light	585 – 575 nanometers (nm) wavelength

Green light	575 – 500 nanometers (nm) wavelength
Blue light	500 – 445 nanometers (nm) wavelength
Indigo light	445 – 426 nanometers (nm) wavelength
Violet light	426 – 390 nanometers (nm) wavelength
Ultra-Violet light	Less than 390 nanometers (nm) wavelength
Limit of resolution of the light microscope	260 nanometers (nm) approximately
Diameter of a polio virus	26 nanometers (nm)
Width of a DNA molecule	2 nanometers (nm)
Wavelength of an electron beam at 10,000 volts acceleration	0.012 nanometers (nm)
Wavelength of an electron beam at 60,000 volts acceleration	0.005 nanometers (nm)

SI units of measurement used in microscopy - The micrometer (um) was formerly known as the micron (u). The nanometer (nm) was formerly known as the milli-micron (mu). The **ANGSTROM** (Å) is a redundant unit, equal to 0.1 nanometers.

Simple Microscopes and Single Lenses – Curry (1982, p. 12) introduces a discussion of microscopes and lenses by explaining "lenses are of two basic types. Those that converge or concentrate light are known as positive lenses and are thicker in the center than at the edges. Those that diverge or spread light are known as negative lenses and are thinner in the center than at the edges. The human eye contains a positive bi-convex lens that unlike a glass lens can be squeezed or stretched to alter its curvature. This means it can focus as easily upon a near object as it can upon a distant one.

Simple microscopes consist of a single lens only, or a single group of lenses. Such simplicity restricts the useful magnification of this microscope to about 20 times (X20). Generally, such a lens is called a hand lens, for the term simple microscope has faded from everyday use.

How can a single lens manage to magnify a small object? Although light can be thought of as travelling in waves, it can also be thought of as travelling in the form of rays; thus light from the object travels outwards in all directions in a straight line at a steady speed through the air. Until that is rays enter a substance that has a higher optical density than the air. The light rays temporarily slow down until they pass through the object of higher optical density and escape back into the air, where their original higher speed is resumed. Glass is a good example of an optically dense substance. If a horizontal light beam hits a vertical sheet of plane glass its speed will be slowed down while passing through the glass, but the beam will continue along its original direction undeviated. If, however, it should hit the glass at any other angle the beam will be bent, both upon entering the glass and upon leaving. This is due to one part of the incident beam striking the glass and slowing down before another part of the same beam strikes the glass. An analogy may be helpful here. Think of a bulldozer, which moves on two tracks. If both tracks move at the same speed, then the bulldozer will travel in a straight line. If however, one track slows down, then the bulldozer will turn in the direction of that track. This bending phenomenon in an optically dense substance is termed refraction. Notice that, as the angle of incidence (i) increases, so too does the angle of refraction ®. Close study reveals a mathematical relationship of great value in optics:

$$\frac{sine\ of\ the\ angle\ of\ incidence}{sine\ of\ the\ angle\ of\ refraction} = a\ constant\ value$$

This constant value for light travelling from air into an optically dense substance is known as the refractive index of that substance."

Many limitations in magnification in simple microscopes are overcome by a more complex arrangement of lenses known as the compound microscope.

On page 15, Curry (1982) continues his discussion of lenses and microscopes by explaining "The compound microscope consists of two main lens groups that magnify the object in two stages. The first group, known as the objective, produces the primary image, which is further magnified by the second group, known as the eyepiece or ocular. The objective and eyepiece consist of two or more individual lenses . . .

Magnification in Compound Microscopes - The total magnification of a compound microscope can be calculated by multiplying the magnifying power of the eyepiece by the magnifying power of the objective, which gives the overall magnification. Both the eyepiece and objective usually have their respective magnifying powers engraved upon them. The calculation is based upon the assumption that the eyepiece and the objective, when in focus, are the distance apart that was intended by the manufacturer. The distance is known as the mechanical tube length, or body tube length, and is the distance between the shoulder of the objective and the flange of the eyepiece. When using lenses intended for a different tube length, the new magnification can be calculated using the following simple equation:

$$\text{Magnification} = \frac{mechanical\ tube\ length\ (mm)}{focal\ length\ of\ objective\ (mm)}\ X\ eyepiece\ magnification$$

The two commonest tube lengths used nowadays are 160mm and 170mm."

Curry (1982) indicates "For the compound microscope, the limit of resolution, R, in um is given by the formula:

$$R = \frac{1.22\lambda}{2n\sin\theta} = \frac{0.61\lambda}{NA}$$

The symbol λ represents the wavelength of the light used, and "NA" the numerical aperture of the objective. And n·sinθ is an expression for the numerical aperture." (p. 18)

The Objective - The objective is the most important part of the optical system of a compound microscope. It provides the primary image with its resolved detail and much of the ultimate magnification seen by the eye. No single objective fulfills all the needs of a microscopist, so several different objectives are used. The clearance between the lowest point of the objective and the specimen in focus is called the working distance. The working distance becomes smaller as the magnification of the objective lens increases. It is obvious that there is very little margin of error with high magnification objectives and great care must be exercised to avoid ramming the objective into the specimen. Most high magnification objective lenses are spring-loaded so that accidental touching of objective and specimen does not result in damage to either. (Curry, 1982; pp. 16-17)

The Eyepiece or Ocular - The eyepiece of a compound microscope enlarges the primary image formed by the objective, so that it can be recorded by the eye of the microscopist or by the film of a camera. As with objectives, there are several types of eyepiece. The Huygenian eyepiece is the most common type and consists of a metal cylinder with Plano-convex lenses (flat on one side and convex on the other), with their convex surfaces toward the objective; they have a circular opening between them called the field stop. An engraved magnification is marked on the metal surrounding the upper lens.

Compensating eyepieces are required for apochromatic objectives, as it is not practicable to correct all the aberrations within the apochromatic objective itself. Some of the aberrations are compensated for in the eyepiece. Such eyepieces are usually engraved with the word 'compensating.' (Curry, 1982; pp. 23-24)

Beyond a certain limit there is no point in further magnifying an object. This is because no more information or fine detail can be seen. Magnifying an object beyond the point where no further resolution is available is called empty magnification. Consider a photograph in a newspaper that is composed of dots of varying density. When viewed normally, the photograph conveys information, but look at the photograph again using a magnifying glass. Now the dots are visible by it is not possible to extract any more detail.

A good quality light microscope, correctly adjusted, can resolve detail in an object if it is greater than a quarter of a micrometer in size. Obviously there is no point in using magnifications greater than that required to make the separation of an objects details visible to the eye. Thus a useful rule is not to let the total magnification exceed 1000 times the NA of the objective in use. For example, using an X40 objective with an NA of 0.65 (1000X0.65 = 650), rather than an X10 eyepiece (10 X 40 = 400) would be more appropriate than an X20 eyepiece (20 X 40 = 800). In practice, field of view, working distance and depth of focus all play a part in the choice of lenses, which with experience becomes automatic. (Curry, 1982; pp. 28 - 29)

Measuring Objects under the Compound Microscope - It is often necessary to measure the size of microscopic objects. Measuring the size of an object, in a photomicrograph, and dividing by the total magnification of the image of that object can achieve this. If a measurement is required more quickly and accurately, use an eyepiece micrometer graticule, which has been calibrated using a micrometer slide. These items are available from some suppliers. The eyepiece graticule is marked, but these marks have no particular measurement value. The eyepiece graticule has to be placed inside the eyepiece so that it is in focus at the same time as the specimen. The eyepiece usually consists of two lenses that have a fixed diaphragm between them. Unscrew top lens and carefully drop eyepiece graticule down the inside of eyepiece tube so it rests on the field stop. Replace top lens of the eyepiece. Look through eyepiece to check that the scale is in focus to your eye. If not, the field stop can sometimes be carefully pushed up and down the eyepiece tube until the desired focus of the graticule is accomplished. Otherwise a focusing eyepiece is required. The micrometer slide has a scale on it divided into 0.01 or 0.1 mm units. The micrometer slide is placed on the specimen stage and brought into sharp focus. The eyepiece scale, or divisions on this scale, can now be compared to the known distances marked on the micrometer slide. Thus if a specimen is examined at the same magnification as the micrometer slide, the size of features on the specimen can easily be calculated. The scale in the eyepiece must be calculated again for each of the objectives of the microscope. (Curry, 1982; pp. 29-30)

For those who cannot afford a good quality microscope or who have been unable to find a good second hand microscope, there is a third alternative. It is possible to construct a microscope of good optical quality with only moderate manual skills. The quality of the objective is the most important part of any compound microscope, but many of the cheap microscopes on sale have objective lenses possessing unacceptable chromatic and spherical aberrations, and such poor quality can easily dampen enthusiasm for microscopy. Money would be better spent in buying a couple of objectives and a Huygenian eyepiece from one of the microscope suppliers listed in appendix II, or from colleges or industry if they are disposing of old equipment. These would then form the basis of a homemade microscope. The maximum usable magnification of such a microscope is X400 so appropriate lenses to purchase would be a Huygenian eyepiece of X10 magnification, and X10, X20, and X40 achromatic objectives. These items can be bought singly or all together, depending on available finances. Once suitable eyepieces and objectives have been purchased, the microscope can be easily and cheaply constructed, and a great deal of satisfaction can be obtained from building it.

This do-it-yourself microscope is capable of producing excellent service and is easily repaired or modified. Its optical performance depends on the quality of construction and the quality of the lenses purchased. It also makes an ideal microscope for photography through the microscope. (Curry, 1982; p. 31) The construction process described below is the work of Curry (1982, pages 31 – 40). It is important to read and understand these instructions prior to building the microscope and students will develop a much deeper understanding of this process if they create (and use as an instruction manual) a series of drawings to illustrate important steps in the process. Student designs should reflect updates in technology such as inexpensive LED lights they may incorporate in their classroom microscope construction project.

> **Making the Body Tube -** This is made from a length of domestic copper pipe; approximately 28mm external diameter. The mechanical or body tube length of many microscopes is 160 mm and is measured from the base of the screw thread of the objective to the lower lip of the eyepiece lens. Cut the copper

pipe to precisely 160mm. Some microscope objectives are designed for use with a tube length of 170mm, so check designated length when buying lenses.

Copper piping varies in its wall thickness, so the internal diameter may allow the eyepiece to fit without modification. If this is not the case, because the internal diameter is too great, a suitable diameter piece of wood (approximately 25 mm (1 inch) long) can be glued into the top of the tube. Araldite or similar epoxy resin is ideal for this purpose. When dry, a 23 mm (9/10 in) diameter hole for the eyepiece should be bored through the center of the wooden insert using a brace and a bit. The eyepiece should slide into place without being too loose, and its sides must be parallel to the copper tube.

Inserting the objective is a little more difficult. The objective lens has a screw fitting approximately 20mm in diameter. The easiest way to fit it into the tube is by (again) gluing a suitable piece of wood into the tube and when dry, boring a central hole parallel to the sides of the tube so that the threads of the objective can be pushed into the hole, while further insertion is stopped by the shoulder surrounding the lens. Now cut a slot approximately 20mm long through the center of the tube, parallel to its long axis. Different objectives can now be secured in the tube by using a suitably sized, adjustable clip over the slot to constrict the tube. Tightening the clip closes the slot in the tube, so that the wooden insert just grips the threads of the objective lens and holds it securely in place. This method of fixing the objective has a number of advantages, in that the lens can easily be changed and misalignment of the lens can be corrected before tightening the retaining clip.

Making Stage Micrometers - Stage micrometers are relatively expensive items. Instead of purchasing one it is possible to make a micrometer photographically. Photograph a 0 – 100 mm scale on a transparent ruler. The negative produced can be further reduced in a photographic enlarger and printed on another piece of negative film material (thus producing a transparent positive) so that the 0 – 100mm scale is reduced to 10mm. This 'positive' can be mounted on a glass microscope slide beneath a coverslip with Canada balsam and used as a stage micrometer.

Making the Stand and Illumination System - The illumination system comprises a lamp holder and a small 15-watt opal lamp contained in a box. The box is constructed from 3/8 inch plywood and should be about 5 inches high and five inches square. Ideally, the box should be nailed and glued together. The front of the box (that part opposite the upright stand) should be detachable for centering the illumination and for changing the lamp. Ventilation holes must be drilled into the box to allow heat dissipation when the lamp is on. It is desirable to line the inside surface of the top of the box with foil to reflect the heat. A hole of ½ inch should be drilled through the center of the top of the box and the light fitting centered below this hole. This is important in order to achieve even illumination of the specimen.

The upright stand, which supports the body tube, must be substantial and accurately made. The stand should be constructed of hardwood not less than 2.5 X 1 inch in section. The upright should run from the base of the box (approximately 5 inches) for a secure attachment and protrude 5.5 inches above it. A horizontal beam 5 inches long is attached at right angles to the top of this upright. This must have a hole the diameter of the body tube drilled into it, approximately 2 inches from one end, in which the body tube fits. A vertical slit is cut through the center of the horizontal beam from the front to about 0.5 inches beyond the body tube hole. A bolt hole is drilled at right angles through the front beam through the vertical slit. A suitable bolt is pushed through this hole and a washer and wing nut screwed on to the protruding end. When the body tube is pushed through its hole, the wing nut can be tightened to secure the tube in place, but can be slackened to allow for vertical movement to accomplish coarse focusing. A second adjustable clip can be attached around the top of the body tube to act as a stop in case the tube drops toward the specimen slide. The horizontal beam must be accurately fitted to the upright stand, so that the objective is centered symmetrically on the hole in the top of the illumination box, so that it is evenly lit. The horizontal beam should be glued and screwed into position. A supporting block must be attached into the right angle formed between the vertical stand and the horizontal beam.

Making the Fine Focusing Mechanism - Rough focusing is accomplished by, raising, or lowering the body tube then clamping it with the wing nut on the horizontal beam.

Fine focusing is essential, and this was our most difficult design problem as it was necessary to have a mechanism that was in the ability of someone without extensive workshop facilities. The most precisely machined, readily available component is a nut and bolt, and this is the basis of the fine focusing mechanism. Use a bolt with a fine thread pattern, approximately 30mm long X 7mm diameter. A central hole is drilled through the front edge of the top of the illumination box about 4 inches from the rear upright. The bolt is pushed through this hole from below and firmly attached by a nit. A further nut is required, to which is glued a large washer. The kind of body washer used to attach rear seat belts in cars is ideal, and these can be obtained from a car accessory shop. A suitable glue is cyanoacrylate ester

(Loctite 'super-glue – 3'), which bonds in seconds. This particular nut is loosely screwed on to the protruding bolt with the large washer on the undersurface of the nut.

The specimen stage -plate comprises a piece of 3mm acrylic sheet – approximately 4.5 X 4 inches. The size of this acrylic sheet should be measured from the front of the upright stand to the front of the bolt on the illumination box. In the microscope's simplest form, this transparent acrylic sheet should be made translucent by scrubbing one surface with household scouring powder. The translucent sheet acts as a diffuser to spread the light from below and thus evenly illuminates the specimen. The acrylic sheet is attached by a suitably sized hinge to the upright stand, in such a way that the sheet is parallel to the top of the illumination box and about 1/8 inch above the top of the fine focusing bolt. The nut with the large washer should be unscrewed until the nut protrudes above the end of the bolt. By rotating the large washer, the stage plate should move up and down.

It was found necessary to attach a rubber band around the upright stand, through the rear corners of the acrylic sheet, to minimize the play in the hinge. If this is not used, the image can shift too far while being viewed under high magnification.

Grease or petroleum jelly should be smeared on the fine focusing bolt. This acts as a lubricant and also takes up the play between the focusing nut and the threads of the bolt. This is necessary when using the microscope at high magnification.

The microscope should be set up with the stage capable of up and down movement (the stage will not move downwards if the nut with the large washer is screwed down below the end of the bolt). The specimen on a glass microscope slide is placed on the specimen stage-plate, and the specimen roughly focused, by slackening off the wing nut on the horizontal beam and moving the body tube carefully up and down until the image can be accomplished by slight rotation of the fine focusing nut, using the large washer.

The basic microscope described above makes an ideal instrument for beginning microscopy. A mechanical stage can be added to it to allow more precise specimen movements; and a rudimentary condenser system, with iris diaphragm, can be added to increase resolution and alter the illumination intensity.

Making the Mechanical Stage – Students may be encouraged to make a cardboard model. The mechanical specimen stage is constructed from acrylic sheet, cyanoacrylate ester glue, two fine-threaded nylon bolts and four nuts. The dimensions of the parts are only for use with a 112mm X 100mm acrylic specimen stage-plate and with the use of 3 inch by 1 inch (76mm x 25mm) glass microscope slides. Mark out the parts listed on the acrylic sheet and accurately cut them out, use a saw fitted with a blade suitable for woodwork. Accuracy in both cutting out and assembling these pieces is essential, in order to produce a mechanical stage which works smoothly and precisely without "backlash." Before final assembly, all parts should be assembled without glue, so that any potential mistakes can be avoided. Incorrect assembly can usually be corrected by tapping incorrectly glued parts with a wooden mallet. The parts may then be reassembled after removing the old glue by sanding and filing as necessary.

Mechanical Stage (shown in illustration – next page) Pieces and Dimensions	
Part 1	***Slide Carrier*: 94mm x 39mm rectangular "cut out" to hold a 76mm x 25mm slide**
Part 2	**112mm x 60mm in which to fit part 1 (rectangular shape two 9mm wide legs & one 10mm wide top)**
Part 3	***Retaining Strips*: 40mm x 10mm - (two of these pieces are needed)**
Part 4	***Front Lip Rail*: 112mm x 6mm**
Part 5	***Skinny Rail*: 45mm x 3mm (unlabeled rail parallel with part 4 along edge of stage plate in the image on the next page)**
Part 6	***Retaining Plates*: 16mm x 10mm - (two of these pieces are needed)**
Part 7	**15mm x 10mm - (THREE of these pieces are needed) drill a 5mm hole centered 9mm from each base. A similar fourth piece should be labeled Part 9. See directions on next page.**
Part 8	**35mm x 20mm**
Part 9	**30mm x 20mm – drill a 5mm hole centered 18mm from the base and 8mm from an edge**
Part 10	***Specimen Stage Plate:***

-

Assembly of the Mechanical Stage (Steps 1-9)

1. Glue *part 4* to the front edge of *specimen stage-plate*, flush with the bottom surface, so that a lip is formed above the surface of the plate.
2. Place (do not glue) the front of part 2 against *the lip formed by part 4*, and then glue *part 5* into position so that *part 2* is free to slide from side to side. *Part 5* must be positioned centrally (on the specimen stage plate) leaving 30 mm on each side, or insufficient lateral movement of the glass specimen slide will result.
3. Glue the retaining plates (*parts 6*) into position on *parts 4 and 5* ***only***. Glue must **not** contact *part 2*.
4. Place the *slide carrier* (part 1) into position *within part 2*, and glue the *retaining strips* (parts 3) into position on *part 2*, so that they overlap *part 1*. Part 1 must be free to move relative to part 2.
5. Glue *part 8* beneath the front right-hand corner of the specimen stage-plate so that 20mm protrudes to the right.
6. Glue one of the THREE *parts numbered 7*, vertically to the *right-hand side of part 2* in the position shown.
7. Glue *part 9* vertically to the end of *part 8* so that the hole, in *part 9*, is at the same level as the hole in *part 7* of step six.
8. Push a long bolt through the hole in *part 9* and attach a nut. Then push the bolt through the hole in *part 9* and attach a second nut. This second nut should be glued to *part 7* using cyanoacrylate ester glue. The first nut should be glued to the bolt near to *part 9* so that they both rotate in part 7 but cannot be unscrewed relative to one another. This is best accomplished by tightening the nut so that the bolt is free to move in *part 7* without play. Mark corresponding positions on the bolt head and nut, using a pencil or a felt-tipped marker pen. Unscrew the nut and put a drop of cyanoacrylate ester glue on the threads of the bolt and screw up the nut once more until the marks correspond. After a few seconds the nut will be glued firmly in place. By rotating the threaded end of the bolt, the mechanical stage should move easily both left and right across the specimen stage-plate.
9. The two remaining *parts 7* should be glued to the left side *of parts 1 and 2* in a position **OPPOSITE** to the position shown in the illustration. The remaining long bolt should be pushed through the hole in *part 7*, attached to *part 1*, and a nut glued into position on the bolt as in step 8. The bolt should then be pushed through the *part 7, on part 2*, and a further nut added which should be glued to the *part 7* attached to *part 2.* Parts numbered 7 and 9 can be replaced with small metal brackets to increase strength and durability.

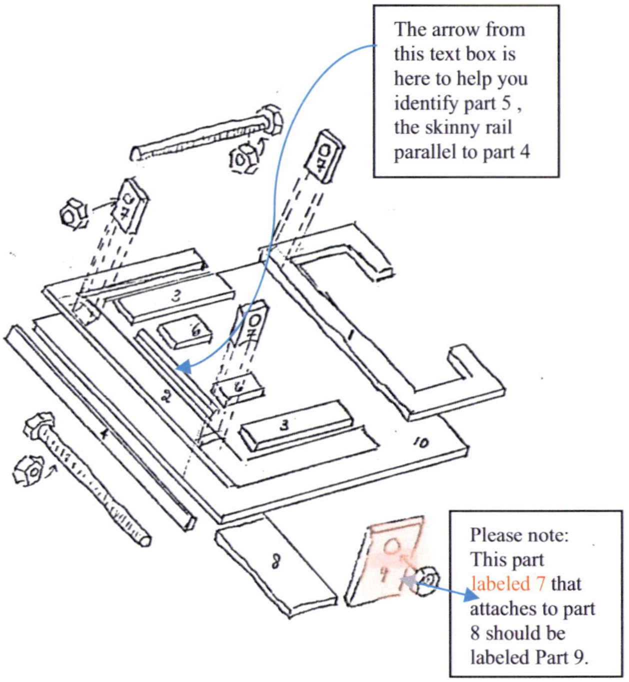

The assembled mechanical stage should be capable of controlled movement, front to rear and left to right of the specimen stage-plate. The specimen stage-plate, with mechanical stage, can now be hinged to the upright stand of the microscope. If the movement of the mechanical stage is not smooth, petroleum jelly can be used as a lubricant.

Adding a Condenser and Iris Diaphragm - To achieve maximum resolution, a condenser lens system can be added to the microscope. All that is needed is a suitable convex (converging) lens of approximately 1-inch focal length. These can be sometimes found by dismantling old redundant camera lenses. A hole of approximately 0.5 inches should be drilled through the acrylic specimen stage-plate beneath the objective lens. The convex lens is placed on top of the illumination box, so that light transmitted from the lamp is converged on to the specimen, through the hole in the specimen stage-plate.

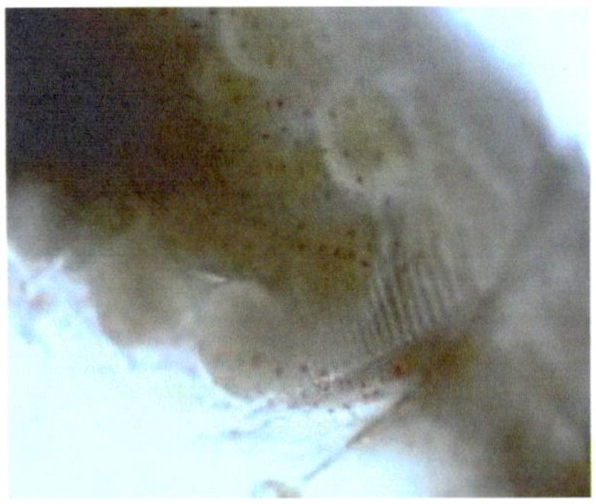

Photo by Robert Jakus

An iris diaphragm can be added to the condenser to control the amount of light reaching the specimen. If the condenser lens was removed from an old camera, then the iris diaphragm from this can be used, with a series of progressively smaller holes drilled in it. This can be moved across the hole in the illumination box, to reduce the intensity of the illumination.

Mounting Specimens for the Microscope - Acquiring a microscope is only the first stage in microscopy. Items for examination under the microscope need to be mounted in a manner suitable for this instrument, and in a form that maximizes the amount of detail that can be seen in the specimen.

The basic requirements for mounting specimens are glass slides, coverslips and a mounted needle. These are essential. The glass slides upon which the specimens are placed are 3 inches by 1 inch. Coverslips are of variable size and are made from extremely thin glass. The coverslip fulfills two major functions. Firstly it helps to flatten the specimen thus ensuring that most of the specimen is in focus, when the depth of focus of the objective is small. Secondly it protects the objective from possible immersion in the fluid in which the specimen is mounted. The mounting needle is used for lowering the coverslip.

Most specimens, particularly those of biological origin, cannot be mounted without some supporting medium to maintain their shape. The most common medium is water, which can be used for rapid preparation of slides, or for viewing small aquatic animals and plants. Water mounted specimens are, however, only temporary. In preparing a water mount, a small drop of water should be placed in the center of a clean glass slide, using either a glass rod or a small dropping pipette. The specimen can then be placed in the water drop, and a coverslip carefully lowered onto it. Dropping the coverslip on to the specimen, or even gently lowering it parallel to the slide surface, should be avoided, as this will inevitably trap numerous air bubbles under the coverslip, thus impeding examination of the specimen. A better result is obtained by lowering the coverslip as if it were hinged to the slide on one edge, by placing a mounted needle under the free edge. If too large a water droplet has been used, then the coverslip will float and the correct focusing of the specimen will be difficult, not only because of the depth of water, but also because even slight movement will jar the floating coverslip. Excess water can be removed from such a mount by touching the edge of the coverslip with absorbent paper

Microscope Activities with Aquatic Organisms

According to Hanes et.al (2002), "tracking growth of algal populations in mass cultures is essential for certain types of aquaculture." Monitoring consists of observing the cultures and counting cells to determine if the culture is healthy and growing. Algal cultures that have stopped growing are either significantly contaminated with microscopic predators or competitors or are approaching the maximum cell density under the existing conditions. In both cases, if the culture is slowing from the log phase of growth, the culture must either be discarded if contaminating organisms are present, fed out if it is approaching a stationary phase of growth, or diluted with nutrient-rich seawater media to stimulate a return to the log phase of growth.

Algal cell cultures are routinely produced as a food for cultured aquatic animals. Many filter-feeding aquatic animals such as bivalve molluscs require specific cell densities to permit the cultured animals to feed optimally. Therefore, another important reason for monitoring the density of cells in the algal culture is to provide and maintain optimal algal densities for filter feeders. This exploration demonstrates how growth of algal cultures is measured.

One method to monitor algal population density and growth is to count the number of algal cells in a fixed volume of culture media. This technique is adapted from the medical techniques for counting red and white blood cells, thus the name of the measuring chamber is a hemocytometer (hemo = blood; cyto = cell; meter = measuring device). The hemocytometer is a fine glass microscope slide that holds a fixed volume of liquid under a special cover slip. Etched into the surface of the counting cell is a grid of squares that are visible under the microscope. The object is to count the number of algal cells within a given number of squares and use the cell numbers observed to calculate total cell density within a milliliter of culture medium, using a given formula. This technique provides an accurate count of cell number within a sample of culture medium. It also allows the culturist to observe the condition of the culture and assess the culture composition and health.

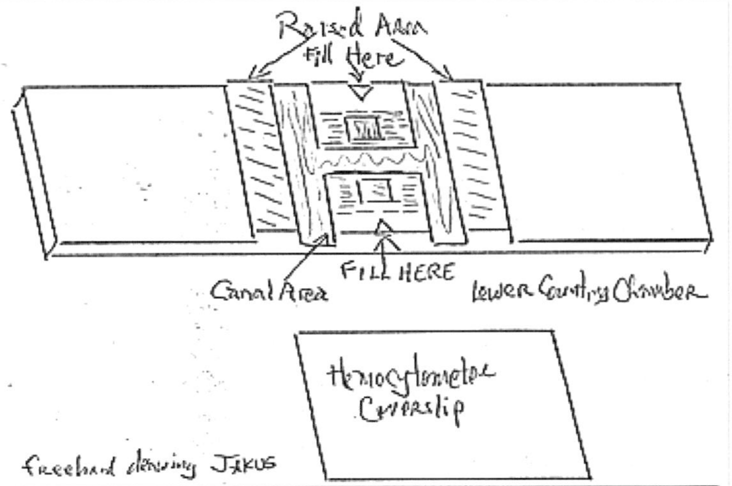

Hemocytometer showing counting chambers and fill points

As educators set-up and introduce this activity to their students they may prepare to guide students to develop an estimate of the number of cells in a classroom algal culture by quantifying algal cells in samples of algae growing in the culture. Depending upon lens magnifications in student manufactured microscopes students may be able to use their own microscopes to develop estimates of cell densities in their algal cultures. Commercially manufactured compound microscopes generally have the necessary lenses to perform this activity. The directions given below are provided in order to explain the proper techniques for using a hemocytometer to count cells in a given area and volume and to then use this data to calculate an estimate of cell density within the entire culture.

Once the aquatic class/lab is properly set-up and all students demonstrate that they conform to the aquatic class/lab safety criteria regarding clothing, jewelry, and hair and are also prepared to properly dispose of potentially dangerous equipment and materials as necessary student teams (or pairs) may perform this activity – under supervision and guidance. Special emphasis may be placed on appreciating the importance of not "contaminating" an algae culture. Cultures that become exposed to other bacteria and/or toxic substances may quickly become useless for the primary purposes of feeding desirable aquatic larvae and feeding the live microscopic food that desirable aquatic larvae must eat in the classroom. The more a culture is handled the higher the risk of contamination through contact with the lab technician or lab equipment used. Ideally, students will be encouraged to use clean/sterile materials while working with algae cultures whenever possible. Some algae species are motile and in order to prevent them from moving – so they may be counted more accurately the sample beaker must be microwaved (just briefly) so the algae cells stop moving and are still recognizable. Microwaving too long can easily cause algae cells to explode or dissolve. Some recommend the use of a drop or more of a weak acid or ethanol to prevent cells in the algae culture from moving.

Numerous techniques for counting cells with a hemocytometer have been developed. Descriptions of counting techniques abound on the internet (http://www.hemocytometer.org/). Teachers should present their preferred technique to the class and ask them all to follow this standard for the class activity. After safely removing enough algae from the culture into one clean sample beaker, all groups in the class may fill their eyedroppers with a drop algae culture from the beaker and with the cover slip already in place atop the hemocytometer allow capillary action to fill in the space in the hemocytometer beneath the cover slip with algae culture from the eye dropper. Now that the hemocytometer is filled with a sample the cells must be given time to settle onto the grid. Using a digital microscope the teacher may project the microscopic image for students to see and demonstrate the counting technique for students to follow. The counting protocol may be introduced by asking students to count independently the number of cells in the top row of four small squares in the upper left box. After completing this count teachers may ask students to share their count with the class. It is common that many cells are double counted (because of their position on a grid line). Students may also miscount because they do not use a reliable and simple tallying method.

To improve counting accuracy (inter-rater reliability) students may now be challenged to develop a technique that will help them count cells accurately (minimizing counting errors because of a cells position on a grid line – "double counting"). Most resources to which you may refer describe the kind of standard counting technique that students may be guided to "invent." With the new technique, students may view the instructor's projected hemocytometer grid and again independently count algae cells in the bottom row of four small squares in the upper left box. After each individual shares the results of this systematic counting technique with the class, it is hoped that there will be much better accuracy in counting cells in a sample. With this standard technique in hand student groups may now fill their hemocytometer with an algae sample (if not done already) and count the algae cells on their own respective hemocytometer according to the following directions. Practicing together is often a quite helpful introductory activity.

Some writers suggest that if it is difficult to count the cells (perhaps the algal density in the culture sample is too high) the class may decide to first dilute the sample before performing the following standard estimate development protocol. When samples are diluted it is important to quantify the volume and percent by which the dilution is being made so that the final calculation as described below can be multiplied by the dilution factor to determine the proper cell density estimate of the culture.

The hemocytometer has four corners, in each corner are16 small boxes. In order to estimate the number of algae cells in the algae culture students must now use their counting technique to count the number of algae cells in the 16 small boxes in each of the four corner squares.

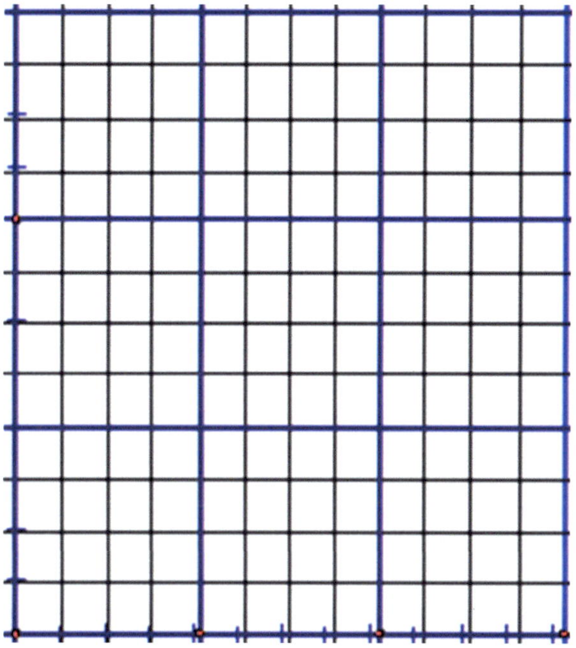

Students working together to count and tally the algae cells on their hemocytometer may find it helpful to draw a grid like the one to the right on a sheet of paper and fill each of the 16 boxes - in the four corners of this grid - with the number of algae cells they counted.

Once all 64 small boxes have been filled in the numbers should be added and divided by 4. This result produces the average number of algae cells per corner square.

Students must next determine the number of algae cells per mL by multiplying the average cells per corner square by 10,000.

Once all students have successfully completed the counting and calculating per their algae sample they may be guided to critique their results, pool their estimates of the number of algae cells per mL and be challenged to devise a way to use this average number of cells per mL to determine the number of algae cells in the classroom culture from which the sample was taken.

Data from regularly performed microscopic observation of algae cultures of this type can help ecosystem managers and analysts remain aware of the growth rate of a culture. This information is useful for deciding when to start new cultures, feed old cultures, or any number of other actions that might be practical.

Write a list of **science-based** activities you have participated in during this project and briefly describe evidence that verifies your interaction with and knowledge of scientific concepts for each item on your project eight list.

Write a list of **technologically-oriented** activities you have participated in during this project and briefly describe evidence that verifies your interaction with and knowledge of technologically-oriented concepts for each item on your project eight list.

Write a list of **engineering-oriented** activities you have participated in during this project and briefly describe evidence that verifies your interaction with and knowledge of engineering-oriented concepts for each item on your project eight list.

Write a list of **mathematical** activities you have participated in during this project and briefly describe evidence that verifies your interaction with and knowledge of mathematical concepts for each item on your project eight list.

Write a brief list of **Non-STEM-related** activities you have participated in during this project and briefly describe evidence that verifies your interaction with and knowledge of Non-STEM related concepts for each item on your project eight list.

Rate your level of involvement in STEM-related activities in this project (1=lowest, 10=highest).

1—2—3—4—5—6—7—8—9—10

Additional Project Eight Assessment Activities and Questions

Use Excel to update and graphically represent the water quality of each system. Create a new descriptive graph and interpretation of the data. Compare it to previous graphs and explain patterns and trends. Create a chart/poster that summarizes your interpretations.

In terms of visual focus and clarity of vision, explain similarities and differences between the way lenses in the eyeball work to make vision clear and how lenses in compound microscopes work to clarify objects.

Describe the essential components of a microscope and your success and difficulties in constructing your own microscope. In order to have a more successful construction experience, what would you do differently?

Use a hemocytometer to sample populations in algal cultures and rotifer cultures (on a daily basis). Record your data and create a graph of the growth or decline of the organisms in the culture.

Sample and record rotifer and algae culture populations every day. Periodically update the data and use this data to develop an updated population growth curve. Enlarge the graph to fit on a poster and label the four stages of the population growth curve as they relate to the culture samples you are reporting on. How does the new data affect the curve? What does it mean about the population trend in the culture? What actions should be taken?

Organize a group of students to read the play by Barrie Stavis (1966) *Lamp at Midnight* and consider presenting it as a fundraising event (about Galileo).

Hanes et.al, (2002) have developed two interesting **spectrophotometer**-oriented cell counting activities that may pique some student's interest while enriching their knowledge of cell counting techniques. The activities are outlined/summarized below.

If possible, obtain a **spectrophotometer** (see Hanes et.al, 2002) to measure indirectly, by light absorption, the cell density of two algal samples. Follow directions for operation of the spectrometer. First, "zero" the spectrophotometer using a sample of sterile culture media that does not contain any algal cells. Then measure light absorbance of an algal sample with unknown cell density. Transfer approximately 100 mL of well-mixed algal culture to a 250 mL beaker. From this beaker create two samples of algae. Record the light absorption of both algal samples. Do a final check of the spectrophotometer by analyzing a second sample of sterile culture media that does not contain any algae cells. If it reports "zero" for this sterile sample, the reading from the algae samples can be used to calculate the average light absorbance from the two algal samples. If not "zero" the procedure must be repeated.

Investigate the meaning of "serial dilution." If it is possible to obtain a **spectrophotometer** (see Hanes et.al, 2002) determine how to use a 100 mL solution of algae from a classroom algae culture system and a supply of sterile algae culture water to create a serial dilution of algae in which each dilution of algae in the series will be 50% less concentrated than the previous sample (from which it was drawn). In series the samples will be 100% pure algae culture, 50% pure algae culture, 25% pure algae culture, 12.5% pure algae culture, 6.25% pure algae culture and 3.12% pure algae culture. For each dilution in the series, take two measurements. First, use a hemocytometer to determine the average number of cells in each dilution. Second, Use a spectrophotometer to determine the average light absorbance for each dilution. Graph ***(hemocytometer cell count, light absorbance)*** data and create a curve. Use a spectrophotometer to determine light absorbance for the culture from which the algae was taken every day for a week. And use the light absorbance data to estimate the daily change of algal cell density in the classroom algae culture. Create a report about this process and explain how knowledge of algal cell densities can b used to inform decisions of the culture manager.

Review activities and information that interested you in this project and determine if this information may fit on the "Research-ability" Potential Topics Form on page 59. If so, add it to the list.

For each adult species in a classroom broodstock larval rearing system, complete the form and include information from it (and the Species Specific Lifecycle Requirements Form) in any presentation or report you give about the system and the organisms in the system.

Species Culture Sheet	**Technique for Collection of Eggs**	**Ideal Hatchery Environmental Conditions** Volume, Temperature, Substrate, pH & Salinity, Light Aeration (circulation & flow rates) , Dissolved Oxygen	**Duration of Larval and Metamorphosis Stages**	**Other Comments and Observations**
Name of Species __________ **and** **Type of Reproductive Adaptation** __________				

Beyond the Classroom Aquarium Daily Journal Guidelines

Entries students place in their daily journals should be:

1. Dated accurately and logged chronologically.
2. Kept in a notebook and written neatly in ink or typed.
3. Written or typed in complete sentences using correct capitalization, spelling, and punctuation.

In their journals, students should keep a daily record of:

1. Who they worked with each day by including the names of school staff, fellow students, community members, and volunteers.
2. What they did each day, such as lecture topic, activities or project design, assembly, or maintenance.
3. What they learned each day by asking questions, following instructions, and/or observing activities.
4. How *Beyond the Classroom Aquarium* experiences relate to past, present, or future life experiences.

For bonus points, students may complete these journal options: include illustrations, diagrams, or creative expression relating to the journal entry. Journal entries will be evaluated on a point system, with four points being the maximum number a student can earn. A half-point will be awarded for each of the first eight journal requirements above. Bonus points will be awarded when journal options are completed. Journal points will be totaled and divided by the number of journal dates. A possible journal grade based on points is included in the project one assessment section.

For project eight, journal writing topics may include answers to the following questions or descriptions of a project activity or assignment. Reflection on these topics may also help students prepare for exams, quizzes, and other assessments.

- What would you like to know about the organisms and systems that you observed?
- What do you want to understand better in order to contribute more to your own productivity as you independently design, assemble and then manage an aquatic ecosystem?
- Create a new drawing picture and create a story about the organism or system that is most interesting to you.
- Create a journal entry that tells what information you learned during project eight. Also write down what you would like to know more about and if you know someone else who might help you learn more if you asked them.
- Select a book or magazine to read that is related to a topic in project eight. Write a report on this book and include your report in your *Beyond the Classroom Aquarium* portfolio.
- Create a poster that illustrates and explains function of the following major parts of a homemade microscope:

Eye Piece	Upright Stand
Body Tube	Objective Lens
Horizontal Beam	Specimen Stage Hinge
Bolt and Wing Nut	Rubber Band
Supporting Block	Specimen Stage
Saw-cut Slit	Ventilation Holes
Objective Lens retaining clip	Illumination Box
Illumination Hole in Illumination Box	Detachable front of Illumination Box
Fine Focusing Mechanism	

<table>
<tr><th colspan="5">Beyond the Classroom Aquarium STEM Self or Peer Assessment Rubric</th></tr>
<tr><td colspan="5">This self-assessment is valid only to the extent to which you can document and display evidence that supports it. For each category in the left hand column, write a STEM-Oriented example in the box beneath 3 if you can display evidence that your STEM-Oriented example represents work that exceeds expectations. If your evidence shows that your STEM-Oriented statement meets but does not exceed expectations, write the statement in the box beneath 2. Similarly, for STEM-Oriented examples that demonstrate some progress toward expectations.</td></tr>
<tr><td colspan="2">Name:
Project Number:
Date Project Work Began:
Date of Project Completion:

List Math Topics and/or Objectives Currently Under Study:</td><td>3
Work Exceeds Expectations</td><td>2
Work Meets Expectations</td><td>1
Work Is Progressing Toward Expectations</td></tr>
<tr><td colspan="2">Write examples of questions and defining problems related to your STEM work.</td><td></td><td></td><td></td></tr>
<tr><td colspan="2">Describe how you have developed and used models in your STEM work.</td><td></td><td></td><td></td></tr>
<tr><td colspan="2">Write examples of STEM investigations you planned and carried out.</td><td></td><td></td><td></td></tr>
<tr><td colspan="2">Describe STEM related data that you analyzed or interpreted.</td><td></td><td></td><td></td></tr>
<tr><td rowspan="5">Give examples of how you used mathematics and computational thinking related to the five standards listed to the right.</td><td>Problem Solving (Math Process Standard One)</td><td></td><td></td><td></td></tr>
<tr><td>Reasoning & Proof (Math Process Standard Two)</td><td></td><td></td><td></td></tr>
<tr><td>Communication (Math Process Standard Three)</td><td></td><td></td><td></td></tr>
<tr><td>Connections (Math Process Standard Four)</td><td></td><td></td><td></td></tr>
<tr><td>Representation (Math Process Standard Five)</td><td></td><td></td><td></td></tr>
<tr><td colspan="2">Describe STEM-Oriented explanations and solutions you designed</td><td></td><td></td><td></td></tr>
<tr><td colspan="2">Explain evidence you used while engaging in a STEM project-based argument.</td><td></td><td></td><td></td></tr>
<tr><td colspan="2">Give examples of STEM information you obtained, evaluated, and/or communicated</td><td></td><td></td><td></td></tr>
<tr><td colspan="2">TOTAL POINTS</td><td></td><td></td><td></td></tr>
</table>

CHAPTER V: Students Design and Execute an Aquatic Ecosystem Research Project (Project 9)

Having engaged in aquatic ecosystem-related design, assembly, and maintenance activities each school day, students have collectively created an aquatic ecosystem classroom/lab environment that offers a multitude of opportunities to produce significant research and to discover or rediscover many unknown ideas about marine life and marine systems. In working with projects that have involved water chemistry, nutrition, food webs, algae, rotifers, black worms, Artemia, aquaponics, magnification, properties of light, aquaponics, coral reefs, aquaculture, and numerous other issues, students have learned to work with most of the basic laboratory/classroom tools and equipment that are used by aquatic scientists. Project nine is a stand-alone project in which students are guided to perform an in depth research project related to aquatic ecosystem skills and knowledge they have been cultivating every day.

By now, most classroom aquatic ecosystems have matured to a point where they are capable of maximum and sustainable productivity. As students proceed through activities in project nine it is imperative that they also continue performing scheduled daily system maintenance and upkeep tasks. Ongoing upkeep is essential in order to make most effective use of the classroom aquatic ecosystems as students design and carry out their research projects over the following three months (generally about February through April).

Following introductory research related discussions, activities, and field trips, students are subsequently guided through the processes of developing an aquatic ecosystem research proposal that addresses specific components of a classroom aquatic ecosystem or an aquatic organism and then of executing their research project. A concluding activity in project nine guides students to recognize their personal membership in a scientific community that extends the classroom and the local community while also setting the stage for an awesome end of year summative assessment activity.

Two introductory field trip experiences perform a number of critical functions in chapter V. Their ultimate purpose is to provide a strong background for students to rely upon as they develop their own research projects. Advance scheduling for student attendance at an aquatic ecosystem conference and prearranging a date for a beach sweep field trip are important first tasks for the teacher. Forms for guiding students through each project nine field trip activity are based on writings by Leedy (1980) and the Center for Marine Conservation (1994).

In the introductory field trip, students become attendees at an aquatic ecosystem-oriented conference. This activity serves a threefold purpose. First of all, students are carefully guided to apply essential research-oriented standards in analyzing and critiquing conference research presentations, thereby reinforcing their awareness of how their own classroom-based aquatic ecosystem research projects reflect these standards. Secondly, as students attend the conference, a portion of the data they are to collect pertains to the business world that provides services and products that support and/or evolve directly from activities performed by conference attendees and presenters. Aquatic ecosystem-related business activities will be studied more fully in chapter VI and in preparing to attend the conference, students will be guided to collect information related to aquatic ecosystem businesses and economics at conference trade booths. Third, the conference field trip sets a standard for conference settings and protocols that will be useful as students are assigned roles and a timeline in setting up their own end of year student conducted conference in chapter VII.

Beyond the Classroom Aquarium educators, by working with local and state members of Sea Grant, NMEA, environmental, agricultural, tradespersons, aquatic hobbyist groups, parent groups, and others, should find that arrangements for student attendance at a locally accessible conference may be handled in many creative ways. In some cases, it may be possible to enlist local help from school district superintendents, university professors, and esteemed secondary school colleagues. Some teachers may prefer to create a virtual conference activity for their students using a series of online research presentation "stations" for groups of students to observe and critique.

During their second introductory field trip activity, students participate as a group in a beach sweep experience. The beach sweep activity is designed to model the experience of direct personal involvement in the process of data collection and in the development of descriptive statistical displays from it. Educators may also use this busy data collection activity to reinforce student awareness of important differences between research and non-research based activities before students embark upon the formulation of their own research project. Necessary beach sweep forms and details are provided in this chapter's text.

Following the two field trips, students are introduced to the twelve week aquatic ecosystem research project schedule. In the first two weeks students perform all preliminary research work, the main purpose of which is to have the instructor approve a research proposal. During the remaining ten weeks, while working independently on setting up their research, documenting their work, collecting data, and writing up a concluding report, students are responsible for continued participation in chapter VI and VII class activities.

In project nine, students are guided through the following sequence of activities in order to ensure that they are successful in developing proposals for independent research and in carrying the investigation to a successful conclusion.

Introductory Activities

- Research discussion and establishment of a basis for critiquing research
- Conference attendance and assessment of research presentations (or a virtual assessment) followed by a critique of student's research assessment skill
- Execution of a beach sweep field trip
- Beach sweep data presentations, conclusions, and critique

Primary Activities

2-weeks preliminary approval;

- Discuss, brainstorm, and select researchable classroom aquatic ecosystem projects
- Preliminary approval of selected topic

2-weeks lit review final approval;

- Literature review/assessment of selected research topic
- Final approval of classroom aquatic ecosystem research project

6 weeks data collection

- Independent Research Activities (individualized assistance with research set-up, execution, and data collection)

2-weeks data analysis and write up

- Independent Research Activities (individualized assistance with data analysis, research project report, and presentation)

Concluding Activity

- Students are guided to organize and operate a conference planning committee for the purpose of developing and conducting an end of school year student-organized and sponsored classroom aquatic ecosystem conference.

PROJECT NINE ...Beyond the Science Fair - Design & Execute an Aquatic Ecosystem Research Project

Introductory Research Related Activities – Project Nine

Mankind has devised only two ways to uncover the unknown; deductive logic and inductive reasoning (the scientific method). Deductive logic is identified with Greek thinkers such as Aristotle. As illustrated in studying classic geometry, deductive logic relies on logical reasoning and, to arrive at valuable conclusions, must begin with an assumption related to the truth of a major premise. History provides many examples of logical but absolutely faulty conclusions that drove cultural customs and behaviors for centuries but that were based on incorrect assumptions about the truth of the major premise (e.g. the Earth is flat). The use of deductive logic to advance knowledge was the only method available until the Renaissance.

The consistent application of inductive reasoning in relation to natural phenomena gave rise to the scientific method. Use of the scientific method leads learners to view knowledge as flexible and improvable. The scientific method eliminates the need for dogmatic absolutism and replaces its use with informed and cautious acceptance of the reliability and validity of any conclusion.

As stated in *Practical Research: Planning and Design* (Leedy and Ormrod, 2010), "the public is often unaware of what researchers do on a day-to-day basis or how their work contributes to people's overall quality of life and general welfare. The word research has been so widely used in everyday speech that few people have any idea of its true meaning. Following are several statements that describe what research is not. Accompanying each statement is an example that illustrates a common misconception about research." The following information is excerpted from *Practical Research* (Leedy & Ormrod, 2010) with the intention of promoting useful classroom discussions of research, and of providing a standard upon which project activities can be based.

1. **Research is not information gathering**. A sixth grader comes home from school and tells her parents, "The teacher sent us to the library today to do research, and I learned a lot about black holes." For this student, research means going to the library to glean a few facts. This may be information discovery, it may be learning reference skills, but it certainly is not, as the teacher labeled it, research

2. **Research is not mere transportation of facts from one location to another**. A college student reads several articles about the mysterious "Dark Lady" in the sonnets of William Shakespeare and then writes a "research paper" describing various scholars' suggestions of who she might have been. Although the student does, indeed, go through certain activities associated with formal collecting of information, organizing it in a certain way for presentation to others, referencing statements properly, and so on, these activities still do not add up to a true research paper. The student has missed the essence of research: the interpretation of data. Nowhere in the paper does the student say, in effect, "These facts that I have gathered seem to indicate this about the Dark Lady." Nowhere does the student interpret and draw conclusions from the facts. This student is approaching genuine research; however, the mere compilation of facts, presented with reference citations and arranged in a logical sequence—no matter how polished and appealing the format—misses genuine research by a hair. A little further, and this student would have travelled from one world to another; from the world of transportation of facts to the world of interpretation of facts. The difference between the two worlds is the distinction between transference of information and genuine research, a distinction that is critical for novice researchers to understand.

3. **Research is not merely rummaging for information**. The house across the street is for sale. You consider buying it, and so you call your realtor to find out for how much your present home would sell. "I'll have to do some research to determine the fair market value of your property," the realtor

tells you. What the realtor calls "doing some research" means, of course, reviewing information about recent sales of properties comparable to yours; this information will help the realtor zero in on a reasonable asking price for your current home. Such an activity involves little more than rummaging through files to discover what the realtor previously did not know. Rummaging, whether through one's personal records or at the public or college library, is not research. It is more accurately called an exercise in self-enlightenment.

4. **Research is not a catchword used to get attention**. The morning mail arrives. You open an envelope and pull out its contents. A statement in colorful type catches your eye: Years of research have produced a new car wash! Give your car a miracle shine with soapy suds! The phrase "Year of Research" catches your attention. The product must be good you reason, because years of research have been spent on developing it. You order the product, and what do you get? Dishwashing detergent! No research, merely the clever use of a catchword that, indeed, fulfilled its purpose: to grab your attention. "Years of research" - what an attention getting phrase, yet how misleading.

> "Research is a systematic process of collecting, analyzing, and interpreting information (data) in order to increase our understanding of a phenomenon about which we are interested or concerned. People often use a systematic approach when they collect and interpret information to solve the small problems of daily living. Here, however, we focus on formal research, research in which we intentionally set out to enhance our understanding of a phenomenon and expect to communicate what we discover to the larger scientific community. The essential nature and characteristics of research are described below" (Leedy & Ormrod, 2010).

1. **Research originates with a question or problem**. The world is filled with unanswered questions and unresolved problems. Everywhere we look, we see things that cause us to wonder, to speculate, to ask questions. And by asking questions, we strike the first spark igniting a chain reaction that leads to the research process. An inquisitive mind is the beginning of research, as one popular tabloid puts it, "inquiring minds want to know."

Look around you. Consider the unresolved situations that evoke these questions: What is such-and-such a situation like? Why does such-and-such a phenomenon occur? What does it all mean? These are everyday questions. With questions like these, research begins.

2. **Research requires clear articulation of a goal**. Are your objectives in clear, concrete terms demonstrating that you have a good idea of what you need to accomplish? A clear, unambiguous statement of the problem is critical. This statement is an exercise in intellectual honesty: The ultimate goal of the research must be set forth in a grammatically complete sentence that specifically and precisely answers the question, "What problem do you intend to solve?" When you describe your objective in clear, concrete terms, you have a good idea of what you need to accomplish and can direct your efforts accordingly.

3. **Research requires a specific plan for proceeding**. Research is not a blind excursion into the unknown, with the hope that the data necessary to answer the question at hand will somehow fortuitously turn up. It is, instead, a carefully planned itinerary of the route you intend to take in order to reach your final destination – your research goal. . . . Researchers plan their overall research design and specific research methods in a purposeful way so that they can acquire data relevant to their research problem. Depending on the research question, different designs and methods will be more or less appropriate.

Therefore, in addition to identifying the specific goal of your research, you must also identify how you propose to reach your goal. You cannot wait until you are chin deep in the project to plan and design your strategy. In the formative stages of a research project, much can be decided: Where are the data? Do any existing data address themselves to the research problem? If the data exist, are you likely to have access to them? And if you have access to the data, what will you do with them after they are in your possession? Each of the questions just listed—and many more—must have an answer early in the research process.

4. **Research usually divides the principal problem into more manageable sub problems**. From a design standpoint, it is often helpful to break a main research problem into several sub-problems that, when solved will resolve the main problem. Breaking down principal problems into small, easily solvable sub-problems is a strategy we use in everyday living. For example, suppose you want to get from your hometown to a town 50 miles away. Your principal goal is to get from one location to the other as expeditiously as possible. You soon realize, however, that the problem involves several sub-problems:

   Main problem: How do I get from town A to town B?
   Sub-Problems: 1. which route is the most direct?
   2. How far do I travel on the highway?
   3. Which exit should I take to leave the highway?

5. **Research is guided by the specific research problem, question, or hypothesis**. Having stated the problem and its attendant sub-problems, the researcher usually forms one or more hypotheses bout what he or she may discover. A hypothesis is a logical supposition, a reasonable guess, an educated conjecture. It provides a tentative explanation for a phenomenon under investigation. It may direct your thinking to possible sources of information that will aid in resolving one or more sub-problems and, in the process, the principal research problem. . . . Even with the best of data, however, hypotheses in research are rarely proved or disproved beyond the shadow of a doubt. Instead they are either supported or unsupported by the data. If the data are consistent with a particular hypothesis, the researcher can make a case that the hypothesis probably has some merit and should be taken seriously. In contrast, if the data run contrary to a hypothesis, the researcher rejects the hypothesis and turns to others as being more likely explanations of the phenomenon in question. Hypotheses are an essential ingredient in certain kinds of research, especially experimental research. To a lesser degree they guide most other forms of research as well.

6. **Research accepts certain critical assumptions**. In research, assumptions are equivalent to axioms in geometry—self-evident truths. . . . The assumptions must be valid or else the research is meaningless. For this reason, careful researchers—certainly those conducting research in an academic environment—set forth a statement of their assumptions as the bedrock upon which heir study must rest. In your own research, it is essential that others know what you assume to be true with respect to your project. If one is to judge the quality of your study, the knowledge of what you assume as basic to the very existence of your study is vitally important.

7. **Research requires the collection and interpretation of data in an attempt to resolve the problem that initiated the research**. After a researcher has isolated the problem, divided it into appropriate sub-problems, posited reasonable questions or hypotheses, and identified the assumptions that are basic too the entire effort, the next step is to collect whatever data seem appropriate and organize them in meaningful ways so that they can be interpreted.

Data demand interpretation. But no rule, formula, or algorithm can lead the researcher to a correct interpretation. Interpretation is inevitably subjective: It depends entirely on the researchers' hypotheses, assumptions, and logical reasoning processes. A number of potentially useful methods of organizing and interpreting data are available from among which researchers must carefully determine the method most appropriate to the data in their project.

8. **Research is, by its nature, cyclical, or, more exactly, helical**. The research process follows a cycle and begins simply. It follows logical, developmental steps:

a. A questioning mind observes a particular situation and asks: Why? What caused that? Home come? (This is the subjective origin of research.)

b. One question becomes formally stated as a problem. (This is the overt beginning of research.)

c. The problem is divided into several simpler, more specific sub-problems.

d. Preliminary data are gathered that appear to bear on the problem.

e. The data seem to point to a tentative solution of the problem. A guess is made; a hypothesis or guiding question is formed.

f. Data are collected more systematically.

g. The body of data is processes and interpreted.

h. A discovery is made; a conclusion is reached.

i. The tentative hypothesis is either supported by the data or is not supported; the question is either answered (partially or completely) or not answered.

j. The cycle is complete. Every researcher soon learns that genuine research yields as many problems as it resolves. Such is the nature of the acquisition of knowledge.

Distribute a Checklist for Evaluation of Research Form to each student in class (provided below). Briefly go over its main points. Explain that students will view a video of a research report and write a response to each question on the form. Following the video, the entire class conducts a critique of the research report based upon the questions on the form. This class critique should be used to clarify important components of research.

A second research presentation video is then presented to the class. Student groups (four per group) evaluate the research and contribute to a team research critique that reflects the opinions of the members of the group. In time, each group presents its critique of the presentation. Similarities and differences between group assessments are compared. Research and presentation of results should strive to attain a standard as described on the form. When students recognize important standards that should be a part of every research presentation, they are prepared to perform the following activity.

Each group of four to five students now views a different research presentation. They develop a group critique of the research presentation they viewed and each group reports their critique to the class. After each group has reported on the research they have assessed, students should be able to identify important similarities between each report despite the fact that each presentation reported on was unique from the others.

On the day before the field trip to the conference, students are given a schedule packet to follow at the conference. In each packet students are provided the name of the person who will present the welcoming speech at the conference (which the entire class will attend). Each student's packet will contain a form to use to help them collect information to use in assessing this introduction. The packet will also contain a Checklist for Evaluation of Research Form for each of the three conference presentations they are assigned to attend (during concurrent sessions). Although a few students from class may be assigned to the same presentation, each student must individually analyze each presentation they attend. The schedule packet will also detail a period of time during which each student is to attend two poster presentations at the conference (and an information collection sheet for each) and a scheduled time to interview people hosting two tradeshow booths in the conference exhibit hall (and another set of information collection forms to fill out for each interview). Students are to carry these packets with them throughout the day and fill out the information forms completely. By the end of the day-long conference field trip, each student will have a total of eight completed presentation forms one from the welcome speech, three from conference presenters, two from poster sessions, and two from tradeshow booths). If possible to arrange during lunch, students will benefit greatly by visiting with individuals from each of the conference's committees for 10 or 15 minutes.

In class on the day following the field trip, groups of four to five students summarize their conference experience on five dimensions (A the welcoming speech, B conference presentations, C poster sessions, D tradeshow booths, and E lunch with committee members) with one another. Strengths and weaknesses of presentations and information shared are decided upon and then each group in turn reports to the whole class on their overall assessment of each component of the conference experience.

CHECKLIST FOR EVALUATION OF RESEARCH (Leedy, 1980 p. 241)

Author______________________________

Last Name **First Name** **Middle Initial**

"______________________________"

Title of research report/thesis/dissertation

Appearing in

____

Journal (title) **Volume** **Month** **Year Pages**

Directions: Place a check mark in the appropriate column after carefully inspecting the research report to see if it contains the item designated.

Factor	***Yes***	***No***	***Comment***
Is the central problem for research clearly stated?			
Are the sub-problems clearly stated?			
Does the research show planning and organization?			
Has the researcher stated one or more hypotheses?			
Are the hypotheses related to the principal problem or sub-problems of the research?			
Is the research methodology which has been employed clearly described?			
Did the researcher interpret the data (i.e. tell what the facts mean)?			
Are the conclusions which the researcher presents justified by the facts presented?			
Is there any indication whether the hypotheses are supported or rejected?			
Is there any reference to or discussion of related literature or other studies by other researchers?			
Number of tally makes in each column (Enter these here) - - - - - - →			
Multiply the total of each column as follows	X 10	X 0	
Total score (Column 1 only)		0	

Conference/Trade Show Interview Questions (Distributed on Day Prior to Conference Field Trip)

Two Interviews Booths - Businesses Sales & Marketing	**Two Interviews Poster Session Research Summaries**	**Attend Three 45-60 Minute Workshops/Presentations**	**Introductory Welcoming Speaker**
Describe products featured	Provide a brief summary of each poster	State the title of each presentation attended	Was the value of this conference explained on a broader scale? What did the introduction lead you to expect?
Describe Primary Purposes of Featured Products	Explain to whom each topic is of most importance	Explain who may find each topic most interesting	Was the need for ongoing study and future work addressed?
Provide a brief history of the product	How were the project developers aware of the value of studying their respective topic?	Describe three strong points of each presentation	What major themes and concerns were presented?
Explain briefly how this product began and how it has grown over time	Report on the hypotheses that were tested	Describe one or two weaknesses of each presentation	Explain how graphs and charts were used
Describe/Explain how charts and tables were used at the booths	Describe/Explain how charts and tables were used in each poster	Describe/Explain how charts and tables were used in each presentation	Other?
What percent of income for the business sponsoring the booth is generated by attendance at trade shows and conferences?	Describe how the data was collected and the skills needed to be able to collect the data	Would you be interested in using information of this type in your own future?	
Describe skills first needed to develop the product (models and prototypes)	Describe how the data was analyzed	Other?	
What skills were needed to standardize the product for the public?	Explain the interest or lack of interest you personally felt about each project		
Describe the complexity of the economy needed to satisfy the public market	Other?		
How has the market changed and how is it growing?			
What directions are being explored for future growth and development for this product?			
Does this product interest you?			
Describe other details about financial business plans related to each booths primary product			

After completing their field trip six assessment activity (see page 21) and performing upkeep and maintenance activities on classroom aquatic ecosystems, the class should assemble and the two following forms should be distributed and explained: the Items Collected Form and the Data Sheet. Students should be grouped in pairs for field trip seven and be prepared to enter data for half the sweep collection time and to collect trash and report it to their partner to record on the data sheet during the other half of the collection time. A map of the site to be cleared of debris is mapped and each team is assigned a space in that area.

A brief walk through of this field trip activity can be conducted in the school yard. On the day preceding the second field trip, trash gatherers will be provided with a glove to keep the collector's hands from becoming contaminated. The trash collector should also carry the bag into which the trash is placed after the data collector (with a clipboard and pencil) enters the collected item on the Data Sheet.

At the end of the trash collection time, the trash from each group is weighed and discarded in a trash receptacle. Students wash their hands and then each team's Data Sheets are brought to class to be shared with the entire class so that each team can then create a descriptive statistical presentation of the distribution of the trash collected and propose ways to reduce the trash.

From previous discussions, students should be able to determine what quality of this activity keeps it from being considered a true research activity. Does the activity have value anyway? Can it be useful in designing future research? Process critique is performed upon completion of activity. How might it be turned into research?

ITEMS COLLECTED - (A Center for Marine Conservation Developed Form, 1989)

It is helpful to work with a buddy as you dean the beach; one of you picking up trash and the other, taking notes. An easy way to keep track of the items you find is by making tic marks. The box is for total items. For an example, see the top line in the beach sweep data sheet below.

Beach Cleanup Data Card
Thank you for completing this data card. Please answer the questions and return to your area coordinator. The information will be used in the *Beyond the Classroom Aquarium* database and reported to help develop solutions to stopping marine debris.

Name: ______________________ Affiliation: _________________

School Address: ______________ City: ______________________ State: _____ Zip: ______

Today's Date: _______________ Name of area coordinator: ______________________________

Location of Beach Cleaned: ___________________ Nearby City: _________________________

Number of people working together on this data card: ______ Estimated Area of Beach: ____

Number of Bags Filled: ______________________

Safety Tips

1. Do not go near any large drums
2. Be careful with sharp objects
3. Wear gloves
4. Stay out of dune areas
5. Watch out for wildlife
6. Do not lift anything too heavy

SOURCES OF DEBRIS: Please list all items with foreign labels (such as plastic bleach bottles from Mexico) or other markings that indicate the item's origin (such as cruise line names, military identification, or debris with names and/or address, of shipping/freighting or fishing companies, or oil/gas exploration activities)

SOURCE	ITEM FOUND
Example ABC Shipping Company	Plastic strapping band

STRANDED OR ENTANGLED ANIMALS (Please describe type of animal and type of entangling debris. Be as specific as you can.)

____

____

____

What was the most peculiar item you collected?

____

Comments:

____

Sample Data A	Total	Sample Data B	Total
Egg Cartons ~~1111~~ 111	8	Cups ~~1 1 1 1~~ ~~1 1 1 1~~ 1 1 1	13
PLASTIC			
Bags		Fishing nets	
Food bags		Hard Hats	
Salt		Light Sticks	
Trash		Pieces	
Other bags		Pipe Thread Protector	
Bottles		Rope	
Beverage/Soda		Sheeting	
Bleach/Cleaners		Longer than 2 ft.	
Milk/Water Gallon Jugs		2 ft. or less	
Oil/Lube		6-Pack Holders	
Other Bottles		Strapping Bands	
Buckets		Straws	
Caps/Lids		Syringes	
Cigarette Buts		Tampon Application	
Cigarette Lighters		Toys	
Cups/Utensils		Vegetable Sacks	
Diapers		Write Protection Rings	
Fishing Line		Other Plastic (specify)	
Fishing Lures/ Floats			
STYROFOAM			
Buoys		Packaging Material	
Cups		Pieces	
Egg Cartons		Plates	
Fast Food Containers		Other Styrofoam (specify)	
Meat Trays			
GLASS			
Bottles/Jars		Fluorescent Light Bulbs	
Beverage Bottles		Light Bulb	
Food Jars		Pieces	
Other Bottles/Jars		Other Glass (specify	
RUBBER			
Balloons		Tires	
Condoms		Other Rubber (specify)	
Gloves			
METAL			
Bottle Caps		55-Gallon Drum	
Cans		Rusty	
Aerosol		New	
Beverage		Pieces	
Food		Pull Tabs	
Other		Wire	
Crab/Lobster Trap		Other Metal (specify)	
PAPER			
Bags		Newspapers/Magazines	
Cardboard		Pieces	
Cartons		Plates	
Cups		Other Paper (specify)	
WOOD			
Crab/Lobster Traps		Pallets	
Crates		Other Wood (specify)	
Lumber Pieces			
CLOTH			
Clothing			

One of the foundational components of the scientific method is the idea of reproducibility (Popper 1959). In order for an experiment to be considered valid, it must be replicated. This process begins with the scientists who originally performed the experiment publishing the details of the experiment. This description of the experiment is then read by another group of scientists who carry out the experiment and ascertain whether the results of the new experiment are similar to the original experiment. If the results are similar enough then the experiment has been replicated. This process validates the fact that the experiment was not dependent on local conditions and that the written description of the experiment satisfactorily records the knowledge gained through the experiment. A successful replication of a physical experiment advances scientific knowledge because it demonstrates that the model outputs can be repeatedly generated and thus the original results were not an exceptional case. In addition, after a replication, the knowledge and data embodied by that experiment can be utilized by the model replicator as a tool to advance their own research agenda.

Based on a combination of instructor preference, available classroom space, and student skill and interest, students may perform research individually or in teams. With intense effort some students may complete the process of gaining preliminary approval of their research topic (following a conference with their teacher) within one week's time. At the preliminary research project approval conference, students present an introductory write-up of their research's purpose, describing their data and the process of collecting it and the hypotheses of their research. Other students may require another week before they successfully complete their preliminary research project approval conference. From this point (generally the end of the research project's second week) forward, students carry out their classroom research projects independently. While the instructor will certainly continue to provide guidance and support to student research efforts, this support must be scheduled around ongoing *Beyond the Classroom Aquarium* activities requiring their participation in project nine's concluding conference-oriented activity and the five remaining *Beyond the Classroom Aquarium* projects. More information of the conference-oriented activity that also occurs at the beginning of week three will be presented soon.

After students have identified their research topic and received preliminary approval, students are assigned a library assignment to identify and review a set of articles related to their research topic. This library assignment is designed to guide students to develop a fuller appreciation of ways data may be collected and analyzed to formulate a decision about the value of a hypothesis. Another two weeks may be allowed for reviewing literature related to the research project and submitting a concise summary. Using the now familiar research checklist form (Leedy, 1980) students identify key elements of each research project related article they review and include this information in a report on the current state of topic-related knowledge and describe how information uncovered in this review influences them to revise/improve their research design.

In addition, the library assignment write-up must include important details that describe how students plan to analyze the data they collect. A blank data form that must be filled out and included with this report is provided below for this purpose. The library assignment will also include a signed non-human vertebrate ethics form (page 129) and a student-developed (and printed out) inventory spreadsheet of space, materials, and equipment needed to collect data and perform other research tasks. The inventory of space is an important factor for the research team and teacher to consider before a research project is granted final approval because, it is expected that a number of research projects will be carried out within different components of the same classroom aquatic ecosystems, and it is important that different research studies being carried out in any one system do not interfere with one another in any way. Each write-up must also include a proposed timeline that specifies the dates during which students may be involved in data collection. By signing the timeline form, students are admitting their awareness that by the end of the research project's tenth week, all data collection must be concluded. After successfully presenting this report to the instructor, the project is given final approval status.

Once a student's research project has received final approval from the instructor (generally by the end of the research project's fourth week if not sooner), students are encouraged to work independently on setting up the experiment and conducting the procedures to complete it. Following the tenth week, students will then have a final two weeks in which to analyze their data, perform repeated critiques of their work, complete a final draft of their written report, and prepare a presentation. During non-scheduled class times, students who ask for assistance will be guided in reviewing, organizing, and drawing conclusions from their data. In order to complete their work on time, students should be encouraged from the earliest days of the project to complete much of their written report in advance. Finally, students develop and present a report documenting their research process, hypotheses, data, and the conclusions that may be drawn from the data and submit an electronic PowerPoint presentation to the conference presentation committee chairperson.

Many educators successfully use a brainstorming activity to initiate the generation of researchable topics. In this process each student puts forth an idea and passes on to the next student who puts forth another idea and so on. All topics/subjects are recorded without judgment. By this point in the year, students should be able to put forth many more ideas than are contained on the following list. Yet if the group has difficulty getting started, as novice researchers, it may be well worth the effort for students to use information from page 56 in combination with suggestions of topics below to create a replicability spreadsheet to help them identify a research topic that might best fit their interests and abilities. Many students choose to replicate microbiological research that has been reported upon in other areas of *Beyond the Classroom Aquarium*. For example, many potential research topics (related to aspects of planktonic culture) may involve replication of assertions presented in the Plankton Culture Manual (Hoff and Snell, 1987). Students working with pairs of reproducing organisms often prefer to develop a research project to help them better understand some aspect of their lab work with these organisms. Which of these topics might you be interested in studying?

- Microalgae's role in balancing pH by removal of excess CO_2 and addition of oxygen
- Macroalgae ability to reduce nitrates in a system from 30 to 40 ppm to 0 to 5 ppm (in three months)
- Therapeutic and antibiotic effects of microalgae on aquatic larvae
- Using microalgae to pack rotifers with essential fatty acids, vitamins, and other nutrients
- Show how trace elements nitrogen or phosphorus are limiting microalgae growth factors
- Changes in temperature and light intensity (and duration) are related to microalgae growth rates
- Assess impact of container size and color on larval growth
- Assess optimal temperatures and light intensities, on larval, planktonic, or algal growth
- Assess impacts of varied ionic water composition upon . . . *(student identified variable)*
- Assess impacts of aeration and pH on . . . *(student identified variable)*
- Variation in germination requirements of seeds (temp, light, pH . . . or other variable)
- Demonstrate the development and operation of PLC aquatic ecosystem applications/ancillary equipment beyond the automated copepod plankton culture an collection apparatus
- Crustaceans In the Classroom (Peppermint Shrimp, Lobsters, Copepods)
- Molluscs In the Classroom (Conch)
- Echinoderms in the Classroom
- Hanging Gardens of Babylon in the Classroom
- Mangroves in the Classroom
- Fish in the Classroom (Salmon, Trout, Tilapia, Perch, Ornamentals or other specific species)
- Energy Project: Wind, Solar, Diesel Fuel from Cooking Oil
- Vermiculure in the Schoolyard Garden

- Hoop House, Pond, Local Waterway Environmental Data
- Demonstrate (according to Soderberg, 1995) that:
    - At constant temperature, fish length increases linearly until sexual maturity is approached.
    - When growth rate at one temp is known, one may predict the proportional growth rate at another.
    - Feeding rates can be rationally calculated based on estimated food conversion, metabolic characteristic, and the anticipated growth rate.
    - The maximum possible weight of fish that can be supported in a rearing unit is determined by the depletion of oxygen and the accumulation of metabolic wastes.
    - Oxygen consumption and metabolite production progress in proportion to the amount of food fed.
- Students may hypothesize that incorporating both broodstock and larval rearing components in the same system may reduce stress on larvae by providing consistent water conditions throughout the period of egg formation and incubation.
- Preview project activities in chapter VI for potential research project themes.
- Review end of chapter assessment questions and suggestions for potential research project themes.

Guide to the Management of Data (Leedy, 1980 pp. 255 - 257)

Students may fill out the following guide related to the consideration and management of data in their research project in order to help them present this information to their teacher during the research project approval conferences. It is derived from the work of Leedy.

The Characteristics of the Data

1. Are the data discrete or continuous?

2. What are the characteristics of the data? Are they
   - □ Nominal
   - □ Ordinal
   - □ Interval
   - □ Ratio

3. What do you want to do with the data?
   - □ Find a measure of central tendency? If so, which one?
   - □ Find a measure of dispersion? If so, which one?
   - □ Find a coefficient of correlation? If so, which one?
   - □ Estimate parameters? If so, which ones?
   - □ Test the null hypothesis? if so, at what confidence level?
   - □ Test of significance? If so, which?
   - □ other—specify?

4. With respect to correlational techniques, what are the characteristics of the variables?
   - □ Independent variable: (identify) 'x'
   - □ Nature of the data
   - □ Dependent variable (identify) 'y'
   - □ Nature of the data
   - □ Correlational technique for dealing with these types of data

5. State your rationale for planning to process the data as you have indicated in sections three and four.

Interpretation of the Data

6. After you have treated the data statistically to analyze it characteristics, what will you then have? (Explain clearly and succinctly)

7. Of what will your interpretation of the data from a research standpoint consist? (What has the statistical analysis done in terms of solving any part of your research problem? (Explain precisely.)

8. What yet remains to be done before your problem (or any one of its sub problems) can be resolved? In other words, what further needs to be done so that the data will result in a resolution of the problem or any of its integral parts?

9. What is your plan for further carrying out this interpretation of the data?

A Research Report Format Form

To help students meet standards the written report should contain, in order:

A. Abstract: three paragraphs one each on purpose, procedure, and conclusion; this serves as a cover page.

B. Non-Human Vertebrate Animal Endorsement required when vertebrates are used a subjects of a research project.

C. Title Page: states the research topic in the form of a question.

D. Table of Contents.

E. Acknowledgements: recognizes individuals who have been helpful with this project.

F. Purpose: includes reason for choosing topic and the hypotheses.

G. Review of Literature: background information on research topic, including work done in the past on the topic. All resources quoted directly or cited indirectly must be cited in the text and identified in the concluding Literature Cited section of the paper.

H. Materials and Methods of Procedure: describes a clear, detailed chronological account of how the project was done and what materials and equipment were used.

I. Results: tables, charts, and/or graphs of quantitative data collected as well as observational logs. All measurements must be in standard units.

J. Conclusion: includes the student's evaluation and interpretation of the results obtained in the project, particularly whether the results support the student's hypotheses and possible applications of the knowledge gained from the study.

At the beginning of the third week of the research project, before most students can get very far on their research project's library assignment, students are given another long-term assignment. This assignment is a necessary first step toward an end of year summative assessment activity. The assignment is valuable because it provides a framework from within which students will be able to provide intensive and highly interactive demonstrations of ways they have gained and applied knowledge during the school year.

During the assignment's introduction, students are provided with the first steps they will need to take to design and operate a *Beyond the Classroom Aquarium* Aquatic Ecosystem Conference. In addition to organizing and running the conference, students will present results of their research at this conference. In fact, featured speakers and presenters at this conference are to consist primarily of the students themselves. And yes, parents and the public will be invited. A *Beyond the Classroom Aquarium* Conference Schedule (provided below) is distributed to the students. A brief discussion of what this might look like and a guiding organizational structure (based on Fortner, 1986) is described. From their own earlier conference experience, most students realize that holding a successful conference requires a lot of cooperation and unrelenting work by dozens of people. In this case those people are them (the students).

The purpose of this activity is to inform students of the teamwork and responsibilities involved in hosting a conference. An overview of the five conference committees is explained and methods to recruit students to one of these committees are considered. The class instructor may wish to engage the students in a brainstorming exercise (taking turns) sharing what they believe would need to be done and organized in advance for a student-run conference like the one indicated on the potential conference schedule. Reviewing notes from the conference they attended may be helpful. One purpose of this brainstorming activity is to help students consider all the specific tasks that must be performed. As the brainstorming session ends, students are asked to categorize the enormous number of tasks into categories of similar jobs that may all be assigned to the same committee. Perhaps this will be completed during class on the following day as they report their ideas. Once all ideas have been posted, the Conference Committee Form below is distributed to students and they are to determine which tasks should be included or deleted from the list.

At this point in project nine, as students continue working on their research project library assignment, they may create a job description sheet for each of the five (more or less) committee chairperson jobs. The teacher may then provide a copy of these five jobs to the class and post a sign-up sheet for students who wish to work on a given committee. Students wanting to work in some critical areas may even need to develop a resume and be interviewed in order to be selected for a given position. The instructor and other staff may wish to select students from each job list to interview for selection of conference committee chairperson jobs by the end of the week. To begin the process of creating the best conference, it is important to fill these jobs as quickly as possible with student workers who may be most well suited to a particular job/role. Every student will be asked to serve on at least one committee.

Once committee chairperson jobs have been filled (this chairperson group - executive committee) creates a conference timeline and is asked to hold weekly meetings and work with their committee members to:

- Create a list of tasks for which their committee will be responsible (a checklist on Word/Excel)
- Orient their committee members to the importance of the committee's work
- Establish a timeline to guide and synchronize their work with that of other committees
- Create a schedule for meetings to update one another on progress and what needs more work
- Estimate the number of workers that will be necessary for the committee to function successfully
    - Estimate budget for committee to perform
- Select committee members (perhaps again by job descriptions and resumes)
- Prepare to report their progress to the executive committee during the posted meeting time (within two days of filling the estimated number of committee jobs).

Beyond the Classroom Aquarium – Year End Conference (for class discussion)

7:30 – 8:00 Refreshments

8:00 – 8:30 Informal Viewing of Classroom Aquatic Ecosystems and Other Bells & Whistles (Exhibit Hall)

8:35 – 8:45 Student Board and Committee Chair/Moderator for Whole Group Sessions
Welcome and Thanks – Parent Recognition (Meeting Hall)

8:45 – 9:15 Introduction of *Beyond the Classroom Aquarium* Supporters: (Meeting Hall)
Board of Ed./ Superintendent Office Representative;
School Principal;
Business and Trade Reps (Field Trip Supporters, Technicians & Staff)
Beyond the Classroom Aquarium Educator

9:15 – 9:30 Break

9:30 – 9:55 *Beyond the Classroom Aquarium* - Research Presentation Sessions

Aquaponics Panel	Room 1
Coral Propagation Panel	Room 2
Broodstock and Larval Rearing Panel	Room 3
Algae/Phytoplankton Culture Panel	Room 4
Zooplankton Panel	Room 5
Technology Panel	Room 6

10:00 - 10:25 Tradeshow Booths and Field Trip Sponsor Booths (Exhibit Hall)

10:30 – 10:55 *Beyond the Classroom Aquarium* Presentations (Continued Rooms 1 – 6)
Juvenile Growout Panel
Aquatic Ecosystem Opportunities Panel (Camps, Internships, Employment, Business Plans)
Feedback Cycles – Gerald Marten – Apo Island – Tipping Points and Related Community Service Project Panel
Advanced & Future Technology Apps Panel (Examples of Working Models or Models in progress)

11:00 – 11:25 Poster Sessions STEM Applications in Aquatic Ecosystems (Exhibit Hall)

11:30 – 12:15 Lunch (Exhibit Hall)

12:15 – 12:30 Break

12:35 – 12:45 Welcome Back – Review of Morning – Schedule for Afternoon (Meeting Hall)

12:45 – 1:15 Honorary Guest Speaker (Meeting Hall)

1:15 – 1-30 Break

1:30 – 1:55 *Beyond the Classroom Aquarium* Presentations (Continued; Rooms 1 – 6)

2:00 – 2:25 *Beyond the Classroom Aquarium* Presentations (Continued; Rooms 1 – 6)

2:30 – 3:15 Student Portfolios (Exhibit Hall)

3:15 – 3:30 Break

3:35 – 4:00 Closing Guest Speaker (Meeting Hall)

4:00 – 5:00 Student Conference Committee Conclusion – Recognitions – Summarization – (Meeting Hall)

4:25 – 5:30 Light Reception (Exhibit Hall)

Proposed Conference Committees

A. Conference Program Committee (Parents, Teachers, Staff, Community, Other Volunteers, and Beyond)
B. Conference site liaison committee
C. Conference schedule coordinator
D. Special events coordinator
E. Public Information coordinator

A. Conference Program Committee (Parents, Teachers, Staff, Community, Other Volunteers, and Beyond)
    a. Reviewer of STEM content of Presentations to facilitate opportunities to demonstrate Science Technology, Engineering, and Mathematical aspects of project design and assembly
    b. Guest Invitation committee
        i. Create final attendance roster for registration table
        ii. Set up Excel spreadsheet or other program to collect guest names
        iii. Create final attendance list with updated contact information for file
        iv. Assemble name tags
        v. Assemble registration listing of participants.
        vi. Send thank you notes to all who helped with the conference
    c. Prizes, Awards, Scholarships, Internships, Camps, Businesses, and Special Recognition Committee
        i. Assign members to identify and solicit sponsors, in kind and auction donations, and determine sponsorship levels and benefits
        ii. Determine the types of giveaways, if any
    d. Journal editor and publisher
    e. Be prepared to host any news media personnel who may attend
    f. Develop event evaluation survey and perform evaluation analysis
B. Conference site liaison committee
    a. Select venue
    b. Review, sign, and return contract
    c. Estimate attendance
    d. Meet with catering, facilities, media services, and conference services to go over the final details
    e. Conference presentation room and electronic equipment reservationist & schedule updater
    f. Contact speakers and panels to determine AV and other equipment needs; place order (example: podium, microphones, Internet access, and so on)
    g. Finalize A/V equipment needs with media services, including PA systems and microphones
    h. Exhibit Hall, Meeting Hall, Presentation Rooms coordinator
    i. Arrange for delivery of meeting materials and supplies to be delivered to event site
    j. Assign set up assignments to committee members
    k. Arrange set up and spaces for:
        i. Registration area table
        ii. Seating style (theater, classroom, rounds, etc.)
        iii. *Beyond the Classroom Aquarium* information material
        iv. Accessibility for disabled persons
        v. Schedule all table and chairs needed
        vi. equipment carts, loading/unloading assistants
    l. aquatic ecosystem assembly/disassembly
C. Conference schedule coordinator
    a. Time date location
    b. Event schedule timing, room locations, etc.
    c. Emcee and podium transitions

d. Speaker/room hosts
e. Speaker/s confirmation letter
f. Poster session coordinator
g. Trade show booth coordinator
h. Liaison with local businesses, volunteer tradespersons, and others

D. Special events coordinator
   a. Snack and meal coordinator
      i. Menu selection
      ii. Special dietary needs
      iii. Confirm catering, facilities
   b. Entertainment coordinator
   c. Confirm volunteer assignments for program details
   d. Assign hosts for speaker/s and VIP guests
      i. Pick up signage or create own
      ii. Provide notebook or folder for Speaker, Emcee, Hosts

E. Public Information coordinator
   a. Contact invited and scheduled speakers and request vita, address and phone number, biography, and head shot photo for publicity
   b. Arrange photography and presentation videographers/recorders
   c. Online conference coordinator/webmaster
      i. Create website
      ii. Website Links (District, etc. local media - broadcast, newspapers, business journals, civic, chambers, etc.)
      iii. Online Presentation up-loader
      iv. Registration link for parents and the community (other schools?)
   d. Invitations and promotional brochure
      i. Compile invitation lists, target audience
      ii. Identify VIP guests and send letter invitation.
      iii. Develop Save-the-Date notice (options e-mail, print)
      iv. Develop first draft/s of brochure/postcard/invitation. Distribute for approvals
   e. Draft program agenda
      i. Finalize brochure/postcard/invitation material and set mail date, details
      ii. Arrange for printing of invitations
      iii. Finalize event sponsors; include with promotional materials
      iv. Produce printed program and get schedule to PR
   f. Produce conference poster (a contest?), if appropriate
   g. Confirm event name/spelling etc. for venue marquee
   h. Order signage (registration, directional, etc.)
   i. publicity/advertising schedule - advertising alternatives
   j. Send Event Report to District Pubic Info Officer
   k. Update all event and registration information at all calendar of events sites

Write a list of **science-based** activities you have participated in during this project and briefly describe evidence that verifies your interaction with and knowledge of scientific concepts for each item on your project nine list.

Write a list of **technologically-oriented** activities you have participated in during this project and briefly describe evidence that verifies your interaction with and knowledge of technologically-oriented concepts for each item on your project nine list.

Write a list of **engineering-oriented** activities you have participated in during this project and briefly describe evidence that verifies your interaction with and knowledge of engineering-oriented concepts for each item on your project nine list.

Write a list of **mathematical** activities you have participated in during this project and briefly describe evidence that verifies your interaction with and knowledge of mathematical concepts for each item on your project nine list.

Write a brief list of **Non-STEM-related** activities you have participated in during this project and briefly describe evidence that verifies your interaction with and knowledge of Non-STEM related concepts for each item on your project nine list.

Rate your level of involvement in STEM-related activities in this project (1=lowest, 10=highest).

1—2—3—4—5—6—7—8—9—10

Beyond the Classroom Aquarium **Daily Journal Guidelines**

Entries students place in their daily journals should be:

1. Dated accurately and logged chronologically.
2. Kept in a notebook and written neatly in ink or typed.
3. Written or typed in complete sentences using correct capitalization, spelling, and punctuation.

In their journals, students should keep a daily record of:

1. Who they worked with each day by including the names of school staff, fellow students, community members, and volunteers.
2. What they did each day, such as lecture topic, activities or project design, assembly, or maintenance.
3. What they learned each day by asking questions, following instructions, and/or observing activities.
4. How *Beyond the Classroom Aquarium* experiences relate to past, present, or future life experiences.

For bonus points, students may complete these journal options:

* Include illustrations, diagrams or creative expression relating to the journal entry.

Journal entries will be evaluated on a point system, with four points being the maximum number a student can earn. A half-point will be awarded for each of the first eight journal requirements above. Bonus points will be awarded when journal options are completed. Journal points will be totaled and divided by the number of journal dates. A possible journal grade based on points is included in the project one assessment section.

For project nine, journal writing topics may include answers to the following questions or descriptions of a project activity or assignment. Reflection on these topics may also help students prepare for exams, quizzes, and other assessments.

- Create posters and other graphical displays, and include these in a presentation of your research to inform the audience about aquatic ecosystem designs that were necessary for your research.
- In what ways do you think these systems involve science, technology, engineering, and math?
- What would you like to know about the organisms and systems that you observed?
- What do you want to understand better in order to contribute more to your team's productivity as it independently designs, assembles, and then manages an aquatic ecosystem?
- Create a new drawing and story about an organism or system that is most interesting to you.
- Create a journal entry that tells what information you learned during project nine. What you would like to know more about? Do you know someone who might help you if you asked them?
- Select a book or magazine to read that is related to a topic in project nine. Write a report on this book and include your report in your *Beyond the Classroom Aquarium* portfolio.
- Design, propose for critique and execute a research project that you may perform to replicate an "action spectrum" curve for oxygen production/photosynthesis of the freshwater seaweed elodea (Anacharis) using data collected from elodea illuminated with different intensities and wavelengths of light.
- Explore effectiveness of filtration resulting from the use of various macroalgae species, such as Chaetomorpha linum, in an algae filter component of a compound filter.

Develop a field trip report that illustrates which field trip steps performed which function in the design and execution of a research project. Prepare this report for class critique.

Some aquatic organisms are used as bio-indicators, their presence indicating quality of the water that contains them. In this project, student teams may choose to do independent research related to bio-indicator organisms in aquatic ecosystems. Such research incorporates a great deal of data collection and analysis.

Photo by Robert Jakus

Use an excel spreadsheet to chart and update algae rotifer system growth rates determined from microscopic exploration of cultures.

In virtually every subject area, our knowledge is incomplete and problems are waiting to be solved. We can address the holes in our knowledge and those unresolved problems by asking relevant questions and then seeking answers through systematic research. Select a career practiced by someone you know or in which you might one day be working and review literature of a topic in that area and describe how the research in that field has influenced and improved the outcome of the product that persons who work in that field are capable of producing.

Students are challenged to consider the many mysteries yet to be discovered by studying species that have yet to be studied, innovative system designs that may be utilized, and many important ecological applications.
Do a self-assessment of your preparedness to present a data summary using descriptive statistics using Excel. This knowledge is an essential component of your research project assessment and presentation.
Learn to use Excel to create histograms and other graphs so that data samples can be compared.

<table>
<tr><td colspan="5">Beyond the Classroom Aquarium STEM Self or Peer Assessment Rubric
This self-assessment is valid only to the extent to which you can document and display evidence that supports it. For each category in the left hand column, write a STEM-Oriented example in the box beneath 3 if you can display evidence that your STEM-Oriented example represents work that exceeds expectations. If your evidence shows that your STEM-Oriented statement meets but does not exceed expectations, write the statement in the box beneath 2. Similarly, for STEM-Oriented examples that demonstrate some progress toward expectations.</td></tr>
<tr><td colspan="2">Name:
Project Number:
Date Project Work Began:
Date of Project Completion:

List Math Topics and/or Objectives Currently Under Study:</td><td>3
Work Exceeds Expectations</td><td>2
Work Meets Expectations</td><td>1
Work Is Progressing Toward Expectations</td></tr>
<tr><td colspan="2">Write examples of questions and defining problems related to your STEM work.</td><td></td><td></td><td></td></tr>
<tr><td colspan="2">Describe how you have developed and used models in your STEM work.</td><td></td><td></td><td></td></tr>
<tr><td colspan="2">Write examples of STEM investigations you planned and carried out.</td><td></td><td></td><td></td></tr>
<tr><td colspan="2">Describe STEM related data that you analyzed or interpreted.</td><td></td><td></td><td></td></tr>
<tr><td rowspan="5">Give examples of how you used mathematics and computational thinking related to the five standards listed to the right.</td><td>Problem Solving (Math Process Standard One)</td><td></td><td></td><td></td></tr>
<tr><td>Reasoning & Proof (Math Process Standard Two)</td><td></td><td></td><td></td></tr>
<tr><td>Communication (Math Process Standard Three)</td><td></td><td></td><td></td></tr>
<tr><td>Connections (Math Process Standard Four)</td><td></td><td></td><td></td></tr>
<tr><td>Representation (Math Process Standard Five)</td><td></td><td></td><td></td></tr>
<tr><td colspan="2">Describe STEM-Oriented explanations and solutions you designed</td><td></td><td></td><td></td></tr>
<tr><td colspan="2">Explain evidence you used while engaging in a STEM project-based argument.</td><td></td><td></td><td></td></tr>
<tr><td colspan="2">Give examples of STEM information you obtained, evaluated, and/or communicated</td><td></td><td></td><td></td></tr>
<tr><td colspan="2">TOTAL POINTS</td><td></td><td></td><td></td></tr>
</table>

CHAPTER VI: Application of Aquatic Ecosys-STEM Knowledge and Skills to Local, Regional, and Global Issues (Projects 10 – 13)

The remaining projects in *Beyond the Classroom Aquarium* are designed so students have opportunities to apply their aquatic ecosystem knowledge and skills beyond the classroom. They are designed to help students better understand and explain universal concerns about natural aquatic ecosystems. In chapter VI, students are guided to assess the ecological sustainability of watersheds in their own community and to develop opportunities to expand and share their knowledge beyond the classroom in space and time!

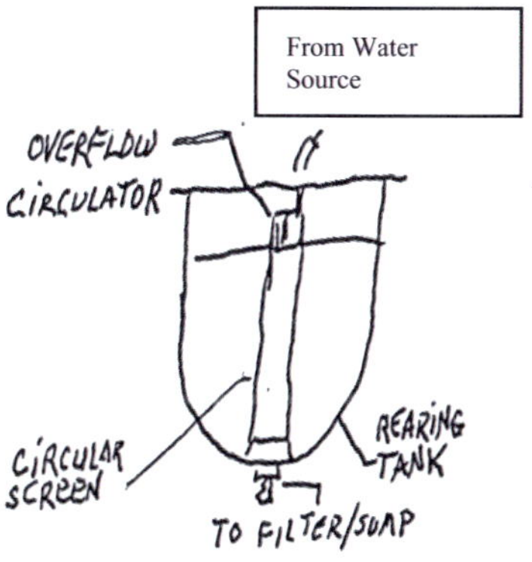

Having successfully established classroom-based aquaponic, coral propagation, and broodstock larval rearing systems in chapters II and III, it is reasonable to expect that adult organisms in a number of these aquatic ecosystems have acclimated to their new homes and are engaging in natural courtship and reproduction-related behaviors. According to the curricular calendar, it is usually possible to provide newly hatched aquatic larvae with at least three months to metamorphosize to a juvenile stage of development between the end of project nine and the end of the school year.

In project ten, as students continue to work diligently upon previously assigned research projects and conference committee-related duties, they are guided through species specific studies of the metamorphosis of larval aquatic organisms into juvenile and adult stages. One of the primary goals of project ten is guiding students to use their knowledge of aquatic ecosystem design to develop systems capable of housing aquatic organisms beyond the larval-stage in their life cycle and into adulthood. It is important for both teachers and students to prepare for the eventuality that a larval rearing project will be wildly successful. In such cases, it will be important to share this information in a timely way with others (locally and beyond) in order to most efficiently pass along a surplus abundance of organisms to others who are prepared to adopt and care for them, potentially raising funds for future *Beyond the Classroom Aquarium* projects.

Depending upon student interest, ability, and the nature of the development of classroom aquatic ecosystems within the time frame of the school year, educators may choose among three additional project ten aquatic ecosystem activities through which to guide groups of students or the entire class. The three additional activities described in project ten include: **1.**) an extended sea urchin embryology lab in which students attempt to raise sea urchins to adulthood {based on the work of Hanes et. al., (2002), Tamaru et. al. from the Hawai'i Institute of Marine Biology (2009), The Sea Urchin Embryology SUE - CORE LAB [at Stanford], and as presented online at: classic.sidwell.edu/us/science/vlb5/Labs/Urchin_Lab/, as well as on the current work of researchers such as Martin A. Moe Jr. (*Diadema, antillarum*), David Cohen (*Tripneustes, gratilla*), and others}; **2.**) development of an upweller nursery assembly for shellfish—students who assemble an upweller system are then guided to stock and maintain the system with fertilized oyster eggs/larvae (based on writings of Hanes et. al., [2002], and Dale Leavitt's work (2009) on FLUPSY – a floating dock/upweller system used for the cultivation of shellfish; and **3.**) a spherical cylindrical kreisel system of the sort used to raise lobster and other crustaceans (based on the work of Hughes et. al., 1974).

Photo by Robert Jakus

Many of the natural aquatic ecosystems explored in projects eleven and beyond are found thousands of miles from most classrooms. Videos and Internet resources are important components of project eleven and may be considered virtual field trip excursions. Project eleven activities are designed to lead student teams through a process that results in the production of an ecologically sustainable service-learning program proposal.

The introductory activity in project eleven brings students to the Chesapeake Bay. Students are guided to identify and analyze a diversity of public policy platforms advocated by various Chesapeake

Bay interest groups. Based on an originally published (1980) role-playing curriculum, this briefer version of the activity allows time to expose students to critical tools beyond those introduced in the Chesapeake Bay activity and essential to the completion of the service-learning program proposal.

The abbreviated Chesapeake Bay activity in *Beyond the Classroom Aquarium* represents a recent collaborative effort between the National Aeronautics and Space Administration (NASA) and the Geography Education National Implementation Project (GENIP) and its constituent organizations on a project called Mission Geography. Support materials that focus on the development of key grade-level appropriate geography skills including remote sensing and map/image interpretation are contained within the Mission Geography publication. The activity *"What's up with Earth's Water Resources?"* uses existing NASA data and images where possible to engage students in active, "hands-on" inquiry, modeling the scientific method, and developing students' understandings of environment-society relations and earth science. The text of this activity may be accessed from the Mission Geography CD-ROM containing all relevant PDF files and live links to the Mission Geography Web site. Teachers can use the CD-ROM as a stand-alone product or capitalize on the Internet connections. Mission Geography is available from the Central Operation of Resources for Educators and the Chesapeake Bay module may be accessed at: http://er.jsc.nasa.gov/seh/Mission_Geography/9-12/Module_1/III-1-1.pdf.

Following the Chesapeake Bay activity, students are next guided through a structured series of human ecology techniques. These techniques come directly from the work of Gerald Marten (*Human Ecology - Basic Concepts for Sustainable Development*, 2001). Materials he developed related to Apo Island (in the Philippines) provide students with practical information about ways to approach issues so as to positively influence environmental policy and change. According to Strong (2001), Marten's examples are "drawn from actual situations that cover a wide range of topics and take us to diverse parts of the world. They allow us to see how to make the jump from theory to practice." After reviewing examples of successfully implemented ecologically sustainable development programs adopted by governments, the private sector, and civil society to develop healthier relationships with the environment, the message that well-informed change is possible—that students can make a difference—is powerfully reinforced. As a result of information gained in these introductory activities, students become better prepared to design a service-learning program proposal that is most likely to generate positive outcomes.

A list of themes is provided for review in project eleven to suggest how students might design the class service-learning proposal as an extension of activities related to classroom aquatic ecosystems or from other applications and resources. Many students and teachers may find resources such as *No Student Left Indoors* (2007) and *The Good Food Revolution* (2012) helpful for providing additional program proposal ideas. Using these resources, student teams are finally guided to present to the class for critique and subsequently develop their own ecologically sustainable service-learning program proposal.

In project twelve, student teams deploy their ecologically sustainable service-learning programs. This is done in a structured way by following the daily implementation schedule included in the proposals they were guided to develop and critique in project eleven. In the concluding activity in project twelve, students are guided to explore practical connections between their developing aquatic ecosystem skills and vocational aptitudes that may have been highlighted through the experiences of designing and implementing a service-learning program. Students are guided to identify opportunities they feel most compatible with and encouraged to then develop resumes and/or submit applications to such organizations (e.g. aquatic ecosystem themed camps, ecological internships, and summer jobs).

To end the school year so as to further promote student attraction to future aquatic ecosystem-related studies/careers, activities in project thirteen emphasize design, not construction. Following the introductory activity, in which students are guided to design and assemble a laminar flow fountain, general guidelines for six advanced (future oriented) design activities are provided. Students may explore one of the six design issues described in the text or design an invention of their own. In each case, students are challenged to apply the engineering design process to design a tool that will be useful for helping researchers study an issue more successfully. Design drawings and 3-D models for students to include in their design presentation at the end of year *Beyond the Classroom Aquarium* conference must also be created.

After aquatic larvae metamorphosize, they have entered what is commonly considered at a juvenile stage of development in their life cycle. Miniaturized larval hatchery conditions (e.g. minimal container volumes and low rates of water circulation and filtration) were necessary at first to protect the fragile larvae. In this safe environment, as they grew in their first few weeks of life, larvae became much hardier. After this brief period, initial hatchery conditions were no longer sufficient to process the increasing accumulation of toxins (from bacterial breakdown of uneaten food and waste products of larval metabolism). To prevent "die offs" resulting directly from exposure to these toxins, growing larvae/juveniles require a carefully planned transfer into aquatic systems that are capable of processing accumulating toxins much more aggressively. These new juvenile grow-out systems hold maturing juveniles to a point where they are resilient enough to be made publicly-commercially available.

Design and Implement Classroom Relevant Juvenile Grow-out Systems – Project Ten Opening Activity

The Species Specific Ecosystem Parameters – By Life Cycle Stage and System Form provided below is provided so that students can create a file of such forms related to each species in the classroom aquatic ecosystems. In order to complete this species specific library assignment, students must record ideal environmental parameters (or a range) for a given species at each particular stage in their lifecycle. This information can be used as a guide to follow in developing a juvenile grow-out system for aquatic larvae and for improving environmental conditions to optimize larval survival and the reproductive health of mature specimens. Students may also find this form useful in creating a database that may be searched when they have a reason for tracking and evaluating the continuity of care that has been provided for the organism in a specific system. Having a historical record of data that reflects environmental conditions from the time an organism is introduced to a system until the time it is transferred to a different system may help students in their attempts to analyze and improve maintenance protocols.

Photo by Robert Jakus

As students collect larvae and succeed in nursing them through metamorphosis, the process of providing these larvae with a suitable grow-out system should be second nature to them. Students may explore the possibility of keeping organisms in each grow-out system separated. They may find some species that may actually be kept together. The collection of system and organism data continues to be important. Some students will prefer to use a large glass aquarium for grow-out while others may prefer to create a custom grow-out system. It is possible that students will be inspired to create a grow-out system they were introduced to during a field trip.

Photo by Robert Jakus

The images below may provide students with additional grow-out system design ideas. Whichever system students decide upon for a given juvenile, they should prepare a report with drawings and illustrations that describe the system in detail.

Photo by Robert Jakus

After the grow-out system is approved, set up, and juveniles are carefully transferred to it (and data is being collected), students are ready to move ahead to the next activity in project ten. This next project involves the use of live sea urchins that have been housed in one or more of the teacher's model classroom systems.

Species Specific Ecosystem Parameters – By Life Cycle Stage and System			
Species Name System	Larval Conditioning System	Juvenile Rearing System	Adult Broodstock System
Organism Count			
Mass of Organisms			
Volume of water			
Shape			
Rate of circulation			
Substrate			
Salinity			
Light **Spectrum intensity** **Duration**			
Aeration rate			
Feeding schedule			
Temperature			
Other			

In this activity, students are guided to personally investigate sea urchin aquaculture work that is similar to that currently being conducted by marine biologists such as Russell, M. (2000), Leber et.al. (2009), and Martin A. Moe Jr. - who is deeply involved in establishing protocols to follow to raise the sea urchin *Diadema antillarum*. Students who have completed project two in *Beyond the Classroom Aquarium* are probably already familiar with Mr. Moe's ideas about light in aquatic ecosystems. As students will discover by reading reports the sea urchin work performed by Moe and others, many other species of sea urchin have been successfully raised from larvae to adulthood in captivity so there is an abundance of information on the reproduction and embryology of the sea urchin to use as a starting point for this activity. This sea urchin fertilization lab is intended to help students see beyond the brief (and certainly exciting) moments of cellular fertilization and early cellular division and differentiation to the wider role that sea urchins play in their native aquatic environments and how learning to culture these organisms to adulthood may be of wider service to mankind and the environment.

In Hawaii's Institute of Marine Biology's Sea Urchin Fertilization Lab Education Program developed by Tamaru et.al. (2009) the sea urchin classroom laboratory (p. 6) is introduced by prompting students to conduct their own preliminary investigation about this topic by reviewing the considerable amount of information available and accessible on the internet. They suggest this preliminary investigation include gathering information about morphologies of a sperm cell and an egg, the acrosomal reaction, the bindin protein molecule, the vitelline membrane, cortical reaction and cleavage to the 8-cell stage. In the *Beyond the Classroom Aquarium* version of this laboratory preliminary preparation to carry out the activity requires students to also acquire information necessary to provide larval urchins with adequate food and space to begin growing into adulthood. In other words students must design, critique and assemble one or more larval rearing system before the fertilization lab can commence.

By having participated in system design and assembly protocols in previous *Beyond the Classroom Aquarium* projects students performing this activity should be able to, with minimal guidance, design a viable sea urchin larval rearing system to propose for class critique. Specifically designed larval rearing compartments (designed by separate student groups) placed in dedicated compartments in classroom broodstock larval rearing system waters (holding larvae producing parents) perform well.

Once student designed larval rearing systems have been critiqued, assembled and prepared to receive larval urchins, it will be possible for an entire class (or set of classes) for the day, to share the gametes collected from two reproductively active sea urchins (perhaps a stock of ten should be kept in the class lab to ensure at least one male and one female sea urchin) residing in one or more of the instructor's classroom demonstration systems. In addition to a digital compound microscope and additional cordless digital compound microscopes or student developed compound microscopes (a few fitted with a digital camera) this activity requires the availability of the following materials.

Materials for each student group (according to Tamaru et.al., 2009)	
Seawater from systems in which urchins have been kept	
0.5M potassium chloride (KCl) solution (3.73g of KCl in 100mL of distilled water)	
Petri dishes	500 mL glass beakers
Depression slides	Rulers
Glass slides and cover slips	10 mL test tubes
Small syringes	Test tube racks
Plastic pipettes	Sea salt solutions
100 mL glass beakers	Miracle grow
250 mL glass beakers	Thermometer
Paper on which to draw observations of egg, sperm, the cell wall at the moment of fertilization, and additional embryonic stages - Using 10 minute time intervals, repeat observations of the egg/sperm mixture over an extended period of time. Describe your extended observations.	

Unless students are extremely advanced it is best if the instructor is the only individual participating in this lab who uses a small syringe (according to directions given) to be artificially induce male and female sea urchins to spawn. Students may observe this process or the instructor may perform the initial activity of gamete collection prior to the time students arrive in class. Upon the instructor's discretion, more advanced students may perform this collection process (as described by Tamaru et.al. 2009). Since this process involves use of a syringe extra caution and inventory of syringes is required.

By performing the following procedure students should be able to observe the production of a zygote. Tamaru et.al., (2009) state that "confirmation that this has successfully taken place is recognized when the fertilized egg begins to cleave… meaning that incorporation of genetic material from both the egg and sperm has successfully occurred."

The description of the gamete collection and spawning process provided in the Stanford SUE document suggests that instructors follow a detailed checklist of activities in order to prepare cultures of embryos 1½ and 3 hours before each class period to allow viewing of early cleavage stages.

Note that the needle goes over the top of the teeth to inject KCl into the body cavity and not into the mouth and not into the "test" (shell).

Directions from the SUE fertilization lab direct the teacher to "inject about 0.1-0.2 ml per inch of urchin width in each side of the urchin. A total of two injections are made. It is best to use as small a needle as possible as a larger needle leaves a larger wound, subjecting the urchin to infection. (#25 - #30)." They then remind the teacher to "gently shake the urchin for a few seconds to mix the KCl solution inside the sea urchin. (shaking too hard can kill the urchin by ripping apart the delicate internal membranes)."

Collection and Use of Gametes Place the animal on a clean petri dish or glass plate - oral surface up. According to Hanes et.al, 2002, "The color of the spawn that will appear on the animal's aboral surface will tell you whether the animal is male or female." If the spawn is milky white the animal is a male. Depending upon species the animal is female if the spawn is pale yellow to orange to dark maroon.	
if Male	**if Female**
Leave the male urchin in the petri dish.	Place the female urchin mouth side up onto the lip of a properly sized beaker **filled to the brim** with seawater. Water contact promotes the release of eggs.
For any sperm that will not be used immediately, collect and store it 'dry' on the petri dish, in a pipette or an eyedropper and keep it in a refrigerator – not mixed with seawater at all. Most sperm cells mixed with seawater will become inactive in less than ten minutes.	The eggs will be shed into the seawater and collected at the bottom of the beaker. This can take from 10 to 30 minutes to finish shedding the eggs. There should be a layer of eggs on the bottom of the beaker.
Place male urchins back in a seawater holding tank (for males only) only after collecting the sperm.	After the eggs have stopped flowing, return the female to a separate female holding tank.
For sperm that will be used immediately, add about 2mm to 100ml of sea water (adjust this volume up or down if necessary to get proper fertilization levels without polyspermy). This may be called the sperm diluent or Stock Sperm Suspension.	To store eggs for future use suspend the eggs in a large flask no greater than 1 cm in depth to insure adequate exchange of gases. For eggs to be used immediately the teacher or a student group leader should remove 0.2 mL of eggs from the bottom of the beaker and place them into individual 50 mL glass beakers filled with 10 mL of seawater. Each group of students should receive an egg filled 50 mL beaker.
Important - Once each student group has identified a fertilized urchin egg in their microscope they must immediately place the larval rearing container they designed in the broodstock larval rearing system and deposit all of the eggs remaining in their 50 mL beaker in their respective larval rearing container (with unicellular algae as food). Culture monitoring and maintenance activities may be performed from this day forward.	1. Have the students place a drop of egg suspension on a glass depression slide under their microscope (a less that 1% egg concentration will allow students to see individual eggs). 2. Students should focus on the eggs. 3. Next, while the student focuses on the eggs, add a small drop of stock sperm suspension to the drop of eggs and add a cover slip. Quickly, let the students focus on the eggs and watch the fertilization membranes rise.

Cell division is different for each urchin species and is also temperature dependent (at warmer temperatures under a microscope, division is faster). After fertilization and usually within 90 minutes students may have opportunities to observe three important internal cellular transitions and drawn them. These transitions include the centering of the nucleus, the streaking in the nucleus and cellular metaphase. If students have access to eggs that were fertilized earlier in the day or on previous days they should watch eggs to try to identify the various stages of growth through which eggs pass from first division that might occur in as few as 90 minutes and a second division that might occur in as many as three hours after fertilization to the eventual development a day later of a blastula, and 24 hour later a gastrula and further development into a larval pluteus five days after fertilization.

Time since mixing eggs and sperm (15 - 20 Minute Intervals)	Observations

Photo by Robert Jakus

By participating successfully in the preceding nine projects, students may by now be observing the variety of carefully selected organisms in them maturing and in some cases, a number of aquatic larval organisms may already be thriving within a range of relatively small student designed and maintained classroom larval rearing systems. This project ten activity can help students develop a fuller appreciation for the fact that it is unrealistic to expect one standard larval rearing system design to effectively raise any species of aquatic larvae given the variety of aquatic organisms and the diversity of their life cycles. In the aquaculture community, it is commonly understood that air lift upweller nurseries and variously configured kreisels are effective technologies for producing good larval growth while minimizing the investment of money and effort. In this activity, student teams are guided to consider the introductory larval rearing system information presented here and are also directed to some sources of accessible literature related to whichever of the three design and assembly challenges presented here they might wish to follow (or modify) – the new larval rearing system designs from among which they may choose include either:

Photo by Robert Jakus

1. A classroom airlift upweller shellfish nursery or,
2. One of two types of classroom kreisel:
   a. A spherical kreisel Hughes et.al, (1974) or
   b. A rectangular kreisel (a pseudo-kreisel)

For some species, it may be possible to arrange for the purchase of larvae in advance.

Note: Educators guiding students through development of auxiliary larval rearing systems should be fully aware of the necessity to comply with local regulations regarding species which with classroom larval rearing systems are stocked.

Photo by Robert Jakus

Introductory literature for students who may wish to design and assemble any of the larval rearing systems mentioned above is listed below - followed by a brief summary. As students review this literature (and beyond) they should gather larval rearing design and assembly information relevant to their choice and prepare a design presentation to be critiqued via the previously described class critique method. Student teams should use their presentation to convince others, to whom the presentation is given, that the design they wish to adopt is based on practical and achievable information. Information in these design presentations should include descriptions or explanations of: the species to be stocked (including life cycle information), materials, equipment, and tools needed to assemble the system (unique supplementary technologies to be included), the dimensions of the system, electrical - chemical - mechanical safety issues related system assembly, how the system serves the needs of the stocked species, placement and operation of the system (including recirculation of water and ongoing filtration and the importance of daily maintenance/cleaning of an operational system), protocols for maintaining the system over time, how larval health and growth rates will be determined, and how students will document their assembly process. Student developed illustrations (properly labeled) will also be included.

Once larval rearing system designs have been critiqued and approved for assembly students may gather necessary materials and assemble their chosen system. After assembly each larval rearing system must be set-up and tested for leaks or other problems. Once a system has been shown to perform reliably, larvae may be introduced to it (when they become available).

Students who want to review background literature related to the use, design and assembly of an airlift upweller system may be guided to consider sources of literature listed here. For purposes of classroom installations, airlift upwellers may be placed within bodies of water that are deep enough such as a sump or classroom wet table. Modern technologies that may be incorporated in airlift upweller designs are described by Leavitt (2009). In addition to presenting excellent design ideas that students may apply to the design and assembly of their own classroom airlift upwellers Hanes et.al, (2002) also provide excellent advice about larval rearing system cleanliness and protocols for adhering to a daily schedule for cleaning larval rearing systems.

Students who want to review background literature related to the use, design and assembly of a spherical kreisel system may be guided to consider the following resource written by Hughes et.al, (1974). As Hughes et.al, suggest this kreisel design is intended for use by those interested in designing a larval rearing system for lobsters (or other crustaceans). As described in their article this system contains about ten gallons and water and can be positioned above a sump tank. Water flows gently through the kreisel to a sump and is recycled back to the kreisel. I believe this larval rearing device may be the most complicated of the three choices given.

The concept of pseudo-kreisel is a recent development and I have found little mention of this technology in academic publications. I am including them here because these systems seem to be user friendly and fun. Students who want to review background literature related to the use, design and assembly of a pseudo-kreisel system may be guided to study and take notes from the following online sources of pseudo kreisel information. The selection of video links provided here will provide students with many ideas about pseudo kreisel design, the kinds of larvae that may be stocked in these systems and documentation of the ways these systems perform. By the time you read this many additional sources of information about pseudo kreisels may exist. Please add them to the list below.

- http://youtu.be/thfYNk_hfK8)
- http://youtu.be/iXx03u42jWU
- http://youtu.be/RhQvz_63AK8
- http://youtu.be/DIOBMrgy1Xs
- http://raisingpetjellyfish.blogspot.com/

Daily maintenance and larvae data should be collected. A daily morbidity count should be included in the data collected data in addition to standard water quality analysis data and other system performance parameters. By collecting this data and documenting their ongoing work with their larval rearing systems students will produce find plenty of interesting information to share with the public during the project 14 concluding activity.

The primary emphasis of the remaining *Beyond the Classroom Aquarium* projects (eleven through fourteen) involves students in extending their aquatic ecosystem knowledge and skills to aquatic ecosystem applications that can be performed "beyond" their classroom walls.

Write a list of **science-based** activities you have participated in during this project and briefly describe evidence that verifies your interaction with and knowledge of scientific concepts for each item on your project ten list.

Write a list of **technologically-oriented** activities you have participated in during this project and briefly describe evidence that verifies your interaction with and knowledge of technologically-oriented concepts for each item on your project ten list.

Write a list of **engineering-oriented** activities you have participated in during this project and briefly describe evidence that verifies your interaction with and knowledge of engineering-oriented concepts for each item on your project ten list.

Write a list of **mathematical** activities you have participated in during this project and briefly describe evidence that verifies your interaction with and knowledge of mathematical concepts for each item on your project ten list.

Write a brief list of **Non-STEM-related** activities you have participated in during this project and briefly describe evidence that verifies your interaction with and knowledge of Non-STEM related concepts for each item on your project ten list.

Rate your level of involvement in STEM-related activities in this project (1=lowest, 10=highest).

1—2—3—4—5—6—7—8—9—10

Beyond the Classroom Aquarium Daily Journal Guidelines

Entries students place in their daily journals should be:

1. Dated accurately and logged chronologically.
2. Kept in a notebook and written neatly in ink or typed.
3. Written or typed in complete sentences using correct capitalization, spelling, and punctuation.

In their journals, students should keep a daily record of:

1. Who they worked with each day by including the names of school staff, fellow students, community members, and volunteers.
2. What they did each day, such as lecture topic, activities or project design, assembly, or maintenance.
3. What they learned each day by asking questions, following instructions, and/or observing activities.
4. How *Beyond the Classroom Aquarium* experiences relate to past, present, or future life experiences.

For bonus points, students may complete these journal options:

* Include illustrations, diagrams, or creative expression relating to the journal entry.

Journal entries will be evaluated on a point system, with four points being the maximum number a student can earn. A half-point will be awarded for each of the first eight journal requirements above. Bonus points will be awarded when journal options are completed. Journal points will be totaled and divided by the number of journal dates. A possible journal grade based on points is included in the project one assessment section. For project ten, journal writing topics may include answers to the following questions or descriptions of a project activity or assignment. Reflection on these topics may also help students prepare for exams, quizzes, and other assessments.

- What do you want to understand better in order to contribute more to your team's productivity as it independently designs, assembles, and then manages an aquatic ecosystem?
- Create a journal entry that tells what information you learned during project ten. Also write down what you would like to know more about and if you know someone else who might help you learn more if you asked them.

- Select a book or magazine to read that is related to a topic in project ten. Write a report on this book and include your report in your *Beyond the Classroom Aquarium* portfolio.

- Participate in a *Beyond the Classroom Aquarium* conference committee and report on progress.

- Review Hoff (page 143 and beyond) and write a report that compares his description of juvenile rearing systems with your juvenile rearing (growout system) experience in project ten.

- Actively contribute to conference committee meetings; keep a written record of contributions you have made and responsibilities you have accepted. Summarize all accomplishments that are reported in each committee meeting you attend. A compilation of summaries will be collected at the beginning of project fourteen.

- Read Dale Leavitt's work (2009) on FLUPSY – a floating dock/upweller system used for the cultivation of shellfish and design a prototype system that a student team could place in a classroom recirculating "wet table" of the sort depicted in the top two images on page 259.

- Use a classroom upweller to collect data about the rate of water flowing through an airlift tube in the upweller (in units of gallons per minute) and how flow varies by volume of air being pumped through the airlift tube (in cubic feet per minute). Graph this data (air flow, rate of water flow).

- Shellfish growers have worked to establish standards to follow in stocking their upweller systems. Development of these standards includes considerations of upweller area and rate of water flow through the upweller. Review related literature that provides a table to which students may refer to determine how many juvenile shellfish they should stock in their upweller.

- Construct five identical upwellers stock one at a density suggested by the literature. Stock the other four in increments of 25% of the standard density -50%, -25% +25%, and +50%. Sample shellfish in each upweller each week to estimate growth rate of the population of shellfish in each upweller. After 6 weeks, graph the growth rate of shellfish in each upweller. Write a report that explains how stocking rates affect growth rates and the implications of this on maximizing production of healthy and mature shellfish.

Beyond the Classroom Aquarium STEM Self or Peer Assessment Rubric

This self-assessment is valid only to the extent to which you can document and display evidence that supports it. For each category in the left hand column, write a STEM-Oriented example in the box beneath 3 if you can display evidence that your STEM-Oriented example represents work that exceeds expectations. If your evidence shows that your STEM-Oriented statement meets but does not exceed expectations, write the statement in the box beneath 2. Similarly, for STEM-Oriented examples that demonstrate some progress toward expectations.

<table>
<tr><td colspan="2">Name:
Project Number:
Date Project Work Began:
Date of Project Completion:

List Math Topics and/or Objectives Currently Under Study:</td><td>3
Work Exceeds Expectations</td><td>2
Work Meets Expectations</td><td>1
Work Is Progressing Toward Expectations</td></tr>
<tr><td colspan="2">Write examples of questions and defining problems related to your STEM work.</td><td></td><td></td><td></td></tr>
<tr><td colspan="2">Describe how you have developed and used models in your STEM work.</td><td></td><td></td><td></td></tr>
<tr><td colspan="2">Write examples of STEM investigations you planned and carried out.</td><td></td><td></td><td></td></tr>
<tr><td colspan="2">Describe STEM related data that you analyzed or interpreted.</td><td></td><td></td><td></td></tr>
<tr><td rowspan="5">Give examples of how you used mathematics and computational thinking related to the five standards listed to the right.</td><td>Problem Solving (Math Process Standard One)</td><td></td><td></td><td></td></tr>
<tr><td>Reasoning & Proof (Math Process Standard Two)</td><td></td><td></td><td></td></tr>
<tr><td>Communication (Math Process Standard Three)</td><td></td><td></td><td></td></tr>
<tr><td>Connections (Math Process Standard Four)</td><td></td><td></td><td></td></tr>
<tr><td>Representation (Math Process Standard Five)</td><td></td><td></td><td></td></tr>
<tr><td colspan="2">Describe STEM-Oriented explanations and solutions you designed</td><td></td><td></td><td></td></tr>
<tr><td colspan="2">Explain evidence you used while engaging in a STEM project-based argument.</td><td></td><td></td><td></td></tr>
<tr><td colspan="2">Give examples of STEM information you obtained, evaluated, and/or communicated</td><td></td><td></td><td></td></tr>
<tr><td colspan="2">TOTAL POINTS</td><td></td><td></td><td></td></tr>
</table>

PROJECT ELEVEN Human Ecology – Feedback Diagram Analysis and Beyond: Design an Ecologically Sustainable Service Learning Program Proposal

The opening activity in project eleven takes students (figuratively speaking) to an aquatic ecosystem beyond the classroom known as the Chesapeake Bay. In this activity, students are guided to identify and resolve conflicting Chesapeake Bay issues and public policies. Although based on the (originally published in 1980) role-playing curriculum, this recently revised (by NASA) version of the Chesapeake Bay activity in *Beyond the Classroom Aquarium* is streamlined so that the project eleven schedule provides time to introduce students to the basic techniques of "human ecology" and the influence they had upon the development of a public aquarium exhibit about Apo Island in the Philippines. This exhibit (at the Shedd Aquarium) contains many incredibly designed aquatic ecosystems. Like many others who visit it, when I was guided through the exhibit, I was unaware of the breadth of the ecological footprint this exhibit represented (particularly to people and organizations whose primary daily activities involve effective development and implementation of good public policy). Yet I quickly sensed that the exhibit transcended the aquatic ecosystems and their designs. I quickly became immersed in the experience of seeing the world through the eyes of indigenous Apo Islanders and to understand the critical ways that their aquatic environment defines every aspect of their existence. Students will be guided to use these techniques to design an ecologically sustainable service learning program at the conclusion of project eleven.

Chesapeake Bay Educator's Guide: Project Eleven Opening Activity

Students play roles of concerned citizens, public officials, regional planners, and scientists to explore the Chesapeake Bay and its environs. They use data and satellite images to examine how human actions can degrade, improve, or maintain the quality of this water resource in order to make policy recommendations for improving the resource for future use. Time required: four to five 45-minute sessions

Materials/Resources

- Briefing (one copy per student)
- Handout 1: Roles (one role card for each of 10 students – print out, copy and cut out prior to class)
- Handout 2: Testimony Points (one copy for three to four investigators)
- Handout 3: Regional Planner Instructions (one copy for each student in class)
- Computer station(s) with Internet access (optional)

Content Preview

Human action in the sensitive, highly interdependent Chesapeake Bay ecosystem has led to environmental degradation during the past century. Land use changes related to deforestation, agricultural use of fertilizers, and urbanization have led to an increase in sediment and pollutants flowing into the bay. Natural climate cycles—periods of drought for example—exaggerate the effects of human action on the bay.

Classroom Procedures - Beginning the Investigation

1. Conduct a discussion on water as a resource. Ask these questions:
   - How do we use it? *(Drinking, transportation, recreation, industry*)
   - What are some ways that water quality is affected by natural processes? *(Drought affects flow of freshwater, storms affect amount of sediment and water flow.)*
   - What are some ways water quality is affected by human actions and processes? *(Pollution from industry, run off, and sewage, diversion/reduction of feeder streams/rivers, sedimentation, destruction/development of wetlands*

   Relate these issues to water resources in your region.

1. Explain that water is such a key resource that citizens and scientists often join forces to ensure the quality and supply of water. This investigation focuses on how this has occurred in Chesapeake Bay.

2. Have students locate Chesapeake Bay and the watersheds that drain into the bay. Ask them to note the major cities in the Chesapeake Bay River Basin. Resources for students to use to situate Chesapeake Bay are listed in the Briefing.

Developing the Investigation

3. Distribute the Briefing. Ask students to read the entire Briefing but assign a few specific paragraphs to groups of three or four. Ask each group of students to generate several questions, summarize their assigned reading, and take turns asking and answering questions.

4. Explain the premise of the role playing:
   - A public forum will be broadcast on public television moderated by the League of Women Voters under the auspices of the federal government.
   - The purpose of the forum is to give citizens and scientists the opportunity to voice their opinions on key issues related to Chesapeake Bay, to collect information, and to begin to develop a sustainable policy to preserve and maintain the integrity of this national resource.
   - Students play a role representing individuals involved in exploring the uses and abuses of Chesapeake Bay, one of the United States' most significant water resources.

5. Briefly explain the format for the role playing.
   - Ten *individuals* (concerned citizens, public officials, scientists) testify before the hearing. They are questioned by a team of three or four invited *investigators*. All students are *regional planners*. They may ask questions of witnesses and will write a draft statement recommending future policies based on the testimony they hear.
   - Role summaries for each *individual* are provided (Handout 1). These are in the form of questions that may be asked of each role player during the testimony. Students learn their role and develop answers to the questions by reading and discussing the Handout.
   - The *investigators* (three or four students) receive the list of questions each role player is prepared to answer (Handout 1) and the points that they should have each testifier make (Handout 2). They must sequence the presenters and questions to ensure the points are made effectively and logically.
   - The *regional planners* (all students when not role playing) keep a summary of points made and draft the policy statement and justification for it after hearing the testimony.

1. Seek volunteers for each role and distribute the role cards (Handout 1). Each student must find the answers to the questions from the Briefing and assume the character. Require presenters to prepare a visual (for example, a map, diagram, overhead of images appearing in the Briefing, graphs made from additional data obtained by researching the suggested resources) to support their testimony. For example:
   - Susan Elliott: a diagram showing local, state, and federal agencies that have cooperated to study the bay
   - Fred Kyle: a diagram showing changes in sea grass, fin fish, oysters, and crabs over the past 100 years
   - Pam Gibbons: LANDSAT images and maps to show population growth in the region
   - Chris Sprinski: a graphic to illustrate the causative role excess nutrients (water/air) play in bay degradation
   - Georgina Giovingo: pictures of her husband fishing on the bay
   - Phil Klein: a graphic describing how sediments flow into the bay
   - Kristin Hyche: a graphic illustrating the role climate plays in the bay ecosystem

- LeVar Jenks: a graphic showing relationships between agriculture, industrialization, and urbanization and bay health/productivity
- Evalia Tweedle: see role card
- Steve Sui: copies of the AVIRIS, SeaWiFS, and LANDSAT images

You may wish to have two-person teams for each individual role to facilitate the preparation of the graphic.

8. Distribute the list of roles and questions (Handouts 1 and 2) to the investigators. Allow them to divide the questions among themselves and to coordinate a strategy to ensure they are efficient and logical in their questioning.
9. Distribute directions to the regional planners (Handout 3). Explain that all students will contribute to the preparation of the policy recommendations. While some students prepare for and present the testimony, others will need to listen carefully, take notes, and synthesize and evaluate the testimony to form policy recommendations later. All students may ask questions. Decide on a format for the policy recommendations.
10. Have participants make name tags for themselves. Check that students are comfortable with their roles. Arrange the room in a town forum setting.
11. Begin the role play. You may wish to act as moderator Susan Elliott or ask a student to play the role. Susan Elliott calls the forum to order, explains its purpose, and introduces the team of investigators. The investigators proceed by calling the concerned citizens and scientists to testify in any order they wish. Continue through the testimony. At close, request that the regional planners (all students) draft a policy statement

Concluding the Investigation

12. Debrief the town forum.
    - Review the points presented by the individuals, the major concerns, and key issues.
    - Give students time to meet as regional planners and draft a policy to preserve and maintain the integrity of Chesapeake Bay. When they are finished, ask for a summary.
    - Discuss it. How well does the policy reflect a new understanding of the natural and human processes that affect water quality and use? What would be the long-term impact of such a policy? Who would be most/least affected? Would it work? Ask students to make generalizations about water resources in densely populated regions of the world as well as in your region of the United States.

Handout 1: Roles

Susan Elliott

You are the president of the League of Women Voters. Because of its impartiality, your organization has been asked to sponsor a town forum on Chesapeake Bay. You are moderating the forum.
As moderator you should:

1. state the purpose of the town forum and provide some background (see the Questions listed below);
2. introduce the investigators, citizens, and scientists;
3. ensure that order is maintained and that everyone is polite and has an opportunity to speak; and
4. remind all participants that they will, as regional planners, develop a policy to preserve and maintain the health of Chesapeake Bay.

Questions You May Be Asked

- Why is water important?
- How is water used?
- Why have many local, state, and federal agencies cooperated to study the bay?
- How successful have such partnerships of citizens and scientists been in setting environmental policies?

Fred Kyle

You are 44 years old and a concerned citizen employed by a high technology company outside of Washington, D.C. You and your family have lived near Chesapeake Bay in Virginia for generations and love it for its beauty and recreation opportunities. You enjoy taking your children on the water and someday hope to take your grandchildren but are worried about the degradation of the bay's water you have observed.

Questions You May Be Asked

- What is your name, where do you live, and what is your occupation?
- How has Chesapeake Bay changed in the past 100 years in terms of sea grasses, fin fish, oysters, and crabs?
- When did you first become concerned with the health and quality of the bay?

Pam Gibbons

You are 28 years old and have just completed a master's degree in geography at the University of Maryland. Your first job is working for the state of Maryland doing long-range land use planning. Your research focuses on land use change in the Chesapeake, and you are considered an expert on the topic.

Questions You May Be Asked

- What is your name, where do you live, and what is your occupation?
- What river basins flow into the Chesapeake?
- What are some ways that the Chesapeake has changed over the last few decades?
- What factors have caused these changes?
- What are the likely population trends in this area in the next 20 years?
- How will population affect land use?
- What role does sedimentation play in affecting the health of the bay?
- From your studies, what are the effects of land use and population change on the bay?
- What tools/images can you use to help in planning?

Chris Sprinski

You are in your mid-30s and a lifelong environmentalist. You have worked for several nonprofit organizations devoted to preserving the environment. In college you majored in biology with an emphasis in marine sciences but took environmental science courses as well. Currently you serve as a consultant to the Friends of Chesapeake Bay, a nongovernmental organization funded by private donations.

Questions You May Be Asked

- What is your name, where do you live, and what is your occupation?
- What is the current model for bay ecosystem management?
- From your studies, what are the effects of land use and population change on the bay?
- What is the role that excess nutrients in water and air play in causing degradation in the bay? Please explain what happens step by step, beginning with algal blooms.

Georgina Giovingo

You are in your late 50s and have been a widow for a year. You and your husband made your living fishing the Chesapeake. Most of your limited income was made harvesting blue crabs. Your late husband, Dave, died of a mysterious infection possibly related to contamination from the waters of the bay. You are clearly upset about the degradation of the bay, the decline in fish, oysters, and crabs, and the loss of recreational value of the bay. You cannot contribute scientific information, but you show the forum that many people care passionately about the bay and rely upon it for their livelihood.

Questions You May Be Asked

- What is your name, where do you live, and what is your occupation?
- How important is fishing to the economic well-being of people in the Chesapeake region and the nation?
- From your experiences of fishing the bay for 30 years, how has the bay changed?
- Why do you think it is important to develop a policy to preserve and maintain the bay?

Phil Klein

You are a legislative aide assigned to the governor of the state of Maryland. You have three young children and live close to your office in Annapolis, the capital. You are an expert in the development of environmental policy and environmental law and act as the governor's liaison on Chesapeake issues, which includes attending meetings of the Chesapeake Bay Program. The governor has asked you to make sure that the state of Maryland is not responsible for causing damage to one of its key resources, Chesapeake Bay.

Questions You May Be Asked

- What is your name, where do you live, and what is your occupation?
- What is the mission of the Chesapeake Bay Program?
- What actions have been taken or programs put in place related to the Chesapeake's health and wellbeing?
- How much sediment enters the bay from Maryland?
- What action has the state taken to affect the amount of sedimentation?

Professor Kristin Hyche

You are a professor at the University of Delaware specializing in climatology and hydrology in the Department of Geography. You study bay ecosystems from a geographic perspective, looking at the interactions between human and natural processes. You live in Newark, Delaware.

Questions You May Be Asked

- What is your name, where do you live, and what is your occupation?
- How does the Chesapeake Bay ecosystem function?
- What role does climate play in the ecosystem? What are some of the effects of climate change on the health of the bay?
- How have policies to preserve the bay been helpful? For example, have subaquatic grasses returned in some bay tributaries?
- What are the causes of the changes observed in the bay in discharge, sea grasses, and blue crabs?
- Which of these are caused by natural processes and which by human actions? Explain.

LeVar Jenks

You are a research scientist with the USGS assigned to the Chesapeake Ecosystem Response Project. You grew up in inner city Baltimore and, as a young boy, loved to fish the bay with your grandfather and cousins. That inspired you to study the environment and to pursue a geology degree in college. You specialize in the geochemical analysis of sediments and worked hard to be assigned to the project. You are in your early 40s and live in a Maryland suburb of Washington, D.C. You are also a history buff interested in the way of life of your ancestors, enslaved persons in Virginia and Maryland.

Questions You May Be Asked

- What is your name, where do you live, and what is your occupation?
- What is the Chesapeake Ecosystem Response Project? What is the project investigating?
- What methods are being used?
- Why is the project investigating the impacts of agriculture, industrialization, and urbanization at different periods of time? What role do they play in bay conditions?
- Do you have any research results to report yet?
- In your professional opinion, which do you think is having a greater impact on the bay: human processes or natural processes? Why?

Evalia Tweedle

You're training as a classroom educator prepared you well for your current administrative position with the Public Affairs Office of the EPA in Washington, D.C., where you also live, sharing a house with five other recent college graduates. You primarily work to heighten public awareness and understanding of key scientific and environmental issues. Since you were an educator and work with the public a great deal, you come to the forum prepared with teaching tools to present what you feel are key ideas that will contribute to the forum's success.

Questions You May Be Asked

- What is your name, where do you live, and what is your occupation?
- What is the difference between point and nonpoint source pollution?
- What are some of the key sources of pollution?
- What are some strategies the EPA suggests home owners adopt to lessen their impact on the environment in general?
- How important is this kind of recommendation for people who live in the Chesapeake Bay watershed?
- What tools and resources can help EPA understand and control sources of pollution?

Handout 1: (continued)

Steve Sui

You are employed as a research scientist specializing in remote sensing at NASA's Ames Research Center located near San Jose, California. You live close to San Francisco Bay and thus became interested in the Chesapeake Bay Project. You have worked closely with Phil Klein in the past; he invited you to participate in the forum. Your expertise is in using AVIRIS to remotely sense Earth, particularly its oceans. You are also expert in SeaWiFS, a sensing system to observe the world's oceans and monitor ecosystem change. You think you may be asked to explain two images produced by these sensors and plan to come prepared.

Questions You May Be Asked

- What is your name, where do you live, and what is your occupation?
- How did you become involved with the Chesapeake Bay Project?
- What is AVIRIS?
- How was it used to monitor Chesapeake Bay? What was the end result?
- Can you draw any conclusions about the health of the bay from the map that was produced?
- What is SeaWiFS? What is it being used for?
- What is LANDSAT? What is it being used for and by whom?
- What do the colors on the image of Chesapeake Bay indicate?
- Discuss the usefulness of remote sensing for tracking changes on Earth.

Handout 2: Investigator's Points That Each Witness Should Make

Susan Elliott

- Water is important to human health and all its endeavors. It is equally important to other living creatures in the ecosystem upon which humans depend. It has a variety of uses, including transportation, irrigation, recreation, industry, and fishing.
- Chesapeake Bay is the nation's largest and one of the world's most productive estuaries.
- Chesapeake Bay is an important resource to many people in six states who live in urban, suburban, and rural areas. A watershed is a complex system. It requires the cooperation of many agencies at different scales to address watershed-wide issues because the boundaries of governmental jurisdictions do not match the boundaries of watersheds.

Pam Gibbons

- Nine river systems flow into the bay. They are the Susquehanna, the Potomac, the Patuxent, the Choptank, the Rappahannock, the Mattaponi, the Pamunkey, the James, and the Appomattox.
- Human activities play a role in the degradation of the bay. Changes in how people use the land (land use change) are the major culprits, including deforestation, urban development, and expansion of agricultural land. As population grows, forests are cut and land cleared to build houses, malls, and businesses. Agriculture expands into new areas to feed the growing population. More people produce more sewage and pollution. When it rains, exposed dirt (sediments) disrupted by human activity washes into the bay.
- The population will continue to grow in this already densely populated region of the nation, and land use changes will continue. There are about 18 million people living in the area now.
- Sediments cloud the bay water so much that subaquatic vegetation like sea grasses cannot survive.
- Land cover maps and images like the one from LANDSAT 7 are used to estimate polluted water runoff. Remote sensing is an important tool.

LeVar Jenks

- The U.S. Geological Survey (USGS) Chesapeake Ecosystem Response Project is designed to improve understanding of large-scale environmental changes that influence water quality and living resources in the Chesapeake Bay. In particular, project workers investigate the links among changes in climate, precipitation, discharge, salinity, and dissolved oxygen over different time periods.
- The human factors that most affect these natural factors are industrialization, agricultural practices, and urban development.
- USGS scientists and other researchers are collaborating to study the bay's sediments which capture the history of its water, plants, and animals during the period before monitoring began in the 1980s. Using ecosystem "indicators" (microfossils), geochemical data preserved in the bay's sediment, and historical and reconstructed discharge data, we have reconstructed trends and responses in the bay since 1950 to determine the natural conditions in the bay over the last few millennia, including periods prior to 17th-18th century colonial agriculture and 19th-20th century industrialization and urbanization. The emphasis has been on separating natural versus human causes of and responses to extreme events.

Evalia Tweedle

- Point pollution is pollution coming from a known, identifiable source such as a pipeline spewing waste. Nonpoint source pollution is pollution that enters surface, groundwater, and the oceans from widespread and distant activities—that is, from no one, single point.
- The key sources of nonpoint source pollution are agriculture and livestock, urban runoff, automobiles, land clearing, sewage, factories producing air pollution, and industrial waste.
- The EPA suggests homeowners manage hazardous waste carefully. Householders need to be aware that they have an impact on the environment when they dump chemicals down the drain.
- This advice is especially important for the millions of people that live in the Chesapeake watershed. Something placed onto the ground anywhere in the region ultimately affects the health of the bay.
- Remote sensing is helping us to better pinpoint the sources of polluted runoff.

Handout 2: (continued)

Chris Sprinski

- The model is the system. All elements of the various systems present in the bay are linked, connected, and affect the other elements. Everything is interrelated; managing just one system element will not necessarily cause positive changes throughout the system.
- Land use and population changes increase the amount of agricultural fertilizers and urban sewage treatment plants, which cause increased phosphorus and nitrogen loading in surface and groundwater.
- Excess nutrients from water and air can lead to an increase in algal blooms, reduced dissolved oxygen levels on the bottom, habitat degradation, and depleted living resources.
- Algal blooms can reduce the clarity of the water, which prevents sunlight from penetrating to the bottom and thus inhibiting the growth of sea grasses or subaquatic vegetation (SAV). SAV is important because it helps absorb nutrients, adds oxygen to the water, and provides a sheltered habitat for organisms, especially juvenile blue crabs. It is also a food source for water birds living in the bay.

Georgina Giovingo

- The bay provides 50 percent of the nation's blue crab harvest, worth about $130 million per year. There are many people who make their living by fishing and many more who make their living serving the needs of recreational boaters, birders, and beach goers in the region.
- The bay is not as clean as it once was. There are fewer fish and shell fish. Sometimes there are algal blooms that make people sick. The water is not as clear as it was. More people live around the bay now. There are fewer birds because there are fewer fish.
- People's livelihoods depend on the health of the bay.

Professor Kristin Hyche

- Ecosystem function is very complex, and there are gaps in our understanding of the relationships among river discharge into the bay, the amount of oxygen dissolved in the water, and aquatic grasses.
- We have better tools now, like the data and images obtained through remote sensing, to measure nutrients in the bay.
- The amount of rainfall affects the freshwater discharge entering the bay. In recent years, fluctuating climate patterns have made this relationship especially evident. The amount of freshwater changes the nutrients entering the bay and the salinity of the water. These factors, in turn, influence how bay phytoplankton use the nutrients and thus, oxygen levels in the water.
- Historical trends, for example, comparing the drought years of the 1950s and 1960s to later decades, indicate the importance of climate to bay health.
- Aquatic grasses have returned in some areas and not others. We do not yet know precisely what works and what does not work to repair environmental damage. The factors of habitat change, overharvesting, natural mortality, and climate change all play varying roles.
- The bay ecosystem is very complex, and little is known about the relations among species and their environment and climate change. It is also difficult to separate human and natural effects.

Phil Klein

- The Chesapeake Bay Program (CBP) was formed in 1983 as a regional cooperative effort between the U.S. Environmental Protection Agency and various state and local governments of the bay watershed including the states of Maryland, Pennsylvania, and Virginia, the District of Columbia, and the Chesapeake Bay Commission.
- Its primary mission is restoration of living resources, including fish, shellfish, bay grasses, and other aquatic wildlife of the nation's largest and one of the world's most productive estuaries.
- The Chesapeake Bay Program monitors water quality and biotic resources in order to assess the bay's ecological health on a regular basis and to produce computer models to predict watershed quality and ecosystem response to nutrient loading.
- The CBP routinely conducts an ongoing ship-based monitoring program to collect *in situ* measurements of physical, chemical, and biological indicators of the bay's health. The CBP is evaluating the use of remote sensing as an adjunct to the monitoring program, with the idea that highly accurate point measurements may be used in conjunction with appropriately processed imagery to generate a spatial dataset of water quality.
- About 155 million metric tons of sediment has entered the bay in the last 100 years from the state of Maryland alone. That is a serious problem.
- Biological resource management plans have been developed to restore and preserve key species for commercial and recreational use.

Fred Kyle

- The bay has degraded seriously in the last 100 years. Sea grass acreage has declined. There are fewer fish and shell fish, particularly blue crab for which the bay is famous. The amount of dissolved oxygen has decreased, and more soil is washing into the bay.
- The changes really became noticeable in the 1970s when the commercial fishermen started to notice declines in fish, and recreational users became aware of reduced water quality.

Steve Sui

- A NASA experiment showed the utility of remote sensing in monitoring the Chesapeake Bay watershed. A sensing device known as AVIRIS was flown over the southern part of the bay to measure the amount of chlorophyll and sediments in the waters.
- AVIRIS is an acronym for the Airborne Visible Infra-Red Imaging Spectrometer. It is a remote sensing instrument that flies aboard a NASA ER-2 airplane approximately 20 kilometers above sea level. It is used to identify, measure, and monitor constituents of Earth's surface and atmosphere. Research with AVIRIS is directed toward understanding processes related to the global environment and climate change.
- SeaWiFS is a sensor mounted on a satellite that observes Earth from a noontime sun synchronous orbit, which means that the sensor is always viewing Earth around local noon at an altitude of 705 kilometers. This orbit provides data to detect concentrations of microscopic green plants, called phytoplankton, which live just beneath the ocean surface. These green plants absorb sunlight during photosynthesis, the most basic and essential chemical process necessary for life on Earth.
- The red colors show high concentrations of chlorophyll in the water, the yellows/greens indicate intermediate concentrations of chlorophyll, and the blues/purples show low concentrations of chlorophyll.
- In the image of Chesapeake Bay, ocean patterns are clearly evident, such as plumes of material discharging out of eastern Long Island Sound. Red and yellow areas in Chesapeake Bay indicate turbid waters while the blue hues offshore represent clear oceanic water.
- LANDSAT is a satellite that picks up information particularly related to land cover such as pasture, crop lands, different types of forest and wetlands, and types of residential land uses. It is very useful for local and regional planning agencies.

Handout 3: Instructions to Regional Planners

You are a planner on the Chesapeake Bay regional planning council. Prior to the town forum, summarize the situation in the Chesapeake in the space provided.

The Situation

___

___

___

___

___

___

___

___

___

___

___

___

___

___

___

___

The Town Forum

The purpose of the town forum is to collect information about the bay from different points of view. Everyone at this meeting is in favor of taking action to "save the bay." Your job will be to make recommendations to develop a sustainable policy to preserve and maintain the health of the bay.

During the forum, investigators will call a series of concerned citizens, local and state government officials, and scientists to give testimony about the bay. You will listen to the testimony and take notes summarizing the points each individual makes. As you listen, synthesize and evaluate the testimony. You will need the information to decide policy later. You may ask questions to clarify a point, but in general you should leave that to the investigators.

Writing the Policy

Your work really begins with the end of the forum when you develop the recommendations. Summarize your findings <u>based on the testimony</u>, and then make at least three specific recommendations to preserve and restore the bay.

Background

Water is essential to life on Earth. Three-quarters of Earth's surface is covered by water. Because it is almost everywhere, we sometimes take it for granted, but we do things that affect its quality and quantity. This briefing shows how human actions have combined with natural processes to affect Chesapeake Bay, one of the United States' most significant water resources. Because Chesapeake Bay is such an important resource, many citizens and government agencies are concerned about its present and future status. Several federal, state, and local organizations and agencies have combined their efforts to study the bay in order to provide ideas and programs to ensure its future health.

Figure 1: Location of Chesapeake Bay and the river systems that flow into it
Source: http://www.pamatters.com/wp-content/uploads/2013/04/SmallmouthBass.jpg

NASA, USGS, EPA, and NOAA are federal agencies doing this research. Three NASA missions have contributed data to bay studies: SeaWiFS (Sea-Viewing Wide Field-of-view Sensor), AVIRIS (Airborne Visible Infrared Imaging Spectrometer), and LANDSAT 7. The data gathered from SeaWiFS and AVIRIS have contributed to a better understanding of the ability of the bay to support plant and animal life. The data produced by LANDSAT 7 allow smarter land use planning and better estimates of polluted water runoff across the 110,000 square kilometer Chesapeake Bay watershed.

Chesapeake Bay is located in the northeastern United States in the most densely populated region of the nation. Figure 1 indicates its location and the river systems that flow into the bay. Turn to Figure 6 to see the land use in this highly industrialized, highly urbanized region.

Ecosystem Trends and Response: Chesapeake Bay

The Nation's Largest Estuary

Chesapeake Bay, the nation's largest estuary, has had serious environmental degradation during the past century. The evidence of damage includes declines in sea grass acreage, reduction of fin fish and shellfish (oysters and crab), seasonal depletions of dissolved oxygen, and increases in sedimentation. These changes raised serious concern in the 1970s because they threatened major commercial and recreational activities. Most scientists attribute these changes, at least indirectly, to human activities. Land use changes in the bay watershed (deforestation, agriculture, and urbanization) brought added pollution and sewage. Future stress on bay ecosystems is likely to worsen, as the Chesapeake Bay Commission predicts that the population in the bay watershed will swell to over 18 million by the year 2020.

Chesapeake Bay is a complex natural ecosystem with many interrelated human and natural subsystems. Changes in each subsystem ripple through others, causing various effects. Ecosystems recycle chemicals needed by living things, redistribute wastes, control pests that cause disease in both humans and plants and animals, and offer a huge pool of resources for humans and other living creatures. Ecosystems are affected by natural events such as droughts. But ecosystems are more drastically changed by human activities.

The model used for bay ecosystem management is a system which shows links among land, sea, and living creatures that result in changes in the watershed affecting the bay. For example:

- Land use and population changes increase the amount of agricultural fertilizers and urban sewage treatment plants which cause increased phosphorus and nitrogen loading in surface and groundwater.
- Excess nutrients from water and air can lead to an increase in algal blooms, reduced dissolved oxygen levels on the bottom, habitat degradation, and depleted living resources.
- Algal blooms can reduce the clarity of the water, which prevents sunlight from penetrating to the bottom and can inhibit the growth of sea grasses or subaquatic vegetation (SAV).
- Subaquatic vegetation helps absorb nutrients, adds oxygen to the water, and provides a sheltered habitat for organisms, especially juvenile blue crabs. It is also a food source for water birds living in the bay.
- Sedimentation is another critical problem. Over the last 100 years, 155 million metric tons of sediment were deposited in the Maryland portion of the bay. Sedimentation rates have increased since colonial times because of land use changes. Sediment can cloud the water so much that SAV cannot survive.

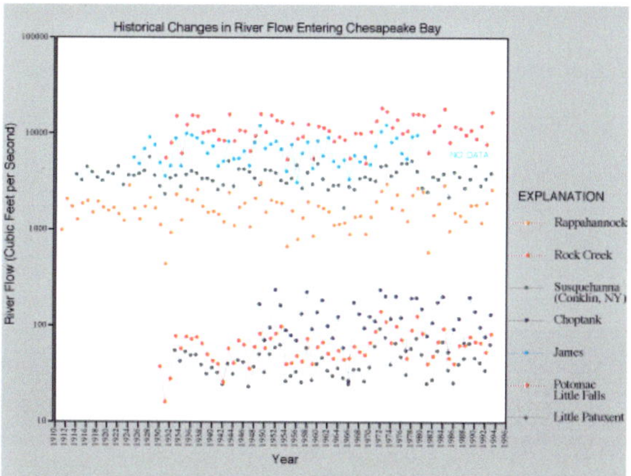

Interannual and decadal variations in river flow are obtained from seven tributary monitoring stations, unimpeded by dams and other structures. River flow strongly influences nutrient input and dissovled oxygen in Chesapeake Bay. Climatic factors in turn, control to a great extent interannual and decadal flow variations.

Figure 2

Source: http://md.water.usgs.gov/publications/fs-213-96/graph.gif

Causes of Recent Trends in Dissolved Oxygen and SAV

Fluctuations in the amount of freshwater entering the bay from its tributary rivers (termed discharge), result from changing precipitation patterns over the last few decades (Figure 2). Recently, these fluctuations have become extreme. Trend data show the influence of discharge on the total nutrients entering the bay and on bay salinity. These influence phytoplankton growth and oxygen levels. Climatic factors play a role in controlling water quality. For example, the drought years of the 1950s and 1960s caused low tributary discharge into the bay. Bay restoration efforts began in the early 1980s. Surveys have indicated that sea grasses have begun to grow again in several tributaries of the bay. In theory, this reflects a response to improved water quality. But there were areas where, despite improved water quality, subaquatic vegetation (SAV) did not return. This continued absence of SAV concerns scientists and resource managers, in part because SAV provides habitat for young blue crabs. The bay provides 50 percent of the nation's total blue crab harvest. The bay crab harvest was worth $126.6 million in 1993. Chesapeake Bay is a significant source of fish and shellfish.

A Complex System

What do recent trends in discharge, anoxia (too little oxygen in the water), sea grasses, and blue crabs mean? Are they caused by habitat change, overharvesting, or natural mortality related to long term climatic factors? These questions remind us of how complex the bay ecosystem is and of how little we know about the relations among individual species, their environment, and climate. They also illustrate how difficult it is for scientists to separate natural versus human effects on critical species and their habitats based solely on monitoring. Detection of a trend can be easy, but assigning cause is much more difficult.

Cross-country Sources Side Effects Spread from Land to Sea

Identifying a water polluter is easy when the source of the pollution is a single point, such as a single pipeline spewing wastes. But it's not so simple when you consider *nonpoint source pollution*, which comes from nearly everything we do on land that contributes to polluted runoff that enters surface and groundwater, as well as oceans. It can lead to beach and shellfish-bed closings and spoiled habitats for fish and other aquatic life. Some of the sources of pollution entering the bay are listed below.

Agriculture and Livestock

Runoff from barnyards, feedlots, and cropland contributes nutrients from manure and fertilizers, as well as pesticides and eroded soil.

Urban Runoff

Urban runoff from buildings and paved surfaces carries sediment, nutrients, bacteria, oil, trace metals, chemicals, road salt, pet droppings, and litter.

Automobiles

Leaking oil and motor fluids run off roadways and into waterways. Emissions send nitrogen and other contaminants into the atmosphere that eventually settle into coastal waters.

Land Clearing

Construction, clearing land, and logging often lead to soil erosion, putting more sediment in rivers and coastal waters. Filling in wetlands takes away vast natural water filters that can break down many common pollutants before they reach other water bodies.

Sewage

Sewage leached from faulty septic systems or dumped directly overboard—instead of emptied at boat pump-out stations—contributes nutrients and disease-causing organisms.

Air Pollution

Airborne pollutants, chiefly from factories and automobiles, are responsible for almost a third of all contaminants and nutrients entering marine waters.

Industrial Waste

Industrial runoff brings heavy metals and other compounds into marine waters from industrial waste landfills, mining, and storm water draining off of industrial sites.

Warning Signs

Too Many Nutrients Leads to Too Little Oxygen

Too much nitrogen (from fertilizers, sewage, feedlot runoff, or air pollution), or too much phosphorus (from the same sources, as well as detergents or water-treatment chemicals), can set off explosive growth of algae and aquatic plants. As the overpopulated plants and algae die off, bacteria can deplete oxygen from the water as they decompose the dead plants. Lack of oxygen kills fish and other animals.

Managing Hazardous Waste Hits Home

Many products for home and garden can burn, explode, corrode, or poison. Dumped down the drain, onto the ground, or into the trash, they can pollute water, pose health risks, and damage water-treatment systems. Since 1980, thousands of community programs have begun to collect household hazardous waste.

Get a Handle on Household Hazards

To cut your "pollution contribution," use fewer hazardous or unsafe products if you can. If you can't, use only as much as you need; share leftovers with neighbors, businesses, or charities and dispose of leftovers safely.

Chesapeake Bay's Health Depends on the Kindness of Many Strangers

The bay's watershed covers six states and the District of Columbia. It drains 150 rivers and streams—an enormous catchall for urban, suburban, and agricultural pollution. Environmental problems often occur over large areas that fall into the control of more than one governmental unit. For this reason, a landmark 1983 agreement formed the Chesapeake Bay Program (CBP).

The Chesapeake Bay Program is a partnership between the U.S. Environmental Protection Agency, Maryland, Pennsylvania, Virginia, the District of Columbia, the Chesapeake Bay Commission, and local governments of the bay watershed. The program has helped reduce nutrient pollution from farming and livestock, banned phosphate detergents and tributyl-tin boat paints, and protected ecologically sensitive shorelines.

The Chesapeake Bay Program

- aims to restore living resources, including fish, shellfish, bay grasses, and other aquatic wildlife;
- monitors water quality and biotic resources;
- produces computer models to predict watershed quality and ecosystem response to nutrient loading;
- used modeling studies to establish a goal of a 40 percent reduction from 1985 levels of nutrient input into the bay by 2000;
- developed biotic management plans to restore and preserve key species for commercial use and recreation; and
- conducts ship-based monitoring to collect measurements of physical, chemical, and biological indicators

Taking Action: Research, Response, and Management

The U.S. Geological Survey's (USGS) Chesapeake Ecosystem Response Project is another effort to understand large environmental changes influencing the water quality and living resources in the bay. In particular, the project scientists

- investigate links among changes in climate, discharge, salinity, and dissolved oxygen over different time periods and geographical scales;
- study effects of past events recorded in the layers of sediment deposited on the floor of the bay over time;
- collaborate with researchers from the University of Maryland, the Maryland Geological Survey, and the Virginia Institute of Marine Science to study the bay's sediments;
- reconstructed trends and responses in the bay since 1950 to determine the natural conditions in the bay over the last few millennia, including periods prior to 17th-18th century colonial agriculture and 19th-20th century industrialization and urbanization;
- found a linkage between climate and land quality and habitat loss in the bay;
- continues studying trends in oxygen, nitrogen, and phosphorus levels, phytoplankton, benthic invertebrates, sedimentation, and biodiversity with emphasis on separating natural versus human causes of and responses to extreme events

Using Remote Sensing as a Research Tool: Three Examples from NASA

Using the unique perspective from space, NASA observes, monitors, and assesses large-scale environmental processes such as the oceans' productivity.

AVIRIS Analysis of Chesapeake Bay

During 1997-1998, NASA's Ames Research Center performed a remote-sensing demonstration project in collaboration with the Chesapeake Bay Program (CBP). The CBP wanted to evaluate the use of remote sensing in its monitoring program, hoping that point measurements could be used with satellite imagery to develop water quality data.

Previously, University of Maryland scientists had explored the use of airborne remote sensing to map chlorophyll concentration in the bay. NASA's AVIRIS (Airborne Visible Infrared Imaging Spectrometer) was chosen for the demonstration. The objective of AVIRIS is to identify, measure, and monitor Earth's surface and atmosphere. Research with AVIRIS seeks to understand global environmental processes such as climate change. The instrument flies aboard a NASA ER-2 airplane at an altitude of about 20 kilometers. It flew over the bay on August 17, 1997, and July 3, 1998. The data it collected allowed scientists to estimate chlorophyll and suspended sediment concentrations. These data were compared with CBP shipboard measurements. The 1998 flight identified submerged aquatic vegetation (SAV). Scientists compared the satellite images with aerial photography and found that imagery analysis was useful in determining sea grass distribution.

SeaWiFS

Figure 3, 4, and 5 show views of Chesapeake Bay from the SeaWiFS (Sea-Viewing Wide Field-of-View Sensor). Figure 3 shows data taken on September 16, 1997, and Figures 4 and 5 record data from September 19, 1997. SeaWiFS data allow assessment of global vegetation patterns, both land and ocean, needed to understand ecosystems and global change. The SeaWiFS instrument observes the oceans from space to measure "ocean color." SeaWiFS is essential to NASA's efforts to study how the global environment is changing.

SeaWiFS observes Earth from a noontime sun synchronous orbit, which means that the sensor always views Earth around local noon at an altitude of 705 kilometers. This orbit provides data at the maximum solar illumination, the most desirable for detecting concentrations of phytoplankton, which live just beneath the ocean surface. These green plants absorb sunlight during photosynthesis

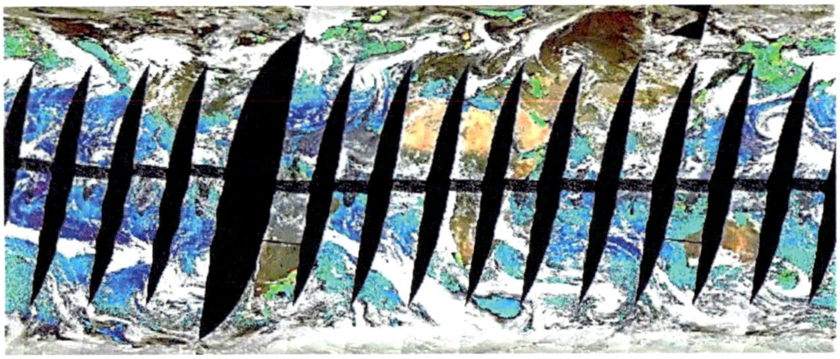

Figure 3: SeaWiFS data from September 16, 1997
Source: http://oceancolor.gsfc.nasa.gov/SeaWiFS/Gallery_Images/revised_ocean_color_and_land_4096x1024.jpg

In Figure 3, the red colors show high concentrations of chlorophyll in the water, the yellows/greens indicate intermediate concentrations of chlorophyll, and the blues/purples show low concentrations of chlorophyll. The black swaths indicate no data because of gaps between the orbits. On Earth, coverage is every two days.

The next two images of Chesapeake Bay represent SeaWiFS high-resolution data obtained over the mid-Atlantic on September 19, 1997. Figure 4 highlights vegetation associated with the coastal plain and mountain ridges and valleys. White areas are clouds and dense aerosols. Note that ocean features are not noticeable. In Figure 5, ocean patterns are evident, such as plumes of material discharging out of Delaware Bay. Red and yellow areas in Chesapeake Bay indicate turbid (sediment filled) waters, while the blue hues offshore represent clear oceanic water. Black areas, such as in the right bottom corner, are locations where the processing could not be completed

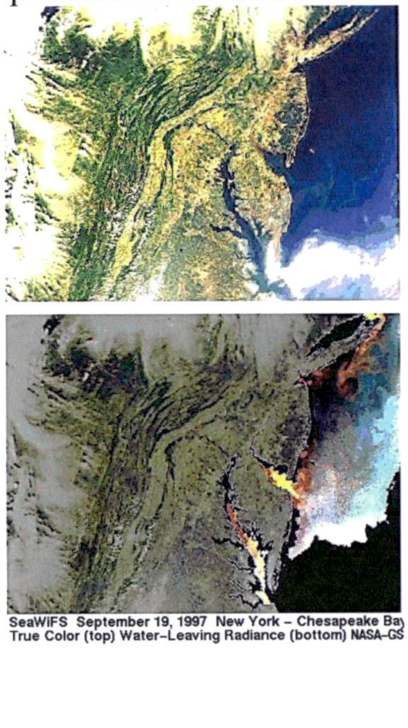

Figure 4: SeaWiFS, September 19, 1997, New York—Chesapeake Bay, true color contrasted with Figure 5: SeaWiFS, September 19, 1997, New York—Chesapeake Bay showing ocean radiance
Source: http://geography.tamu.edu/class/sbednarz/websites/sbednarz/missgeog/912Mod1Inv1/pg6_ches_combined.jpg

LANDSAT (Land Remote Sensing Satellite) is designed to gather data on Earth's resources in a regular and systematic manner. It collects information related to land use, geological and mineralogical exploration, crop and forestry assessment, and cartography. The image in Figure 6 was produced by LANDSAT 7 to assess the amount of different land cover types in the bay region, including residential development, wetlands, forests, and crop lands.

The maps produced from LANDSAT will help make estimates of polluted river runoff flowing into the bay by identifying pasture land and different types of crops. The amount of nutrient pollution entering the bay can be measured by knowing the area of a type of land cover and estimating the average water quality of runoff from that type of land. Heavily fertilized agricultural fields, for example, produce higher levels of nutrients in runoff than the same acreage of pasture land. The image and maps also distinguish low and high-density residential development from wetlands and different types of forest. These images will allow smarter land use planning and better estimates of polluted water runoff across the watershed.

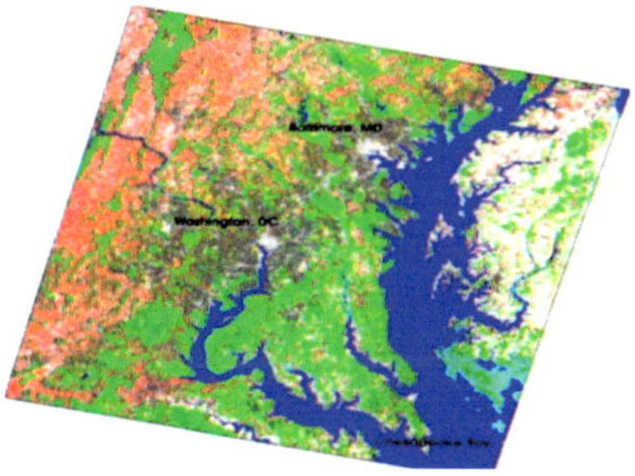

USGS Factsheet FS-116-00. September 2000. "Effects of Climate Variability and Human Activity on Chesapeake Bay and the Implications for Ecosystem Restoration

Figure 6: LANDSAT 7 image showing land cover types in Chesapeake Bay area including residential development, wetlands, forests, and croplands

How Success Works

In this second activity in project eleven, teachers guide student exploration of a feedback diagram model (presented below) to introduce them to an ecological analysis process, as used by Apo Island residents and as displayed in the Apo Island exhibit at the Shedd Aquarium. By the end of the activity, students are expected to see what it takes for an EcoTipping Point lever to overcome a vicious cycle (that drives decline of ecological productivity), thereby transforming a vicious cycle into a virtuous cycle that drives restoration. After participating in this Apo Island activity, students will be prepared to take their first steps in applying their understanding of this model to local community-related environmental conditions beyond the immediate classroom.

How Success Works is an ideal lesson to teach what EcoTipping points are and how they work. Marten (2004) reports this lesson is based on a flagship case related to the Apo Island Fishery. Students watch a short video about Apo Island to introduce the idea of decline and restoration. They then read a short narrative for this case and are challenged to identify ingredients for success in that story. They review the ingredients with a teacher-led PowerPoint, or they review independently using the respective Student-Centered PowerPoint. All of the materials are easy to modify for the needs of individual classrooms. Below is one suggested way to approach this multi-faceted lesson. The Apo island instructional materials include the following:

- Suggested Procedure (Lesson Plan)
- Narrative Handouts
- Ingredients for Success Handouts
- Feedback Diagrams (Vicious Cycles)
- Student Worksheets
- Teacher Keys
- PowerPoint Presentations
- Short Video

Depending on the grade level and the unit of study that provides the context for a lesson, teachers can pick and choose from the steps below, tailoring the style of the lesson to the circumstances and changing the order of some of the steps, if they wish. For example, if a lesson emphasizes "systems thinking," it might proceed directly from a video or narrative handout (Steps 3-4) to feedback diagrams for that success story (Steps 8-9), and then create a "negative tip" feedback diagram for an issue that students identify from their local scene (Step 11), covering Ingredients for Success (Steps 5-7) later if desired. A very simple lesson, appropriate for primary school, might proceed directly from a story (Step 3 or 4), and reviewing the story, to exploring briefly what might be done about a local problem.

Step 1: Have students write a journal responding to the following prompt:

> "Think of a time when something you or your family cared about was falling apart or falling into decline (You can give examples: friendships, a project, a class, a job). What did you do to try turning things around? Did it work? How or how not? Do you think doing something else might have worked better?"

Step 2: After students have written their journals and shared back with the class, introduce the idea of communities being in decline or falling apart because of their relationship with their environment. Explain how these communities must change in order to sustain themselves, but knowing what to change is rarely clear or simple. Tell students that you will use the experience of a community that turned things around, from decline to restoration and sustainability, to learn lessons about what it takes to achieve that kind of success.

Step 3: Show a short video on the Apo Island success story to quickly introduce students to the basic concept of EcoTipping Points. The Apo Island case is a compelling story with a simplicity that clearly reveals the basic concept. An effective way to briefly introduce the video is to:

- Explain the story setting and what will happen during the "negative tip" portion of the video, emphasizing what set decline in motion and what changed during decline.
- Ask students to watch for what the people in the story did to turn things around (what they did first, what they did next, what they did after that, and so on).
- After seeing the Apo Island video, divide students into small groups to list what the people in the story did, first to turn things around, then what they did next, and so on. The entire class can then go over it together.
- To continue in further detail, students will now use what they understand about an EcoTipping Point to perform an analysis of the Apo Island success story to identify the ingredients for success in that story and determine how decline was reversed.

Step 4: A narrative handout can also be used for the Apo Island story. There are two levels of the narrative available: a short version and an extended version, depending on the level of your course. The student worksheet comes with frontloaded vocabulary. Review the vocabulary with the students to improve comprehension and introduce key concepts of human ecology. It may be useful for students to underline those words when they read them. The vocabulary is also useful for assessment.

2. Narrative Handout (Short Version) - Apo Island's Marine Sanctuary: Restoring a Coral Reef Fishery

Among the thousands of small islands in the Philippines is a fishing village of 145 households known as Apo Island. The men use outrigger canoes to fish on the coral reef that extends around the island to a distance of about 500 meters from shore. In the past, they used hook-and-line, gill nets, and bamboo fish traps to catch all the fish they needed to feed their families and earn cash income selling the extra fish. The fishing was sustainable, but about fifty years ago an unexpected series of events set in motion serious destruction of the reef. Within a few decades, the fish supply was collapsing and the very survival of the village was threatened.

The problems began when a growing human population and expanding markets for fish encouraged fishermen to catch more fish. They began using destructive fishing methods such as dynamite, cyanide, and even chasing fish into nets by pounding on the coral with rocks. These methods were very effective for catching fish, but they were not sustainable because they damaged the fish's coral habitat. The fishery in many parts of the Philippines descended into a vicious cycle. Habitat destruction and overfishing led to dwindling fish stocks, which in turn forced fishermen to increase their use of destructive fishing methods to catch anything at all. Fishermen had to travel to distant fishing grounds, working long hours in search of fish, and using any means they could to catch them. The national government reacted with laws banning destructive fishing, but it could not police thousands of islands. Fishermen felt that they had to ignore the ban if they were going to feed their families.

But Apo Island broke out of this downward spiral. In 1979, a marine scientist from a nearby university showed Apo Island fishermen an experiment that he had done on another island in the area. He had set aside part of the coral reef on that island as an area where no fishing was allowed, to see if fish stocks would rebound under the protection, serving as a source of fish for the surrounding area where fishing was allowed. And it was working. The protected area displayed an impressive diversity and abundance of fish.

At first only a handful of Apo families thought it was worth the trouble to try the same thing at their own island. This kind of protection was a new idea for them. But in 1982, with strong support of the island government's leaders, they set aside 10 percent of the island's fishing grounds (450 meters of shoreline) as a no-fishing marine sanctuary. Families rotated the responsibility of standing guard by watching the sanctuary from the beach. Their dedication and patience paid off as nature used its healing powers to restore the coral and the fish populations in the protected area, a process that was strengthened by the rich diversity of life in the coral ecosystem. Within three years, the sanctuary was overflowing with fish, and fishing in the surrounding area was much better than before.

Inspired by this success, the Islanders made some decisions that took their gains even further. They agreed that no one should be allowed to use destructive fishing methods around their island. They would return to the methods they had used in the past, which they knew to be safe. In addition, to help ensure that their recovering coral ecosystem was not overfished, they agreed on one more rule: No outsiders could use their fishing grounds. To Apo Islanders the sanctuary was a sacred site that filled them with pride and offered a better future. They knew that outsiders would not have the same respectful attitude toward Apo's marine resources. They organized a local 'marine guard' to enforce the rules, but they were able to go ahead only after intense political negotiations with higher levels of government for permission to take charge of their local ecosystem and fishery. It wasn't easy, but with persistence and the supportive help of a nongovernmental organization (NGO), they pulled it off!

Apo Island's coral ecosystem has been restored, and fishermen can now catch all they need with a short day's work right around the island. There have been other benefits as well—spin-offs that reinforce and "lock in" sustainability. The revived reef brought eco-tourists, making it even more important to maintain a healthy coral ecosystem. The new economic opportunities that scuba divers and snorkelers brought, such as two small hotels, boating services, and T-shirt sales, gave them a wider range of means to earn their livelihood. Because of their past experience, islanders were quick to see when the reef was being overused by tourists, and imposed necessary restrictions.

The island's primary school added a marine ecology curriculum, and the islanders used part of the tourist income to create scholarships for their children to attend high school and college on the adjacent mainland where some study marine ecosystem management. This growing interaction with scientists and policy makers outside of Apo continues to enable the islanders to manage their ecosystem sustainably. The Islanders also started a very successful family planning program to ensure that their future population does not exceed what their fishing grounds can provide. And using Apo Island as a model of success, 700 villages in the Philippines now have marine sanctuaries.

Step 5: Pass out a copy of the "Ingredients for Success" handout and review the meaning of each ingredient. There are two levels available: a short version and an extended version, depending on the level of your course. They come with frontloaded vocabulary for student comprehension. It may be useful, as you go over each ingredient, to ask students to share real life examples of what that ingredient might look like. This will help them, later, when they try to identify those ingredients in other case studies.

Vocabulary

1. Stimulation: To excite activity or growth.
2. Social System: Everything about a human society, including its organization, knowledge, technology, language, culture, and values.
3. Ecosystem: All the living things (plants, animals, microorganisms) and their environment at a particular place.
4. Restoration: A return of something to its original, unharmed condition.
5. Facilitation: Helping to make something easier or happen successfully.
6. Institution: An established pattern of behavior or relationships in a particular society.
7. Co-adaptation: Two or more things adjusting to each other and fitting together so they function well.
8. Obstacles: The people, things, or events that can block our way.
9. Diversity: Variety (many things different from each other).
10. Resilience: The ability to return to an original form after severe stress or disturbance.

4. Ingredients for Success in EcoTipping Point Case Studies (Short Handout)

1. Outside stimulation and facilitation. An EcoTipping Point story typically begins when people or information from outside a community bring on a shared awareness about a problem (how the situation is changing and what seems to be responsible) and provide fresh ideas for possible actions to deal with it.
2. Strong democratic local institutions and enduring commitment of local leadership. The community moves forward with its own decisions, manpower, and financial resources, so everyone feels a sense of ownership for their community action with the commitment to persist in the face of difficulties.
3. Coadaptation between social system and ecosystem. Social system and ecosystem fit together, functioning as a sustainable whole. When one gains, so does the other.
4. "Letting nature do the work." EcoTipping Points create the conditions for an ecosystem to restore itself by drawing on nature's healing powers.
5. Rapid results. Quick "payback" and something that can stand as a symbol of success help communities stay committed to change.
6. A powerful Symbol. Something that serves as inspiration or stands for success in a way that helps communities stay committed to change.
7. Overcoming social obstacles. Overcoming social, political, and economic obstacles that could block positive change.
8. Social and ecological diversity. Greater diversity of people, ideas, experiences, and natural elements provide more choices and opportunities, and therefore better chances that some of the choices will be good.
9. Social and ecological memory. Learning from the past adds to diversity and often points to choices that were once sustainable. Nature has some of this memory built in at an evolutionary level
10. Building resilience. "Locking in" sustainability by creating the ability to adapt and deal with new (and often unexpected) conditions that threaten sustainability.

Step 6: Either individually, or in groups, have students read a narrative handout for the story and write bullet points onto their Ingredients for Success worksheet, listing examples of events, actions, or conditions in the story that they associate with each ingredient for success.

Instructions: This worksheet will help you to identify the Ingredients for Success in a real situation. First review the Vocabulary needed for this success story. Second, read the success story, underlining the new vocabulary as you go. Finally, write on your worksheet a bullet-point list of elements in the story that are examples of each ingredient for success. For instance, if someone from another city introduced a new idea or helped get a project going, you could put a phrase or sentence about that under 'outside stimulation and facilitation.' Be as specific as possible. You should have at least one example for each ingredient, though for some you may have many.

Vocabulary

1. Fishery: Area where fish are caught.
2. Expanding markets: when there are more people wanting to buy something for sale.
3. Coral Habitat: The reefs of coral that fish use as their homes.
4. Overfishing: When fishermen catch too much for the fish to keep up their population.
5. Ban: To make certain behavior against the rules, usually by law or regulations.
6. Marine Sanctuary: An area where people are not allowed to disturb the marine life.
7. Interdependent Relationship: When two or more things rely upon one another for particular benefits.
8. Persistent: When something or someone continues even if it is difficult.
9. Non-Governmental Organization (NGO): A nonprofit organization that addresses human concerns not covered adequately by government. NGOs are an important part of "civil society."

Student Responses

1. Outside stimulation and facilitation
2. Strong local democratic institutions and enduring commitment of local leadership
3. Co-adaption between social system and ecosystem
4. "Letting nature do the work"
5. Rapid results
6. A powerful symbol
7. Overcoming social obstacles
8. Social and ecological diversity
9. Social and ecological memory
10. Building resilience

Step 7: When students have completed their work, review it as a whole class with the teacher-led Ingredients for Success PowerPoint presentation. The PowerPoint presentation has comprehensive bullet-points that can be used as is or modified to meet your needs. On each slide there are also several photographs to help illustrate the information. There are also additional background notes for the instructor on the notes section of each slide, as well as the Ingredients for Success Teacher Key, which describes examples of each ingredient in the selected case. If you prefer to use a PowerPoint without any pre-written notes on the slides as your point of departure, an editable slide show (with photo captions in the notes section but no text on the slides themselves) is available for your use.

Step 8: After students have seen success in action, they can learn systems thinking to understand what drives decline, how turnabouts from decline to restoration work, and what drives restoration. Explain the following ideas:

- Tipping point – A "lever" (i.e., an action that sets dramatic change in motion).
- Negative tip – The downward spiral of decline.
- Negative tipping point – The action or event that sets a negative tip in motion.
- Positive tip – The upward spiral of restoration and sustainability.
- Positive tipping point (also called EcoTipping Point) – The action that sets a positive tip in motion by leveraging the reversal of decline. An EcoTipping Point is typically an "eco-technology" (in the broadest sense of the word, such as the marine sanctuary at Apo Island) combined with the social organization to put that "eco-technology" effectively into use.

Step 9: Pass out a feedback-diagrams worksheet for the students to map out the vicious cycles driving the negative tip in the story and the virtuous cycles driving the positive tip. There is also a Teacher Key for feedback diagrams with notes for teachers. Fill out the Negative Tip diagram together, noting the negative tipping point, drawing arrows between boxes, and writing the direction of change (increasing or decreasing) in each box. Then, turn to the Positive Tip diagram. After noting the EcoTipping Point as a starting point, let the students fill out the Positive Tip diagram on their own and review the results as a class.

Feedback Diagrams: Apo Island (Student Worksheet) - Diagram for students to fill in

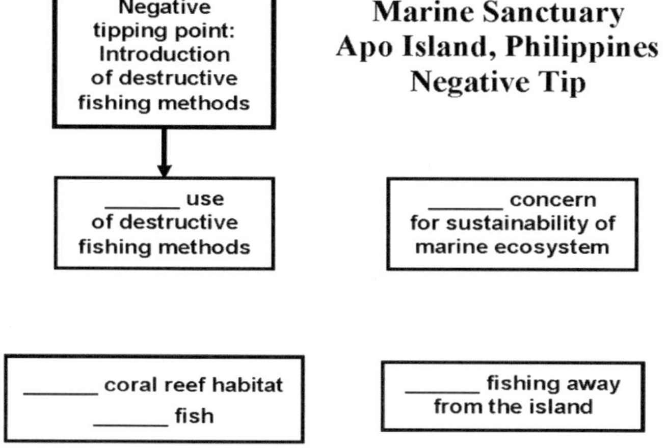

Instructions: Fill in the blanks in the boxes, writing "more" to indicate an increase during the story's period of decline, or "less" to indicate a decrease during that period. Draw arrows between boxes to show which factors were affecting other factors strongly enough to cause their increase or decrease. There should be at least one arrow pointing away from each box and at least one arrow pointing into each box. Finally, trace circular patterns of the arrows that represent "vicious cycles" that were set in motion by the negative tipping point and driving decline.

Instructions: Fill in the blanks in the boxes below writing "more" to indicate an increase during the story's period of restoration, or "less" to indicate a decrease during that period. Draw arrows between boxes to show which factors were affecting other factors strongly enough to cause their increase or decrease. There should be at least one arrow pointing away from each box and at least one arrow pointing into each box. Finally, trace circular patterns of the arrows that represent "virtuous cycles" that were set in motion by the positive tipping point and helped to drive restoration.

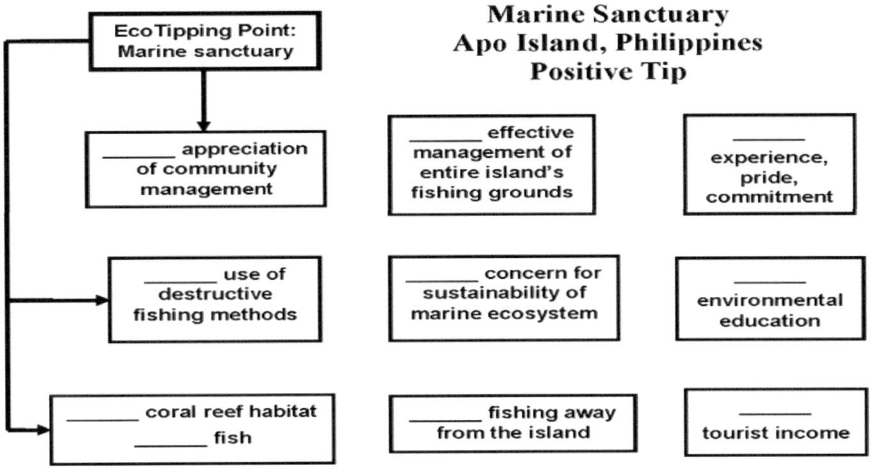

Students should understand that they are not expected to come up with diagrams identical to the Teacher Key. While most of the arrows showing "what affects what" are obvious, others are a matter of interpretation. When students compare their Negative Tip and Positive Tip diagrams, they will discover that:

- One portion of the Positive Tip diagram is identical to the Negative Tip diagram, except change is in the opposite direction (i.e., transformation of vicious cycles to virtuous cycles).
- The rest of the Positive Tip diagram is new virtuous cycles created by the positive tipping point. Those virtuous cycles help to lock in the gains.

Feedback diagrams look complicated at first glance because most of us are less accustomed to looking at cause and effect cyclically. But students of all ages catch on very quickly. After you have gone through one worksheet together, most students find it no more difficult to understand than the basic cause and effect diagrams they already know. This diagram is just cyclical instead of linear!

Class Preparation for Design/Delivery of an Ecologically Sustainable Community Service Program

This third activity in project eleven guides students through the process of developing ecologically sustainable, service learning program proposals. This program, when implemented, is designed to enhance the natural local environment and to simultaneously allow students to demonstrate (and put to practical use) knowledge they have been acquiring throughout the year. Metaphorically speaking, every school community has one or more of its own Chesapeake Bays, Apo Islands, or threatened ecosystems. This activity helps students identify and enrich such places in their own communities.

Teachers may initiate the service learning program design process by engaging the class in a discussion for the purpose of formulating steps that students must follow to be able to perform a service learning activity (such as the coral reef restoration project described in project five) and referred to, as an example, in *Going Blue: A Teen Guide to Saving Our Oceans, Lakes, Rivers, & Wetlands (Kaye, Cousteau, et. al., 2010*). By reviewing examples in this book and other resources, teachers will have enough information to easily guide the students through a process of designing a local service learning program proposal and implementing it. In *Going Blue*, students are guided through a process of five stages. These stages include: 1. Investigate, 2. Prepare, 3. Act, 4. Reflect, and 5. Demonstrate. In this first stage of your service learning journey, you will take steps to find out more about yourself and your skills and the community you are trying to help. You will learn about the issues and identify an ecologically oriented aquatic ecosystem-related need that you want to address. In stage two, it is time to prepare yourself for action by gaining an in-depth understanding of the larger ecological aquatic ecosystem-related system you aim to help. In stage three, students implement their plan. The fourth stage of the service learning project is reflection. It gives students a chance to pause and think about their experience from many angles and to learn from their slipups and successes. During the fifth stage of the service learning project, students inform the world of what they did, how they did it, who was involved, what you learned, and what the results were and directions for future programs. Ask student groups to compile a one day library review of information in this book, or others like it, so that they can expand on each of these steps and help all students in the class follow the same standards as the community service project is designed.

In subsequent class discussions and critiques students will develop a full understanding of the service learning design process. As they establish firm protocols to follow in providing a schedule of service learning events in behalf of the local community, it is reasonable to expect that they will recognize that although every service learning project may not require an in-depth feedback diagram to justify judging it as being of value. However it is not unreasonable to expect that they also come to appreciate more fully how such a diagram might help convince a hesitant member of the school board or administration to approve a proposed service learning activity. And furthermore, diagrams and data related to service learning activities are useful tools for supporting requests for financial support from agencies that provide grant money.

While working through the steps in the service learning design process, students evaluate the community's needs. One technique for guiding students to gather information about local environmental concerns/needs is to help them explore what information can be collected from a survey of students and the community.

As a result of their work in this curriculum, most students have the ability to comfortably perform many of the activities on the Service Learning Skill Assessment Form list below. Although each activity on the list differs from the others, a common theme between them is that each activity requires personal interaction with nature and has the ability to impact the ecology of the local watershed. Students often discover a number of appropriate service learning themes from among which they may choose. Students may also add new service learning themes to the list when appropriate.

After going through the forms and narrowing down the possible service learning programs the class is considering providing to the community, the teacher may help students make a final selection by giving them the following form to fill out. On this form, students can include all pertinent pieces of information relating to community need, practicality of the activity, and value to the students themselves.

Ecologically Sustainable Community Service Learning Program – Design Process (Examples & Forms)

A list of themes upon which students may develop an ecologically sustainable service learning program is provided here. Design themes which students may choose include but are not limited to:

A. establish and maintain riparian buffer zones (on personal or public properties)
B. recycle plastics and paper; compost soils and kitchen scraps, soil enrichment, vermiculture (lunchroom); solicit wood chips and grocery store food waste
C. eradicate non-native plant species and re-establish native plant species
D. develop home or public aquaculture or aquaponics system
E. install, set-up, and maintain a hoop house; seed germination and ground preparation services
F. establish watershed monitoring, data collection, and reporting services that performs local stream cleanup
G. perform prairie restoration support services

H. develop a nature trail and lead nature hikes, canoe/kayak, fishing trips
I. design and beautify school grounds and gardens; install a water garden, pond (e.g. Freidman, 1991), rain garden, butterfly garden, vegetable garden; install and maintain rain barrels
J. assist librarians and other staff with the installation of a fully operational aquatic ecosystem display, aquarium, touch tank, constructed wetland, living machine, algal scrubber
K. develop a program to assist the DNR in installing and maintaining trash/liter/doggie and fisherman waste receptacles along local shorelines/boating areas; rinse boats exiting infested wasters on public ramps

Service Learning Program by Skill Assessment			
Student Name:			
Community Service	**Knowledge & Skills Needed**	**Skills Possessed?** (Yes or No)	**Verification of Skills** (per Self-Assessment Folder) (Adult Supervisor signature)
establish and maintain riparian buffer zones (on personal or public property)			
recycle plastics and paper; compost soils and kitchen scraps, soil enrichment, vermiculture (lunchroom); solicit wood chips and grocery store food waste			
eradicate non-native plant species and re-establish native plant species			
develop home or public aquaculture or aquaponics system			
install, set-up, and maintain a hoop house; seed germination and ground preparation services			
establish a watershed monitoring, data collection, and reporting service that also performs local stream cleanup			

Community Service	Knowledge & Skills Needed	**Skills Possessed?** (Yes or No)	**Verification of Skills** (per Self-Assessment Folder) (Adult Supervisor signature)
perform prairie restoration support services			
develop a nature trail and lead nature hikes, canoe/kayak, fishing trips			
design and beautify school grounds and gardens; install a water garden, pond (e.g. Freidman, 1991), rain garden, butterfly garden, vegetable garden; install and maintain rain barrels			
assist librarians and other staff with the installation of a fully operational aquatic ecosystem display, aquarium, touch tank, constructed wetland, living machine, algal scrubber			
develop a program to assist the DNR in installing and maintaining trash/liter/doggie and fisherman trash waste supplies and receptacles along local shorelines and boating areas; rinse boats on ramps as they exit infested waters			

After reviewing information and deciding on a theme, students are to work in groups of three and generate a service learning implementation proposal with daily guidelines and goals to document progress. A form for this purpose is provided on the following page. After two days, student teams are to present their proposals to the class for a critique. Following the critique, all teams are to adjust their proposals to adhere to the standards that were established for the service learning project during the critique.

Project Title ____

Team Member Names ______________________________________

Task to Complete	Estimated Time Needed	Target Date for Task Completion	Task Completed (enter completion date)	Comments about the task
1.				
2.				
3.				
4.				
5.				
6.				
7.				
8.				
9.				
10.				

Write a list of **science-based** activities you have participated in during this project and briefly describe evidence that verifies your interaction with and knowledge of scientific concepts for each item on your project eleven list.

Write a list of **technologically-oriented** activities you have participated in during this project and briefly describe evidence that verifies your interaction with and knowledge of technologically-oriented concepts for each item on your project eleven list.

Write a list of **engineering-oriented** activities you have participated in during this project and briefly describe evidence that verifies your interaction with and knowledge of engineering-oriented concepts for each item on your project eleven list.

Write a list of **mathematical** activities you have participated in during this project and briefly describe evidence that verifies your interaction with and knowledge of mathematical concepts for each item on your project eleven list.

Write a brief list of **Non-STEM-related** activities you have participated in during this project and briefly describe evidence that verifies your interaction with and knowledge of Non-STEM related concepts for each item on your project eleven list.

Rate your level of involvement in STEM related activities in this project (1=lowest, 10=highest).

1—2—3—4—5—6—7—8—9—10

- Review a description of a case besides Apo Island in which Marten's "human ecology" principles were applied. Develop a report explaining the new cases vicious cycles, feedback diagrams, intervention, and outcome. Explain similarities and differences between the Apo Island case and the case you selected.
- Watch the video: *Turning the Tide: Sustaining Earth's Large Marine Ecosystems* (***http://www.youtube.com/watch?v=-Cr7QiGmbCQ***) Uploaded on May 15, 2011, produced by Francois Odendaal Productions for GEF IW: LEARN. Develop a report related to the ways that policy positions are being addressed, progress that is being made, and steps that must be taken next.
- Watch the 2008 documentary: *The End of the Line*: *The inconvenient truth about overfishing's impact on our oceans.* Develop a report related to the ways that policy positions are being addressed, progress that is being made, and steps that must be taken next.
- Create a journal entry that summarizes what you learned during project eleven. Also write down what you would like to know more about and if you know someone else who might help you learn more if you asked them.
- Select a book, magazine, or scientific journal article to read that is related to a topic in project eleven. Write a report on this reading and include it in your *Beyond the Classroom Aquarium* portfolio.
- Actively contribute to conference committee meetings; keep a written record of contributions you have made and responsibilities you have accepted. Summarize all accomplishments that are reported in each committee meeting you attend. A compilation of your summaries will be collected at the beginning of project fourteen.

Beyond the Classroom Aquarium STEM Self or Peer Assessment Rubric This self-assessment is valid only to the extent to which you can document and display evidence that supports it. For each category in the left hand column, write a STEM-Oriented example in the box beneath 3 if you can display evidence that your STEM-Oriented example represents work that exceeds expectations. If your evidence shows that your STEM-Oriented statement meets but does not exceed expectations, write the statement in the box beneath 2. Similarly, for STEM-Oriented examples that demonstrate some progress toward expectations.				
Name: **Project Number:** **Date Project Work Began:** **Date of Project Completion:** **List Math Topics and/or Objectives Currently Under Study:**		**3** **Work Exceeds Expectations**	**2** **Work Meets Expectations**	**1** **Work Is Progressing Toward Expectations**
Write examples of questions and defining problems related to your STEM work.				
Describe how you have developed and used models in your STEM work.				
Write examples of STEM investigations you planned and carried out.				
Describe STEM related data that you analyzed or interpreted.				
Give examples of how you used mathematics and computational thinking related to the five standards listed to the right.	Problem Solving (Math Process Standard One)			
	Reasoning & Proof (Math Process Standard Two)			
	Communication (Math Process Standard Three)			
	Connections (Math Process Standard Four)			
	Representation (Math Process Standard Five)			
Describe STEM-Oriented explanations and solutions you designed				
Explain evidence you used while engaging in a STEM project-based argument.				
Give examples of STEM information you obtained, evaluated, and/or communicated				
TOTAL POINTS				

The length of time the class is allowed to implement their service learning project is flexible. It depends upon the number of student teams, the time remaining in the school year, the depth of action required, and the level of additional support and assistance from the community—to list a few influential factors. The schedule of implementation actions that was developed in the proposal should be followed and student project foremen should check off the work (for quality and quantity) to verify its completion (can they use excel to generate an appropriate form?). Before, during, and after photos should be taken every day to document the process. Student teams develop a presentation illustrating some of their own contributions during the implementation of the classes' service learning program. Contained in the body of each team's presentation is a section that discusses flipping the service learning program into a business, potential economies impacted by the business, and employment opportunities that incorporate skills performed by the students. In the team reports, student team members are encouraged to mention accredited service learning programs, aquatic ecosystem summer camps, contests, scholarships, and internship opportunities they have discovered and to which they may apply if they have not yet done so. A list of aquatic ecosystem-related job descriptions is provided to give students information to help them get started.

Due to the flexible nature of the activities in projects ten and eleven it is likely that for most classes at some time (concurrent with the time when students are performing their service learning field work) in project twelve, the eight week time limit for research project data collection will arrive. When this period of time arrives, students must complete the following two activities: 1) the completion and submission of a final copy of each student team's research report (an activity that may take up to two weeks and 2) the disassembly of many auxiliary classroom aquatic ecosystems that were used for research project purposes.

At the conclusion of project twelve, as research projects and the multi-week implementation of the service learning program are concluded, the process of finalizing conference preparations is emphasized. A format for PowerPoint presentations to be shown at the conference is provided and the format for individual presentation of STEM-Oriented experiences by project to be displayed at the student teams trade show booths (poster sessions) is also provided.

Design a Large Scale Aquatic Ecosystem Trade Show Booth - Diorama – Project Twelve Opening Activity

The final project assignment relates to the trade show booths that student teams will sponsor. For this assignment, student teams are to design (as a living diorama - a trade show booth, like a float in a parade) an aquatic ecosystem exhibit that illustrates, in human scale diorama form, an aquatic ecosystem with the team members in costumes representing the organisms and environmental features that are unique to the system. Through the booth's characters and décor, students can share their understandings of the people native to the region, their socio-economic and cultural existence, and important environmental decisions they make in relation to the watersheds in which they reside. Booths designs should highlight the diverse features that make their chosen culture unique (including economics, customs, music, art, stories, dance, language, history, and where possible, students may include a demonstration documenting the preservation of traditional knowledge e.g. with respect to food processing and preparation). Booths will be judged based on the authenticity of the library review information gathered about the representative region and other representative anthropological artifacts and demonstrations portraying human-ecosystem interactions and based on their portrayal of aquatic creatures representative of the chosen unique aquatic ecosystem. Each team's booth design should result from thoughtfully applying human ecology concepts studied in relation to Apo Island.

Each booth may also distribute a team-designed brochure to promote interest in the region by including tasteful art, important facts (history of the area—lifestyle in some cases stepping back in time while in other cases looking into the future), necessary information, how to obtain further information, and useful maps. Brochures may also provide information about unique species, a specific habitat, an organization that operates in the area (or could operate there), and information that promotes ecologically sustainable tourism in the area.

Student teams are to select a watershed theme for their tradeshow booth from those listed in the table below.

Asiatic & Indo-Pacific	**South American**	**African & European**	**North American**	**Caribbean & Gulf of Mexico**
• Ganges River	• Tribal Amazonian Rainforest	• Dead Sea	• Great Lakes	• Turks & Caicos
• Hawaii – Kawai Nui Marsh	• Galapagos	• Lake Victoria	• Mississippi River	• Lake Chapala
• Solomon Islands	• Sea of Cortez	• Nile River	• Chesapeake Bay	• Lake Xochimilco
• Papua New Guinea	• Pantanel	• Na'aman River	• Long Island Sound	• Lerma-Santiago River
• Mahakam	• Lake Atitlan	• Drava & Rinya Rivers	• National Marine Sanctuaries	• Usumacinta River
• Philippine Environments	• Rio de la Plata	• Danube River	• National Estuarine Research Reserves	• Cubuy & Icacos Rivers
• Tanshuei River	• Sao Francisco River	• Lake Nakuru	• Walpole Island	• Rio Tempesque
• Singapore River	• Parana River	• Swaziland	• Alaska	• Rio La Mina
• Johnston Atoll	• Magdalena River	• Rhine River	• Colorado River	• Salt River Bay

Learn what it's like at sea with this collection of interviews from participants in Ocean Explorer research expeditions. NOAA Ocean Explorer staff has organized, on this single page, all interviews conducted over the years to introduce students to professions that may not have been covered in as much detail in the NOAA OceanAGE Careers section. Click on the links for an interview with these ocean explorers.

Assistant Professor **Associate Professor** **Associate Scientist** **Deep-sea Biologist** **Fish Ecologist** **Geophysicist** **Marine Archaeologist** **Marine Biologist** **Marine Ecologist** **Marine Mammal Biologist** **Marine Taxonomist**	**Microbiologist** **Natural Products Biologist** **NOAA Corps Officer** **NOAA Marine Scientist/Educator** **NOAA Physical Scientist/Administrator** **Oceanographer** **Principal Research Scientist** **Research Coordinator** **Senior Scientist** **Submersible Pilot**

Explore the careers above by following each career's podcast link. While viewing each podcast, categorize each career on the basis of the criteria in the career checklist matrix below. Select the three careers that most closely approach the ideal career for you. Write a report that explains how you prioritized the careers and what factors influenced you most in selecting your top three potential jobs. Include information in the report that describes publically available opportunities for young people to gain first hand personal experience related to the top three careers selected; for example, camp programs, youth internships, summer classes. Also describe the process students need to follow in applying for attendance in these programs.

Career Checklist Matrix			
Career Title	Work Setting Environment	Academic Training & Professional Development Including Math Applications	Potential for Advancement

Write a list of **science-based** activities you have participated in during this project and briefly describe evidence that verifies your interaction with and knowledge of scientific concepts for each item on your project twelve list.

Write a list of **technologically-oriented** activities you have participated in during this project and briefly describe evidence that verifies your interaction with and knowledge of technologically-oriented concepts for each item on your project twelve list.

Write a list of **engineering-oriented** activities you have participated in during this project and briefly describe evidence that verifies your interaction with and knowledge of engineering-oriented concepts for each item on your project one list.

Write a list of **mathematical** activities you have participated in during this project and briefly describe evidence that verifies your interaction with and knowledge of mathematical concepts for each item on your project twelve list.

Write a brief list of **Non-STEM-related** activities you have participated in during this project and briefly describe evidence that verifies your interaction with and knowledge of Non-STEM related concepts for each item on your project twelve list.

Rate your level of involvement in STEM related activities in this project (1=lowest, 10=highest).

1—2—3—4—5—6—7—8—9—10

- Create a journal entry that summarizes what you learned during project twelve. Also write down what you would like to know more about and if you know someone else who might help you if you asked them.
- Select a book, magazine, or scientific journal article to read that is related to a topic in project twelve. Write a report on this reading and include it in your *Beyond the Classroom Aquarium* portfolio.
- Actively contribute to conference committee meetings. Keep a written record of contributions you have made and responsibilities you have accepted. Summarize accomplishments reported in each committee meeting you attend. A compilation of your summaries will be collected at the beginning of project fourteen.

Begin preparing to conduct a mock run-through of the conference to complete conference preparations. As conference committees finalize conference plans they should include logistical considerations involved in displaying the three primary aquatic ecosystems at the conference, including:

  - disassembling the classroom ecosystems
  - transporting them to the conference location
  - reassemble them for display
  - disassemble them at the end of the conference
  - return them to storage at the school (unless they are raffled off to another school or interested community member/educator)

- Write a job description of skills knowledge and attitudes you would expect from an employee who would be responsible for successfully operating and maintaining the aquatic ecosystems in the classroom.
- What skills do you not yet have that would make you more valuable to a business or employer?
- Identify aquatic ecosystem-related camps, internships, jobs, and other voluntary opportunities that you consider interesting and fill out and submit an application (for two of these) to be invited to attend.

- Assist students in another class in designing and assembling an anthropologically representative aquatic ecosystem exhibit including geological, geographical, sociological, anthropological, and economic data; artifacts and illustrations that characterize human-ecosystem interactions; incorporate aquatic creatures from the selected ecosystem into the living exhibit when this can be done without harming their native area
- Review the Aquaculture Interactive DVD report on information from each of the seven main units that you found most important and useful. Describe any shortcomings that you noticed about the information provided. Search for an up to date aquaculture management simulation program. Explain if you think it would help you prepare to open and successfully operate an aquaculture business.

Beyond the Classroom Aquarium STEM Self or Peer Assessment Rubric

This self-assessment is valid only to the extent to which you can document and display evidence that supports it. For each category in the left hand column, write a STEM-Oriented example in the box beneath 3 if you can display evidence that your STEM-Oriented example represents work that exceeds expectations. If your evidence shows that your STEM-Oriented statement meets but does not exceed expectations, write the statement in the box beneath 2. Similarly, for STEM-Oriented examples that demonstrate some progress toward expectations.

<table>
<tr><th colspan="2">Name:
Project Number:
Date Project Work Began:
Date of Project Completion:
List Math Topics and/or Objectives Currently Under Study:</th><th>3
Work Exceeds Expectations</th><th>2
Work Meets Expectations</th><th>1
Work Is Progressing Toward Expectations</th></tr>
<tr><td colspan="2">Write examples of questions and defining problems related to your STEM work.</td><td></td><td></td><td></td></tr>
<tr><td colspan="2">Describe how you have developed and used models in your STEM work.</td><td></td><td></td><td></td></tr>
<tr><td colspan="2">Write examples of STEM investigations you planned and carried out.</td><td></td><td></td><td></td></tr>
<tr><td colspan="2">Describe STEM related data that you analyzed or interpreted.</td><td></td><td></td><td></td></tr>
<tr><td rowspan="5">Give examples of how you used mathematics and computational thinking related to the five standards listed to the right.</td><td>Problem Solving (Math Process Standard One)</td><td></td><td></td><td></td></tr>
<tr><td>Reasoning & Proof (Math Process Standard Two)</td><td></td><td></td><td></td></tr>
<tr><td>Communication (Math Process Standard Three)</td><td></td><td></td><td></td></tr>
<tr><td>Connections (Math Process Standard Four)</td><td></td><td></td><td></td></tr>
<tr><td>Representation (Math Process Standard Five)</td><td></td><td></td><td></td></tr>
<tr><td colspan="2">Describe STEM-Oriented explanations and solutions you designed</td><td></td><td></td><td></td></tr>
<tr><td colspan="2">Explain evidence you used while engaging in a STEM project-based argument.</td><td></td><td></td><td></td></tr>
<tr><td colspan="2">Give examples of STEM information you obtained, evaluated, and/or communicated</td><td></td><td></td><td></td></tr>
<tr><td colspan="2">TOTAL POINTS</td><td></td><td></td><td></td></tr>
</table>

PROJECT THIRTEEN …And Beyond! A Prospective Vision of Future Design Projects

Take a look around your classroom; the sights, smells, and sounds. By the end of project thirteen it will be transformed from an aquatic ecosystem jungle back to its pre-project one condition. As systems are drained, organisms transferred to new homes, and system frames are disassembled (properly labeled and stored away), a few tasks must yet be performed. The design activities that conclude project thirteen are designed to raise student awareness of the idea that for many students it is the power of human curiosity—the "what if" questions and the persistence in seeking answers to questions through unique applications of information—that makes them capable of successfully pursuing further studies/careers in aquatic ecosystems.

Design and Construct a Basic Laminar Flow Fountain - Project Thirteen Opening Activity

Photo by Robert Jakus

Since the 1980s "fountains with laminar streams have become relatively common. "Fountains can be useful in teaching engineering. Fountain oriented projects can involve introductory level students in reproducing simple historical fountains while for more advanced students, the design and construction of a special-effects fountain can be an excellent project" (Shakerin, 2005).

The inspiration for the laminar fountain template upon which students will create their own laminar flow fountains was originally motivated while viewing the YouTube video (http://www.youtube.com/watch?v=OV-IazRk0sU). However, the laminar flow fountain design plans students will follow in this project are a bit more high tech than those in this wonderful introductory video. The instructions provided here constitute a restructuring of information provided by Schauer (1997). The first important difference is that the classroom laminar flow fountain, instead of using a garden hose, will use a powerful submersible pump (≥ 1200 gallons per hour) to supply water to the fountain. In the classroom laminar flow fountain described here, water first flows into and through:

- A submersible pump that has been positioned in a reservoir, pool, or pond (if possible, the reservoir should have a radius ≥ 3 feet. For initial development and testing, it is only necessary that the depth of water depth in the reservoir be sufficient to cover the submersible pump)
- through an inlet assembly between the pump and the bottom plate of the fountain cylinder
- upwards through the cylinder's 0.5-inch nozzle opening in the top plate of the land based fountain
- through the air in a laminar stream back into the reservoir, pool, or pond from which it first came

Since this project uses electricity and water students should be monitored when participating in this activity. A GFCI outlet or GFCI circuit breaker should be used. Always keep bare wires apart from each other so they will not short. Always unplug the equipment when connecting wires so no "shorts" develop. And be sure to turn off the submersible pump when it is not in use. It is also suggested that teachers model the behavior of recognizing when assistance is needed and of asking for help from qualified local tradespersons, as necessary.

The first set of detailed instructions below provides steps students may follow to develop the three main components of the laminar fountain's cylinder: the top plate, bottom plate, and internal components of the cylinder itself. The description of the inlet assembly connection between the pump and the cylinder is provided in the second set of detailed instructions because the proper attachment between the cylinder and the pump depends upon the availability of a "ready to use" bottom plate. After following these steps and setting up the fountain, students can carefully conduct field tests on it.

Dimensions of laminar flow fountain cylinders vary. In the interest of minimizing costs while still producing interesting fountains, the development of a fountain from a 4-inch diameter PVC pipe of 11-inch length is described here. Students following this description are guided to fabricate the cylinder's top and bottom plates. The dimensions of the plexiglass (or Lexan) from which the top and bottom plates of the cylinder are formed are 5.5 x 5.5 x 0.25 inches thick. Water will enter the cylinder through the bottom plate. A laminar stream of water will exit the fountain cylinder through a 0.5-inch nozzle in the top plate. To begin making the top and bottom plates, drill six evenly spaced 0.375-inch holes, about 2.25 inches from the center of both pieces of plexiglass.

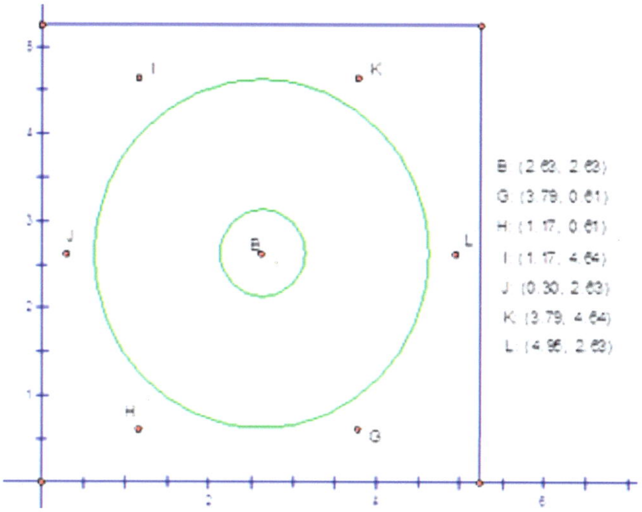

Design and Fabrication of the Top Plate and Nozzle

The first step in preparing a spot in the plexiglass sheet that will serve as the top plate of the fountain drill a two-inch hole in the center of it. Then drill a 0.5-inch hole in the middle of a separate 3-inch diameter piece of 0.25- inch thick plexiglass. This 0.5-inch opening will be called the "nozzle." Next, make the nozzle convex by using a Formica-trimming router bit. Sand the nozzle, first with 100 grit sand paper and then use 220 grit. The nozzle has to be smooth. By carefully heating the nozzle opening with an open flame it is possible to smooth it out even more (depending on the smoothness of the opening, the water flow will have a more or less glasslike consistency). If there is the slightest flaw, it will deform the laminarity of the flow. Lastly, attach this piece of plexiglass with its polished nozzle opening to the top plate with six stainless steel screws (0.5-inch long, 10/32-inch threads). Drill the holes into top plate and the nozzle at the same time, then "tap" the treads into the top plate. Drill the holes one size bigger in the nozzle. Attach this to the top plate with a bead of silicone around the screws.

Design and Fabrication of the Fountain Cylinder

From a separate 4-inch diameter PVC pipe, cut three 1-inch wide rings. Remove a 0.5-inch wide vertical strip (top to bottom) from each ring. The 0.5-inch gap in the three rings makes each ring compressible enough to fit snugly INSIDE the 4-inch diameter 12-inch long cylinder. These compressible sections will serve as retaining rings once all burrs have been filed off.

In order to prepare the 11-inch long piece of PVC that is to become the fountain cylinder, students must cut three 4.5-inch diameter circles out of nonmetallic window screen fabric. One of these circular screen pieces is to be placed on the top of the cylinder. Students are to then place a 1-inch PVC ring on the screen and press it into the cylinder for a distance of 1.5-inch. The next step is to turn the cylinder over and place 200 plus drinking straws into the cylinder so they rest on the screen. Students should keep jamming straws in until they cannot get another one in. After the cylinder is packed with straws, another circular screen cutout should be placed on the edge of the cylinder and a PVC ring placed on top of it. This screen should be slid into the cylinder to the position of the straws. Lastly, two 4-inch diameter scouring pads should be placed on the screen inside the cylinder and a final piece of cutout screen should then be slid down into the cylinder and held in place with the final PVC ring. This completes the preparation of the fountain cylinder.

Design and Fabrication of the Bottom Plate and the Gate Valve and Inlet Screen Assembly
For the bottom of the fountain cylinder, a hole with a diameter that matches the diameter of the outlet fitting on the submersible pump must be drilled in the center of the remaining plexiglass plate. Once a hole of the proper size has been drilled, students will carefully use a torch to heat up a metallic male threaded plumbing fitting (of the same diameter as the outlet fitting on the pump) and use this threaded hot metal fitting to thread the hole that was drilled in the bottom plexiglass plate. The fitting must cool before it is removed. Do not force it. If it does not come out of the hole easily, leave it in and use it. Reinforce the bond between the fitting and the plexiglass with some two part epoxy. Once a fitting is firmly in place in the bottom plate, place silicone between the top and bottom plates and the edges of the PVC pipe. The silicone may act as a gasket to help reduce the chance of fountain leaks. Line up the six pre-drilled holes. Use six 0.375 x 12-inch threaded rods with washers and lock washer and nuts to sandwich the cylinder between the bottom plate and the top plate.

Finally, students must measure the distance between the point where the submersible pump has been placed and the position of the fountain. While plumbing a connection between these two components, it is suggested that a properly sized gate valve and inline hose screen be placed between the pump and gate valve. The screen is intended to prevent the internal components of the fountain cylinder from becoming clogged with debris. Once these steps are completed the fountain is ready for field testing.

This **Alternative Method** is described on the *Beyond the Classroom Aquarium* webpage in the following link: **https://wp.me/p3pycD-hV** What you may notice in this method is the use of an Arduino MEGA, LED lighting, a simple prototype cylinder and contents and a 3-D printed mechanism to fit over a 4 inch diameter PVC pipe for use with a next stage prototype laminar fountain (including a DC stepper motor and motor driver). Many images are provided there; such as the soldering and shrink wrap protection of a RGB LED, testing the LED-Arduino circuit and color fade-in/fade-out code, submerging the prototype fountain and finally running it - in the dark to assess the stream (smoothness, etc.) and light; how many LEDs could be installed in the fountain cylinder? How powerful of a pump? How can a DC pump be used and controlled with Arduino code using pulse width modulation (and another motor driver)?

Photo by Robert Jakus

Photo by Robert Jakus

Photo by Robert Jakus

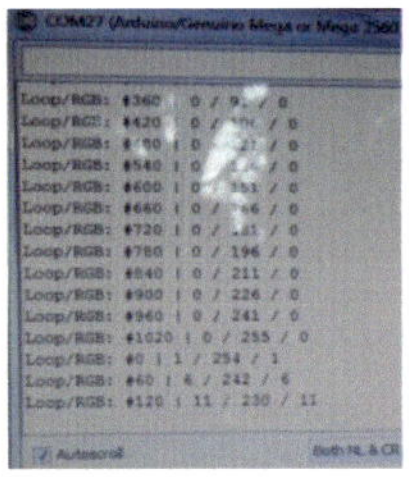
Photo by Robert Jakus

Photo by Robert Jakus

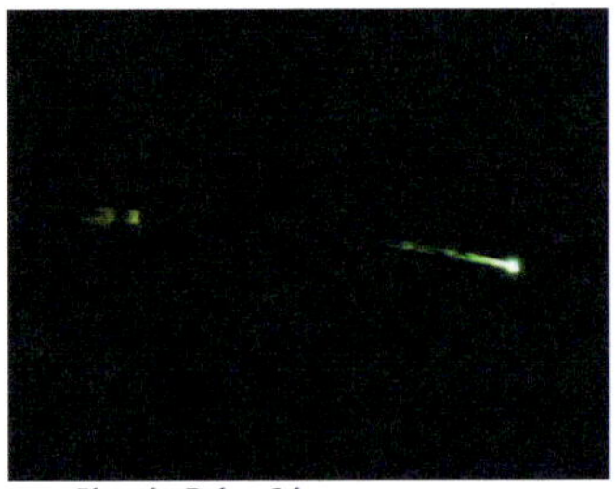
Photo by Robert Jakus

Any of the following seven design project themes may be selected and conducted as a whole class activity. In other classes, student teams may be allowed to independently select and work on one of the six design themes. Ideally, student teams will have a full week to develop their designs. In many cases, but not all, supportive assistance from parents (and local tradespeople) may be encouraged. Information for the purpose of guiding student participation in activities related to each of the six design themes is provided on the pages that follow.

When the time limit designated for this design activity arrives, student teams may be given one additional day to prepare a formal presentation of their design to the class. These presentations are well worth the few additional days of class time because each in each presentation the design team is informing their fellow students of concepts and technologies that they may wish to study for themselves more fully in the future.

Prototype demonstrations models of these projects will also be displayed at the *Beyond the Classroom Aquarium* end of year conference.

1. Design supplementary features to enhance the laminar flow fountains developed in the introductory activity (e.g. computer automation, colored LED lights, optic fibers, manual, and automated flow cutters).
2. Design a mobile video/acoustic system (MVA) to permit real-time synchronous recording of the sounds and behaviors of individual, free-swimming dolphins. Students who are mathematically advanced may also include information in their presentation that explains the mathematics that underlies bio-acoustical sampling methods incorporated in software used in sampling sound.
3. Design a Submersible Remotely Operated Vehicle (ROV) with environmental sensors and mechanical functions (tethered to a controller).
4. Design a wireless remote ecosystem water quality data collection and communication system. Include a feature that permits students on the Internet to access classroom aquatic ecosystem water quality data from anywhere in the world.
5. Design a proposal for supplying aquaponics/greenhouse system electronics renewably (lights, pumps, fans, and other permaculture related applications).
6. Design a futuristic experimental aquarium system or component.
7. Design:
    - an automated floating trash collector (per the provided URL: http://healthyharborbaltimore.org/whats-happening-now/water-wheel), or
    - a mobile, solar powered plastic melting mechanism than can be used to reform trash plastic into seats/stools or other useable items (using 3-D printer technology – follow this URL: http://player.vimeo.com/video/58461689), or

The Design Team Project Form on the following page provides a checklist of organizing steps that each design team may follow as they work on one of the six design activities described above. The steps on this list are designed to guide student teams through their design project from beginning to end.

Design Team Project Form		
Task	**How it is Done**	**Result**
Project Selection	Team members: • Review information on each of the six project themes • Summarize information for fellow group members • Explain the kind of product that work on each theme will require • Critique design options • Select a project theme • Attempt to document (photos and in writing) this design process from beginning to end for inclusion in the final report	
Project Brainstorming	Each team member creates a design overview that represents their vision of the selected design project. Each member reports on their understanding of how each design theme relates to their future academic/career ambitions.	
Team Critique	Each member presents information they believe to be important to the selected design project and all essential elements are discussed and merged into one common design.	
A Detailed Design Project	Team members gather and organize information related to the design they develop: • A written description • Posters of the project with diagrams, graphs, and illustrations • A 3-D representation of the equipment being designed	
Final Written Report	Reports should explain: • The purpose of the piece of equipment being designed • Functions it will be capable of performing • Materials and equipment needed to operate it • Principles upon which the design is based • The purpose of each component in the equipment's design • Potential applications to future studies Resources should be properly cited.	
Conference Presentation (Booth)	Team members set up a display table at one of the trade show booths at the conference. At this table, they lay out the 3-D model of the equipment designed, posters, and documentary information.	
Disassembly	Copies of reports and documentation of project participation are placed in each team members' portfolios.	

The resources listed below are provided to introduce students to key ideas related to an advanced laminar flow fountain design activity. Students who elect to participate in this design activity are encouraged to begin by exploring literature listed below. As they review these resources, most students will discover other relevant resources useful for the purpose of designing an amazing laminar flow fountain show.

> "Although studying patents is not a routine part of engineering curricula, the practice can offer a rich source of design knowledge and can be suitable for any class level. Dave Ayer, Mark Fuller and Lee Sim designed a laminar-flow fountain nozzle while seniors at the University of Utah, and Fuller later patented (4,795,092) a nozzle producing a stream with "substantially no turbulence" in 1989. Fuller and Robinson refined this design by devising a quick diversion method by which the laminar stream can be controllably terminated to give the effect of slicing the stream perpendicularly to its longitudinal axis (patent 4,889,283). Further improvements (patents 4,955,540, 5,078,320 and 5,115,973) included adding a mounting assembly used to change the angle of and reposition the nozzles so that the laminar stream appears to emanate from a fixed location at different angles. This allows varying the characteristics of the arch-like laminar stream in dynamic displays" (Shakerin, 2005).

In this design activity, after reviewing the additional technical features described below and performing an expense analysis of costs to acquire the additional equipment, part two of this design activity requires students to produce and field test a number of identical additional laminar flow cylinders. Much of the laminar fountain information provided here is excerpted and modified from work described at Jerry L. Schauer's website (http://www.hiddenmickeys.org/Imagineering/LeapFrog.html). Based on this information, the total number of cylinders in this activity should not exceed eight. I suggest no more than three for now. Students will document their participation in this activity and develop drawings, diagrams, and results of field tests on each fountain built, including descriptions of modifications they made to the plans as presented here. Besides reviewing the patents listed above, there are also many laminar fountain YouTube videos to check out.

1. Set up a circular reservoir (diameter ≥ 3 feet; depth ≤ 18 inches).
2. Place the submersible pump in the water in the reservoir and route the electric cord to a safe outlet.
3. Build a PVC pipe ring (of a radius matching that of the pumps outlet radius) around the reservoir.
4. Use a PVC tee to connect the ring of pipe surrounding the reservoir to the pump.
5. Carefully position and attach a threaded PVC tee fitting to the PVC ring for each fountain.
6. Connect each tee opening to an electric solenoid valve through which water to the fountain must pass using brass plumbing fittings. Make sure that the input side of the valve is connected to the PVC tee.
7. Aim the output nozzle of the valve toward the middle of the pond.
8. Connect electrical wire to each solenoid and run each wire to a dry, central location.
9. At the fountain control center, connect X-10 Appliance Modules to the wires from the solenoids.
10. Waterproof all electrical connections for safety by covering each connection with silicone.
11. Plug the X-10 Appliance Modules into your GFI outlet.
12. Adjust the X-10 codes to control each solenoid valve separately. X-10 System directions are helpful.
13. The wireless, handheld X-10 remote control unit has 8 buttons to control up to eight jets in this fountain.
14. Adjust plants to hide your PVC pipes and fountains.
15. Field-test this system to identify and remove problems.

Students who want to pursue fountain design even further may explore ways to incorporate optical fibers into laminar fountains, build a few more fountains, and create a laminar flow fountain light show for presentation at the end of year *Beyond the Classroom Aquarium* Conference. . . seriously! A basic introduction to the principles upon which optical fiber light shows in laminar fountains are based is provided in *Lights, Lenses, and Lasers* (Berger, 1987). And then there are musical fountains. . . .

Students who elect to participate in this design activity are to be guided to explore literature listed below along with other relevant resources they find useful for the purpose of designing a mobile video/acoustic system (MVA) to permit real-time synchronous recording of the sounds and behaviors of individual, free-swimming dolphins that research divers may use. Information in the design presentation should incorporate bio-acoustical issues related to cetaceans. Presentations should explain the purpose of the system, materials and equipment needed to operate it, and the principles upon which the design is based. Resources should be properly cited. The purpose of each item in the design should be noted and explained. Students who wish to participate in a mathematical extension of this activity may include a description and illustration of how they might use the sampling theorem (shown beneath the list of five resources on this page) to assist them in determining optimal sampling frequencies for use in sound sampling software.

Photo by Robert Jakus

- Dolphin Communication Project - http://www.dolphincommunicationproject.org/about-dcp/the-mva-and-ecd.html
- Watch the 2011 video: *Sea of Sound: Journey into the Surprising World of Ocean Sound*; produced by the National Oceanographic Partnership Program and the Cornell Lab of Ornithology
- Kraus, Scott; (2010); Whales in the Urban Ocean: http://brctv.org/blog/2010/07/01/whales-in-the-urban-ocean/
- Knowlton, A.R. and S.D. Kraus; 2001; Mortality and serious injury of northern right whales (Eubalaena, glacialis) in the western North Atlantic Ocean; *J. Cetacean Research and Management, Special Issue* 2:193-208.
- Read, A.J., S.D. Kraus, K.D. Bisack, and D. Palka; 1993; Harbor porpoises and gill nets in the Gulf of Maine; *Conservation Biol.* 7(1):189-193

$$F_a = \left| F - \frac{(k+1) \cdot R}{2} \right|, \qquad \frac{k \cdot R}{2} \le F \le \frac{(k+2) \cdot R}{2}$$

Where

- F is the actual frequency in Hz
- R is the sampling rate in Hz
- k is any odd integer which satisfies the inequality
- F_a is the apparent frequency in Hz

Students who elect to participate in this design activity are guided to explore literature listed below along with other relevant resources they find useful for the purpose of designing either:

Photo by Robert Jakus

A. A wireless remote ecosystem water quality data collection and communication system, or
B. A non-autonomous **remotely operated underwater vehicle** (ROVs) – controlled and powered from the surface by an operator/pilot via an umbilical or using remote control.

Designs should explain the purpose of the system, materials and equipment needed to operate it, and the principles upon which the design is based. Resources should be properly cited. The purpose of each item in the design should be noted and explained.

- Introduction to Wiring the Abyss; http://b.vimeocdn.com/ts/437/457/437457014_100.jpg
- Canadian Scientific Submersible Facility; http://www.ropos.com/
- Ocean Observatories Initiative; http://en.wikipedia.org/wiki/Ocean_Observatories_Initiative;
- Victoria Experimental Network Under the Sea; http://venus.uvic.ca/
- NEPTUNE Canada Goes Live - *Dec 15, 2009;* http://www.neptunecanada.ca/news/news-details.dot?id=20604
- Monterey Accelerated Research System; http://en.wikipedia.org/wiki/MARS_(oceanography)
- MIT Sea Grant; (2011); *Sea Perch: Construction Manual*; http://seaperch.mit.edu/docs/seaperch-build-october2011.pdf
- MIT Sea Grant; (2011)Building a Sea Perch – Videos; http://seaperch.mit.edu/build_videos.php

Students who elect to participate in this design activity are guided to explore literature listed below along with other relevant resources they find useful for the purpose of designing renewable energy supply systems to install in a local/classroom aquaponics system or greenhouse. Information included in the design explanation should provide system users with best practices recommendations for energy use. Designs should also explain the purpose of each component of the energy system the materials and equipment needed to operate it and the principles upon which the design is based. Resources should be properly cited. A projected energy budget should also be developed to help system users understand advantages that result from using renewable energy.

Illustrations and diagrams

- Grubinger, Vern; *Renewable Energy for Greenhouse Heat*;
- Bellows, Barbara; updated by K. Adam; NCAT Agriculture Specialists; (2008); *Solar Greenhouses*; NCAT; IP142
- Kid Wind Project Inc.; 2013; http://learn.kidwind.org/teach
- Kindberg, Leif; (2010); NCAT; IP369; *Energy-Efficient Lighting for the Farm*; 12 pages; https://attra.ncat.org/attra-pub/summaries/summary.php?pub=341
- Svejkovsky, Cathy (NCAT Energy Specialist); 2006; *Renewable Energy Opportunities on the Farm*; https://attra.ncat.org/...-pub/summaries/summary.php?pub=306
- Open Aquarium - Aquaponics and Fish Tank - Monitoring for Arduino**;** http://www.cooking-hacks.com/documentation/tutorials/open-aquarium-aquaponics-fish-tank-monitoring-arduino (Libelium Comunicaciones Distribuidas)
- Open Garden - Hydroponics & Garden Plants - Monitoring for Arduino**;** http://www.cooking-hacks.com/documentation/tutorials/open-garden-hydroponics-irrigation-system-sensors-plant-monitoring; (Libelium Comunicaciones Distribuidas)

Students who elect to participate in this design activity are guided to explore literature referenced in the caption along with other relevant resources they find useful for the purpose of designing a calcium reactor. Design a calcium reactor that will be activated/de-activated by a calcium sensor connected to a programmable logic controller). The student developed system information sheet may include descriptions of applications to ocean acidification and species specific larval rearing pulse systems. Thorough and safe directions for connecting, monitoring, and maintaining pressurized carbon dioxide containers to the calcium reactor system and directions for proper and safe placement of a calcium reactor and additional details, equipment used and specifications may be required per the teacher's request. Also see Moe (2014a) pages 181 – 184.

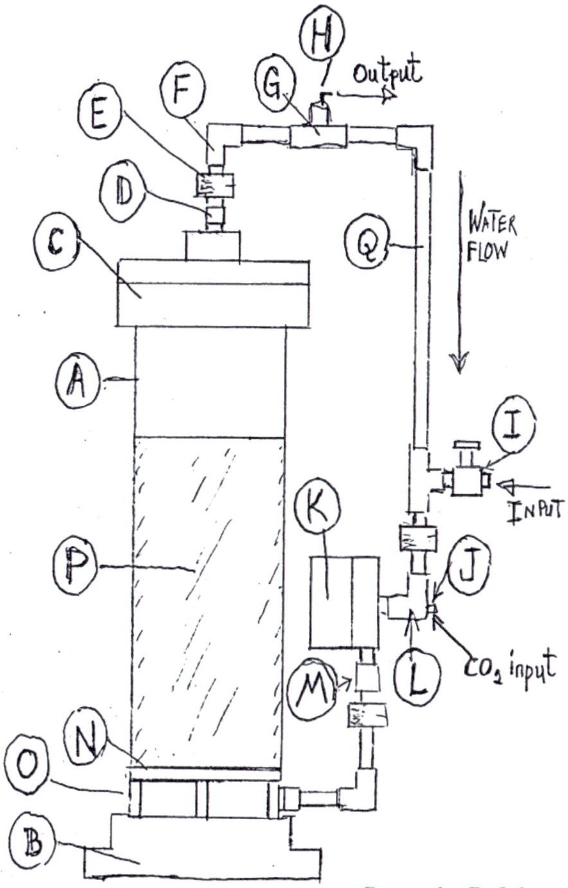

Drawn by R. Jakus

PART	DESCRIPTION	QUANTITY
A	Four and a half inch OD (1/8)" thick 24" long acrylic tube	1
B	Four inch diameter PVC closed - flange	1
C	Four inch PVC threaded female adapter & threaded cap assembly	1
D	Half inch PVC male adapter (MNPT x Slip)	2
E	Half inch PVC union (Slip x Slip)	3
F	Half inch PVC elbow (Slip x Slip)	3
G	Half inch PVC Tee (Slip x Slip x FNPT)	2
H	Tubing adapter elbow 0.5" x 0.25" barb	1
I	Half inch needle valve	1
J	Tubing adapter – 0.125" NPT x 0.25" (or 3/16) barb	1
K	Properly sized water pump	1
L	Half inch PVC elbow (MPT x Slip)	1
M	Half inch PVC female adapter (Slip x FPT)	1
N	Egg crate covered with fiberglass mesh	1
O	Supports for egg crate	3
P	Aragonite sand medium	1 bag
Q	Half inch diameter PVC pipe	2.5 Feet

Students who elect to participate in this design activity are guided to explore literature listed below along with other relevant resources they find useful for the purpose of designing a seawater system for an experimental aquarium. Designs should explain the purpose of the system, materials and equipment needed to operate it, and the principles upon which the design is based. Resources should be properly cited. The purpose of each item in the design should be noted and explained.

Some resources to help students begin their search for experimental aquarium design ideas to develop may be found in the following resources.

- Quesada, Laura; (2011); Automatic Clownfish Hatchery; http://digitalcommons.calpoly.edu/
- Crandall, T.; Busack, C. A.; & Gall, G. A. E.; (1981); An easily constructed recirculating aquarium system for research requiring many small groups of animals; *Aquaculture*, 22, 193-199
- Clark, JR; Clark, RL; (Eds.); (1969); *Sea-water systems for experimental aquariums, A Collection of Papers*, TFH Publications, Jersey City, NJ
- Adey, Walter H.; Loveland, Karen; 1991; *Dynamic Aquaria: Building Living Ecosystems*; pp. 643; Academic Press, Inc.
- Kreisel http://raisingpetjellyfish.blogspot.com/2012/01/how-to-easy-diy-jellyfish-kreisel-tank.html
- http://superpod.ml.duke.edu/cachalot/ - *Mobile Marine Megafauna: A Marine Science Digital Textbook*
- Design of a properly scaled wave flume research tank for studying impacts of surface shore interactions on specific species of planktonic larvae (e.g. echinoderm, mollusc, or crustacean) http://www.wrl.unsw.edu.au/site/resources/facilities/06m-wave-flume/

Write a list of **science-based** activities you have participated in during this project and briefly describe evidence that verifies your interaction with/knowledge of scientific concepts for each item on this project thirteen list.

Write a list of **technologically-oriented** activities you have participated in during this project and briefly describe evidence that verifies your interaction with and knowledge of technologically-oriented concepts for each item on this project thirteen list.

Write a list of **engineering-oriented** activities you have participated in during this project and briefly describe evidence that verifies your interaction with and knowledge of engineering-oriented concepts for each item on this project thirteen list.

Write a list of **mathematical** activities you have participated in during this project and briefly describe evidence that verifies your interaction with and knowledge of mathematical concepts for each item on this project thirteen list.

Write a brief list of **Non-STEM-related** activities you have participated in during this project and briefly describe evidence that verifies your interaction with and knowledge of Non-STEM related concepts for each item on this project thirteen list.

Rate your level of involvement in STEM-related activities in this project (1=lowest, 10=highest).

1—2—3—4—5—6—7—8—9—10

- Actively contribute to conference committee meetings; keep a written record of contributions you have made and responsibilities you have accepted. Summarize all accomplishments that are reported in each committee meeting you attend. A compilation of your summaries will be collected at the beginning of project fourteen.
- Conduct mock run-throughs of the conference to complete conference preparations. As conference committees finalize the conference plans, they are reminded to include logistical considerations required to display the three primary aquatic ecosystems at the conference. These logistics must include arrangements for:
    - disassembling the classroom ecosystems
    - transporting them to the conference location
    - reassemble them for display
    - disassemble them at the end of the conference
    - return them to storage at the school (unless they are raffled off to another school or interested community member/educator)

- Create a journal entry that summarizes what you learned during project thirteen. Also write down what you would like to know more about and if you know someone else who might help you learn more if you asked them.

- Select a book, magazine, or scientific journal article to read that is related to a topic in project thirteen. Write a report on this reading and include it in your *Beyond the Classroom Aquarium* portfolio.

- Assist the teacher in disassembling, cleaning, inventorying, and storing classroom aquatic ecosystems and equipment.

- Design an investigation of the influence of changing the number of fountains and PVC pipe diameter on water pump output.

- Explore patents related to laminar flow fountains. Explain why inventors submit patents.

- Design an electronic system by which students can control their laminar fountain display remotely via smart phone technology and a wireless adapter connection attached to a PLC that is in turn connected to the solenoid valves described in part two of the laminar flow fountain activity.

- Summarize feedback entered on each of your project assessment activity sheets (projects one through thirteen).

Beyond the Classroom Aquarium STEM Self or Peer Assessment Rubric

This self-assessment is valid only to the extent to which you can document and display evidence that supports it. For each category in the left hand column, write a STEM-Oriented example in the box beneath 3 if you can display evidence that your STEM-Oriented example represents work that exceeds expectations. If your evidence shows that your STEM-Oriented statement meets but does not exceed expectations, write the statement in the box beneath 2. Similarly, for STEM-Oriented examples that demonstrate some progress toward expectations.

<table>
<tr><td colspan="2">Name:
Project Number:
Date Project Work Began:
Date of Project Completion:

List Math Topics and/or Objectives Currently Under Study:</td><td>3
Work Exceeds Expectations</td><td>2
Work Meets Expectations</td><td>1
Work Is Progressing Toward Expectations</td></tr>
<tr><td colspan="2">Write examples of questions and defining problems related to your STEM work.</td><td></td><td></td><td></td></tr>
<tr><td colspan="2">Describe how you have developed and used models in your STEM work.</td><td></td><td></td><td></td></tr>
<tr><td colspan="2">Write examples of STEM investigations you planned and carried out.</td><td></td><td></td><td></td></tr>
<tr><td colspan="2">Describe STEM related data that you analyzed or interpreted.</td><td></td><td></td><td></td></tr>
<tr><td rowspan="5">Give examples of how you used mathematics and computational thinking related to the five standards listed to the right.</td><td>Problem Solving (Math Process Standard One)</td><td></td><td></td><td></td></tr>
<tr><td>Reasoning & Proof (Math Process Standard Two)</td><td></td><td></td><td></td></tr>
<tr><td>Communication (Math Process Standard Three)</td><td></td><td></td><td></td></tr>
<tr><td>Connections (Math Process Standard Four)</td><td></td><td></td><td></td></tr>
<tr><td>Representation (Math Process Standard Five)</td><td></td><td></td><td></td></tr>
<tr><td colspan="2">Describe STEM-Oriented explanations and solutions you designed</td><td></td><td></td><td></td></tr>
<tr><td colspan="2">Explain evidence you used while engaging in a STEM project-based argument.</td><td></td><td></td><td></td></tr>
<tr><td colspan="2">Give examples of STEM information you obtained, evaluated, and/or communicated</td><td></td><td></td><td></td></tr>
<tr><td colspan="2">TOTAL POINTS</td><td></td><td></td><td></td></tr>
</table>

CHAPTER VII: A Student Produced *Beyond the Classroom Aquarium* Expo & Conference (Project - 14)

The activities in this chapter are primarily about assessment. The bases for assessment usually vary by instructor, class, and school. In consideration of this variation, summative assessment guidelines must remain general. Academically, instructors may wish to assemble and share end of course review materials with their students and schedule a few exam preparation session days to review these materials with their students to help them prepare for academic final exams. The beginning of project fourteen is also the deadline for students to turn in a paper copy of their research reports.

Instructors who also wish to assess students on any combination of one or more of the dimensions listed below may provide support to their students by informing them of exactly which dimensions will be evaluated and in what ways. A number of matrixes/grids that suggest additional evidence-based summative assessment modalities and structures that teachers may use to help guide students through the process of reflection upon and organization of information (for conference presentation) they have produced throughout the year are provided on the pages that follow.

- A 10 minute documentary PowerPoint/Video displaying participation in the sequence of projects
- A 15 minute presentation of the project nine research project
- A STEM-focused timeline illustrating student specific applications of STEM in project work
- A creative depiction of a watershed ecosystem per their tradeshow booth
- A futuristic design displaying potential future applications of STEM (per trade booths)
- Student contributions to the organization and operation of the conference itself
- A portfolio of book reviews, journal entries, art, posters, accomplishments, struggles, a qualitative essay, and a compilation of self-assessments by project
- A poster that provides a brief summary of student participation in the service learning program
- A way of demonstrating the ability to design and operate ecosystem electronic applications via a programmable logic controller (or Arduino)
- A field trip summary presentation/poster

These assessment criteria should not come as a surprise to any student who has participated in conference committee planning meetings since project nine. After determining which combination of one or more of the assessment dimensions above their instructors intend to base their conference evaluations, students should be given time to assemble and review all evidence of their participation in the class throughout the year. This evidence may include student-produced material that is: academic or empirical, qualitative or quantitative, from portfolios, in videos, PowerPoint presentations, art work, posters, technological designs and models, or in mathematical applications.

While organizing presentation information, another activity students should participate in involves the organization and dress rehearsal run-through of the conference. The conference may be considered the culmination of the entire *Beyond the Classroom Aquarium* year. It is the central activity of project fourteen.

Following the conference, in addition to performing an equipment inventory that the teacher and staff may use to determine budgetary and equipment concerns for the following school year, and in addition to assisting the staff in cleaning the classroom and in helping to put the class in a condition of readiness for the following school year, it is suggested that students conduct a final conference critique and summary.

PROJECT FOURTEEN Back From Beyond: Student-Organized Aquatic Ecosystem Conference
Summative Assessment

Conference Staging and Run-Through Activity

Teachers whose students are presenting an end of year *Beyond the Classroom Aquarium* conference have been guiding their students, throughout the year, to prepare to successfully present the results of their participation in activities related to these nine ancillary summative assessments at the conference. Per the conference schedule that is listed at the end of project nine (and in assessment activities at the conclusion of each succeeding project), each of these summative assessment measures is designed to be an important part of conference planning committee meetings in developing the conference program (including concurrent sessions, poster sessions, tradeshow booths, and lobby and conference hall displays).

It is during the conference that the instructor (and/or additionally selected judges and staff) will perform summative evaluations of student performance. Student committees must include this factor when organizing the conference. It is important that evaluators have the opportunity to view all student work in an unbiased environment. Students should ensure that the conference schedule permits students to be present at each event that will involve evaluation of their work (posters, booths, presentations, videos, and demonstrations). The run-through will help students prepare to represent their work personally during the conference.

In the run-through activity, students are to demonstrate their preparedness to efficiently run the conference by conducting a streamlined run-through of the conference, according to a timeline and plan that they have developed via their committee work. It is hoped that this run-through (dress rehearsal) will help the actual conference run more smoothly. Speeches and other time consuming events will occur only briefly during the run-through. For example, students will not give their full presentations during it. During the run-through, it is important to get the communication channels, teamwork, logistics, and timing of simultaneous tasks to run smoothly. At the dry run, some students may not yet have completed all of their own project-related presentation information that they will need on the day of the actual conference.

Following the dry run, students should be given the remaining time, individually and in teams, to assemble, complete, and refine their own project-related presentation information. Students should use the following checklist to determine work that remains to be done. The list should also help them understand the basis upon which their assessment will be judged successful.

The summative assessment grids that follow the checklist page may be helpful in identifying areas in which students have made progress and in which they skillfully demonstrate their ability to perform a number of tasks. Students may also find the assessment grids and the end of project assessments found in their portfolios (for projects one through thirteen) useful for developing their personalized STEM timeline. Opportunities for creativity and the celebration of student progress abound.

It may be possible to create a huge grid and have all students in the *Beyond the Classroom Aquarium* program write skills and knowledge they personally acquired through participating in each of the projects. Such information about every individual student should be available in the STEM timelines they developed for their respective booths. A large and combined grid might be quite dramatic, artistic, and fun! Perhaps the grid could be on a ground and filled in with colored chalk.

It is suggested that a comprehensive student run critique of the conference be conducted as the final end of year activity.

STUDENT ASSESSMENT CHECKLIST for Conference Preparation

Assessment Modalities	**Evidence**	**Student Self-Assessment**	**Instructor Assessment**
Academic final exams			
10 minute PowerPoint			
Conference research presentation			
Tradeshow booth timeline			
Tradeshow booth design			
Project 13 booth display			
Conference committee work			
Portfolio			
Service learning poster			
Exhibition of knowledge of basic electrician, carpenter, and plumbing skills and safety practices			
Other			

Summative Assessment Grid and Personalized SCIENCE Timeline Guide

Science by Project	Biology	Chemistry	Physics	Geology	Astronomy	Anthropology	Astronomy
Project 1							
Project 2							
Project 3							
Project 4							
Project 5							
Project 6							
Project 7							
Project 8							
Project 9							
Project 10							
Project 11							
Project 12							
Project 13							
Project 14							

Summative Assessment Grid and Personalized TECHNOLOGY Timeline Guide

Technology by Project	Digital Software	Digital Hardware	Sensors and Probes	Pumps	Lights		
Project 1							
Project 2							
Project 3							
Project 4							
Project 5							
Project 6							
Project 7							
Project 8							
Project 9							
Project 10							
Project 11							
Project 12							
Project 13							
Project 14							

Summative Assessment Grid and Personalized ENGINEERING/DESIGN Timeline Guide

Engineering by Project	Design	Electricity	Hydraulics	Structural	Mechanical	Dynamics & Statics	Other
Project 1							
Project 2							
Project 3							
Project 4							
Project 5							
Project 6							
Project 7							
Project 8							
Project 9							
Project 10							
Project 11							
Project 12							
Project 13							
Project 14							

Summative Assessment Grid and Personalized MATH Timeline Guide

Math by Project	Arithmetic	Algebra I	Algebra II	Geometry	Trigonometry	Probability & Statistics	Pre-Calculus & Calculus
Project 1							
Project 2							
Project 3							
Project 4							
Project 5							
Project 6							
Project 7							
Project 8							
Project 9							
Project 10							
Project 11							
Project 12							
Project 13							
Project 14							

Summative Assessment Grid and Personalized Timeline Guide – Other Curricular Areas

Other Curricular Areas by Project	Fine Arts, Music, Drama	Physical Education and Health	Language Arts	Social Studies	Philosophy	Skills Related to Trades and Crafts
Project 1						
Project 2						
Project 3						
Project 4						
Project 5						
Project 6						
Project 7						
Project 8						
Project 9						
Project 10						
Project 11						
Project 12						
Project 13						
Project 14						

Summative Field Trip Assessment Grid

Field Trip	Location & Project	Summary of Journal/Portfolio Field Trip Descriptions	Additional Comments about Skills Developed and Knowledge Gained from Participation in Field Trips
FT-1	**AM: Aquatic Ecosystem Model Sites; PM: Construction Demonstrations**		
FT-2	**Ecosystem specific water-flow and containment demonstrations**		
FT-3	**Introduction to Ecosystem Maintenance Concepts and Techniques (Whole class or division of labor student groups)**		
FT-4	**AM: Plankton tow and algae demo PM: PLC - Computer Automation**		
FT-5	**Planetarium visit/ scope demo**		
FT-6	**Aqua conference a) booths/businesses, b) posters, c) presentations**		
FT-7	**AM - Apo Island exhibit (virtual/real) PM Feedback Diagram – tip points.**		
FT-8	**AM beyond the classroom apps. PM Alternative/green energies**		

Beyond the Classroom Aquarium

A Qualitative Student Summary of the Yearlong Experience

Bibliography

Adey, Walter H.; Loveland, Karen; (1991); *Dynamic Aquaria: Building Living Ecosystems*; pp. 643; Academic Press, Inc.

Alcala, A. C.; (2001); *Marine reserves in the Philippines: historical development, effects and influence on marine conservation policy*; Bookmark, Makati City, Philippines

Allen, D; & McDonald, J; (2003); *the tuning protocol: A process for reflection on teacher and student work;* Coalition of Essential Schools (CES); http://old.essentialschools.org/cs/resources/view/ces_res/54

Allen, F. (1994); The Microbrewery Laboratory Manual – A Practical Guide to Laboratory Techniques and Quality Control Procedures for small scale Brewers, Part 1: Yeast Management: *Brewing Technique Magazine* (http://brewingtechniques.com/library/backissues/issue24.allen.html)

Allen, Will; (2012); *the Good Food Revolution*; *Growing Healthy Food, People, and Communities*; Gotham Books

Amadori, Michael (2012). *Fish, Lettuce and Food Waste Put New Spin on Aquaponics*. http://www.newswise.com/articles/view/578382/?sc=rssn&utm_source=feedburner&utm_medium=feed&utm_campaign=Feed%3A+NewswiseScinews+%28Newswise%3A+SciNews%29 Environmental Science and Forestry (ESF)

American Public Health Association; (2005); Standard methods for the examination of water and wastewater (21st ed.); Washington, D.C.; American Public Health Association, American Water Works Association, and Water Environmental Federation; p 4 - 136 to 4 - 137; http://www.standardmethods.org/.

Ameyaw-Akumfi, C.; The Breeding Biology of two Sympatric Species of Tropical Intertidal Hermit Crabs, *Clibanarius chapini* and *Clibanarius sengalensis*; *Marine Biology*; 29, 15-28 (1975); Springer Verlag 1975

Andrews, Chris; Exell, Adrian; Carrington, Neville; (1988); *The Manual of Fish Health*; pp.208; Tetra Press

Aquaculture Information Center, Public Service Division, Beltsville, MD, 20705; *Aquatopics: Aquaculture for Youth Educators*; ED 328 434

Aquaponics Community; *"Bluegill or Bream Growers;"* http://aquaponicscommunity.com/group/bluegillorbreamgrowers Retrieved December 2011

Aquaponics Community; *Catfish Growers*; http://aquaponicscommunity.com/group/catfishgrowers; Retrieved December 2011

Aquaponics Community; *Goldfish Growers*; http://aquaponicscommunity.com/group/goldfishgrowers; Retrieved December 2011

Aquaponics Community; *Koi Growers*; http://aquaponicscommunity.com/group/koigrowers; Retrieved December 2011

Aquaponics Gardener; (2013); http://www.aquaponicsgardner.com/aquaponic-systems.php**;** Accessed April 29, 2013

Aquarium Guys; (2008); *Are Your Fish Really Suffering from Disease? A guide to non-disease causes of illness;* Aquarium Guys web site: http://www3.sympatico.ca/drosera1/fish/illness.htm; accessed Feb 4, 2013

Arentsen, Cavazini, Imahoko & O'Brien; *Coral Farming Report: For Rutgers GSM course in Marketing Research and Eco Vitality, Inc.;* December 15, 1998; http:www.employees.org/~redfield/reefs/coral2/Coral_doc.html; accessed 2/22/01

Asakura, Akira; Sexual Differences in Life History and Resource Utilization by the Hermit Crab; *Ecology*; Oct. 1995, v76, n7, p2295 (19)

Auerbach, Paul S.; Don't Get Burned: Your Summertime Guide to the Causes, Treatment, and Prevention of Sunburn; July 1993; *Dive Training*

Backyard Aquaponics; *Fish Page: Other Species*; http://backyardaquaponics.com/information/45-fish.html; Retrieved December 2011

Bak, R; (1976); the growth of coral colonies and the importance of crustose coralline algae and burrowing sponges in relation with carbonate accumulation; *Netherlands Journal of Sea Research*; 10: 285-337

Barron, B., & Darling-Hammond, L. (2008); Teaching for Meaningful Learning: A Review of Research on Inquiry-Based and Cooperative Learning (PDF). *Powerful Learning: What We Know About Teaching for Understanding*. San Francisco, CA: Jossey-Bass

Beane, DeAnna Banks; (1985); *Mathematics and Science: Critical Filters for the Future of Minority Students*; the Mid-Atlantic Center for Race Equality; the American University of Washington, DC

Bennett, Lloyd M.; Marine Aquaria -Their Value in Teaching Science in the Elementary School; *School Science and Mathematics*; 72; 1; 57-72; Jan 1972

Bennett, William J.; (1986); *First Lessons: A Report on Elementary Education in America*; Washington, DC; U.S. Dept. of Education; ED 395

Berger, Melvin; (1987); *Lights, Lenses and Lasers*; pp. 85; Putnam's Sons, New York; ISBN 0-399-61214-9

Berger, Ron; (2003); *an ethic of excellence: building a culture of craftsmanship with students*; Heinemann, Portsmouth, NH

Bergese, Elisabeth Mann; (1977); *SeaFarm: The Story of Aquaculture*. Harry N. Abrahms, Inc. 0-8109-1604-5

Bermuda Biological Station for Research, Inc. *Sample Field Course Curriculum for Visiting High Schools: Bermuda's Biodiversity - A Field Course for High School Students*; pp. 7 http://www.bbsr.edu/Visitor_Information/Samplecurriculum/body_samplecurriculum.html

Bernstein, Sylvia; http://theaquaponicsource.com/2010/01/15/fish-to-eat-or-not-to-eat/; accessed 1-25-2013

Birkeland, C; (1997); *Life and Death of Coral Reefs*; Chapman and Hall Publishing Company; 536 pages

Bishop; Bourke; Connolly & Trebic, (2009); *Baird's Village aquaponics project*: AGRI 519/CIVE 519 Sustainable Development Plans; Holetown, Barbados: McGill University

Blosser, Patricia E.; (1985); Meta-Analysis: Research on Science Instruction. ERIC/SMEAC *Science Education Digest*; No. 1

Bocek, A. (2010). *Water harvesting and aquaculture for rural development*; Retrieved December 24, 2010; University of the Virgin Islands, Agricultural Experiment Station [AES]; Aquaculture - Aquaponic Systems

Bonnet, Robert L.; Keen, G. Daniel; (1990); *Computers: 49 Science Fair Projects*; Tab Books Blue Ridge Summit, PA

Boutwell, Juanita (December 15, 2007), "*Aztecs' aquaponics revamped*", *Napa Valley Register* "Space agriculture for habitation on Mars and sustainable civilization on Earth", *Recent Advances in Space Technologies, 2009. RAST '09*: 68–69, 2009

Boyd, C. E; (1990); *Water quality in ponds for aquaculture*

Boyes, Walt; (18 August 2009). *Back to Basics: SCADA;*

Brennan, John; (2013); *What Are Acceptable Levels of Ammonia in Water?* http://www.ehow.com/about_6700519_acceptable-levels-ammonia-water_.html**;** Accessed April 2013

Brodderman, Ted; (1982); Activity Science - The Evidence Shows it Matters; *Science and Children*; Vol. 20; No. 1; 39-41

Brooks, David W.; *Electrolysis of Water: Becker Demos*; http://chemmovies.unl.edu/chemistry/dochem/DoChem044.html; accessed Dec. 2012

Bruckner, A.W.; (2001); Coral health and mortality: Recognizing signs of coral diseases and predators. In: Humann and Deloach (eds.), *Reef Coral Identification*. Jacksonville, FL: Florida Caribbean Bahamas New World Publications, Inc. pp. 240-271

Brylske, Alex; Coral Restoration: Giving Mother Nature a Kick Start; *Dive Training* – December 2009; http://www.fsu.edu/~flserve/resources/Dec09CoralRestor.pdf

Buchheim, Jason; (1998a); *Coral Reef Bleaching*; Odyssey Expeditions- Marine Biology Learning Center Publications; http://www.marinebiology.org/coralbleaching.htm; accessed 3/2/99

Buchheim, Jason; (1998b); Coral Reefs: *Living Corals,* Odyssey Expeditions- Marine Biology Learning Center Publications; http://www.marinebiology.org/coralreefs.htm**;** accessed 3/2/99

Buchheim, Jason; (1998c); Coral Reefs: *Oceanography: Water, Seawater and Ocean Circulation and Dynamics*; Odyssey Expeditions- Marine Biology Learning Center Publications; accessed 3/2/99

Buchheim, Jason; (1998d); Coral Reefs: Odyssey Expeditions - Marine Biology Learning Center Publications http://www.marinebiology.org/fish.htm, and http://www.marinebiology.org/fishecology.htm; accessed 3/2/99

Buckhat, Ron; Miller, Wade W.; *Model Aquaculture Recirculation System (MARS);* 1995 Preliminary Edition; pp. 50; Aquaculture Education Project - National Council for Agricultural Education

Bulloch, David, K.; *The American Littoral Society Handbook for the Marine Naturalist*; 1991a; pp. 153+; Walker Publishing Company; ISBN 0-8027-7353-2; Includes Measuring Salinity by Chloride Titration Measuring Dissolved Oxygen in Seawater by Winkler (Iodometric) Titration

Bulloch, David K.; *Plankton Productivity of the Sea in The Underwater Naturalist: A Layman's Guide to the Vibrant World beneath the Sea*. New York: Lynons & Burford, 1991b; ed. Bulloch, David K

Burgess, Warren; (1987); *a Complete Introduction to Corydoras and Related Catfishes*; pp. 125; TFH Press

Burnett, L. E.; *Respiratory Responses of Grass Shrimp, Juvenile Spot Fish and Killifish to Hypoxia and High CO_2*; http://inlet.geol.sc.edu/chp/burnett.html

Burton, David M.; (1997); *the History of Mathematics: An Introduction*; McGraw Hill Corporation; ISBN 0-07-009465-9

Bybee, Rodger; Scientific and Engineering Practices in K–12 Classrooms: Understanding a Framework for K–12 Science Education; December 2011; *NSTA's Journals*

Cadiz, P. L; and H. P. Calumpong; (2000); Analysis of revenues from ecotourism in Apo Island, Negros Oriental, Philippines; *Proceedings of 9th International Coral Reef Symposium* (Bali, Indonesia, 23-27 October 2000), Volume 2:771-774

Calfo, Anthony; (2003); *Book of coral propagation: Volume one, Reef gardening for aquarists*; Reading trees publication, USA. www.readingtrees.com

Calfo, Anthony; *Coral Fragmentation: Not Just for Beginners*! Reef Central; pp. 15; Accessed 11/15/2006

Callan, C; Laidley, C. et al; *Pacific Threadfin, Polydactylus sexfilis (Moi), Hatchery Manual - 2nd Edition*; Center for Tropical and Subtropical Aquaculture; pp. 133

Cape Fear River Watch; *Lower Cape Fear River Program: Water Quality Sampling Parameters*; http://smec.uncwil.edu/GLAX0/cfrp/cft.erm.htm#waterq

Carlson, Bruce A.; How to Build a Powerful Surge Device; *SeaScope*, summer 1996, Vol. 13

Carner, Don; (1997); *DIY Reef Type Fluorescent Light Hood Plans for Saltwater Aquariums*; http://saltaquarium.about.com/od/diyhoodcanopylighting/ss/sbsdiydonhood_4.htm; accessed July 2012

Carroll, Bob; (2009); *Estuarine Aquarium Keeping for Beginners PDF*; http://www.vims.edu/cbnerr/_docs/education_docs/EstAquKeepwriteup.pdf; Publisher: Virginia Institute of Marine Science

Carroll, David J.; Kempf, Stephen C.; Laboratory Culture of the Aeolid Nudibranch *Berghia verrucicornis* (Mollusca, Opisthobranchia): Some Aspects of its Development and Life History; *The Biological Bulletin*, 179, 243-253, Dec. 1990

Carson, Rachel; (1955); *The Edge of the Sea*; pp. 238; New American Library

Carson, Rachel; (1962); *Silent Spring*; pp. 304; Fawcett

Cassuto, David N.; (1994); *Cold Running River*; pp. 141; University of Michigan Press

Center for Marine Conservation - Teachers and Students; *Searching out Nonpoint Sources of Pollution: A Million Points of Blight*; http://www.cmc-ocean.org/mdio/nonpoint.html

Center for Marine Conservation; *The Ocean Book: Aquarium and Seaside Activities for All Ages*; 1989; 0471-50973-6 or 0-471-62078-5 (paperback)

Chase, Valerie (editor); *Learning Ocean Science through Ocean Exploration: A Curriculum for Grades 6-12*; 3rd Edition; NOAA; http://oceanexplorer.noaa.gov

Chase, Valerie; (1987) *Living in Water Curriculum*; pp. 302; Dept. of Education and Interpretation of The National Aquarium in Baltimore

Cherr, Gary N.; Griffin, Frederick J.; Pillai, Murali C.; Higashi, Richard M.; *Fertilization and Early Development in Marine Organisms and the Effects of Environmental Stressors*; http://www-bml.ucdavis.edu/org4.htm

Cherton, Barbara; Tietjen, John; (1988); *The Facts on File Dictionary of Marine Science*; 0-8160-2369-7 (paper) or 0-8160-1031-5 (hardcover)

Clark, Joe; Picard, Joe; (1989); *Laying down the Law: Joe Clark's Strategy for Saving Our Schools*; Regnery Gateway

Clark, John R; Clark, Roberta L; (1964); *Sea-Water Systems for Experimental Aquariums*; pp. 192; TFH

Cohen, Michael J.; *Reconnecting with Nature: The Restoration of the Missing Link in Western Thinking;* Project Nature Connect; pp. 6; http://www.pacificrim.net/~nature/newbook.html

Conroy, C; Walker, N; (1998); *National Aquaculture Curriculum Evaluation Phase II Final Report*; Ithaca: Cornell University

Crossley, Phil L. (2004), "Sub-irrigation in wetland agriculture," *Agriculture and Human Values* (21): 191–205

Cuddihy, Geri; *Mangrove Replenishment Initiative - Educational Objectives*; http://mangrove.org.education.htm

Curry, Alan; *Under the Microscope*; (1982); pp. 160; Van Nostrand Reinhold Co. Inc.; ISBN 0-442-21598-3

Daily Janakantha; Fish & vegetable culture through aquaponics technology; In Bengali; *The Daily Janakantha*, January 28, 2011

Daily Kalerkantho; Innovation of a BAU researcher, "Aquaponics technology" three times production without any cost; In Bengali; *The Daily Kalerkantho*, January 25, 2011

Daily Sangbad; Some Important Talks on Pest Management; In Bengali; *the Daily Sangbad*, 29 January 2011

Darling-Hammond, L. (1996); *What Matters Most: Teaching for America's Future* (PDF). New York, NY: National Commission on Teaching and America's Future.

Davidson, Osha Gray; (1998); *The Enchanted Braid: Coming to Terms with Nature on the Coral Reef;* published by John Wiley & Sons; http://www.OshaDavidson.com

Davis, M.; A.L. Shawl; (2005); *a Guide for culturing queen conch, S. gigas*; American Fisheries Society Symposia 46:125-142

Day, Robert; The marine aquarium as an educational tool; *Seascope Magazine*, Volume 10; Summer (1993); Published by Aquarium Systems Inc.

DeAlteris, Joseph T.; *Conference Summary: Aquaculture Outreach Activities at the University of Rhode Island*; http://www.mdsg.umd.edu/extensionconf/AQ34.html

De Leon, VB; Ballado Jr., AH; Dychitan, P; Gustilo, R; MCU-Based Aquaculture System; *Proceedings of the World Congress on Engineering and Computer Science 2011*; Vol. II; WCECS 2011, October 19-21, 2011, San Francisco, USA

Deming, H. G.; (1975); *Water: The Fountain of Opportunity*; pp.342; Oxford University Press

Deutschmann, G. V.; (1998) History of hydroponics

Dewey, J.; (1998); *Education and Experience: The 60th Anniversary Edition*; Kappa Delta Pi (original work published 1938)

Dickinson, Aaron; Johnson, Donald M.; A Low Cost Programmable Logic Control (PLC) Trainer for Use in a University Agriculture Electricity Course; *Journal of Agricultural Systems, Technology, and Management*; March 2006, Volume 21; http://jastm.org/legacy/past21.htm accessed June 16, 2012

Dillingham, John; (1990); *A Study of the Aquaculture Industry in Texas to Assist in Establishing Aquaculture as a Course Offering in Agricultural Science and Technology; A Final Report*.; ED33895

Diver, Steve (2006); *Aquaponics - integration of hydroponics with aquaculture, ATTRA - National Sustainable Agriculture Information Service* (National Center for Appropriate Technology), http://www.backyardaquaponics.com/Travis/aquaponic.pdf

Drewes, Charles; Those Wonderful Worms; *Carolina Tips*, August 1996, Vol. 59, No. 3

Drewes, Charles D.; *Connecting Biology Teachers and Students with Living Invertebrates*; http://www.eeob.iastate.edu/faculty/DrewesC/htdocs/NABT-narrative.PDF; accessed Jan. 21, 2003

Ebeling, James M; (1998a); Computer Monitoring and Control Technology: Part 1 - What, where, and how; *Aquaculture Magazine*; 24 (5); 99-104

Ebeling, James M; (1998b); Computer Monitoring and Control Technology: Part 2 - Putting it together; *Aquaculture Magazine*; 24 (6); 81-86

Echeverria, John D.; Barrow, Pope; Roos-Collins, Richard; *Rivers at Risk: The Concerned Citizen's Guide to Hydropower*; 1989; pp. 217; Island Press

Ellis, S; (1999); *Farming Soft Corals for the Marine Aquarium Trade*; Center for Tropical and Subtropical Aquaculture; p. 6

Ellis, S; Ellis E; (2002); *Recent Advances in Lagoon-based Farming Practices for Eight Species of Commercially Valuable Hard and Soft Corals - A Technical Report*; Center for Tropical and Subtropical Aquaculture; pp. 63

Ellis, S. (1997). Spawning and early larval rearing of giant clams (Bivalvia: Tridacnidae) (No. 130); Center for Tropical and Subtropical Aquaculture. pp.55

Ellis, S; Sharron, L; (1999); the culture of soft corals (Order: Alcyonacea) for the marine aquarium trade. Center for Tropical and Subtropical Aquaculture; pp. 77

Elwood, R. W.; Neil, S. J.; (1992); *Assessments and Decisions: A Study of Information Gathering by Hermit Crabs*; pp. 192; Chapman & Hall (London); ISBN 0412391104

Esquith, Rafe; 2007; *Teach like Your Hair's on Fire: the Methods and Madness inside Room 56*; Viking Adult; pp. 256

Esquith, Rafe; 2010; *Lighting Their Fires: How Parents and Teachers Can Raise Extraordinary Kids in a Mixed-up, Muddled-up, Shook-up World*; Penguin; pp. 208

Evans, Lautenbacher, Mahoney & Davidson; *A National Coral Reef Action Strategy: A Report to Congress on Implementation of the Coral Reef Conservation Act of 2000 and the National Action Plan to Conserve Coral Reefs in 2002 - 2003*; http://coris.noaa.gov/activities/actionstrategy/action_reef_final.pdf; NOAA June 2002, pp. 156

Fengqi, Liu; *Production and Application of Rotifers in Aquaculture*;
part one: http://www.netpets.com/fish/reference/reefref/rotifers.html;
part two: http://www.netpets.com/fish/reference/reefref/rotifers2.html

Ferguson, John C.; The Structure and Mode of Function of the Water Vascular System of a Brittlestar, *Ophioderma appressum*; *The Biological Bulletin*, Feb-March 1995, v188, n1, p. 98 (13)

Foley, Allen M.; *A Preliminary Investigation on Some Specific Aspects of Latex Balloon Degradation*; August 1990; Florida Department of Natural Resources Florida Marine Institute

Foley, Allen M.; *Information on the Release of Latex Balloons Filled with a Lighter than Air Gas*, Florida Marine Research Institute - St. Petersburg, FL; August 1990; pp. 6

Fortner, Rosanne W.; Public Information Conferences as Tools for Environmental Communication Training; *Journal of Environmental Education*; Volume 17, Issue 2, 1986; pages 14-19

Foster, S; Vincent, C; 2004; Life history and ecology of seahorses: implications for conservation and management; *Journal of Fish Biology;* (65), 1 – 61

Fotheringham, N.; Shell Utilization and Reproduction in Tropical Hermit Crabs; *Marine Biology*, 55, 287 - 293; 1980

Fowler, P; Baird, D; Bucklin, R; Yerlan, C; Watson, C; and Chapman, F; (April, 1994); Microcontrollers in Recirculating Aquaculture Systems; *Document EES-326*, Florida Energy Extension Service, Florida Cooperative Extension Service, Institute of Food and Agricultural Services, University of Florida

Frakes, Thomas; a Miniature Coral Reef, *Carolina Tips*, Vol. 65, No. 2, spring 2002

Franson, M. A. H. (1995); American public health association American water works association water environment federation. Methods, 6, 84

French, Chris; Gatling-Austin, Bruce; Voith, Melody; Rigby, Chris; *Chemical and Biological Analysis: The Hardware River Watershed, Central Virginia*; pp. 10; http://raraday.clas.virginia.edu/~crf8a/hardware.html not sure of the "hard" in the URL

Friedman, Allan; A Big Lesson in a Small Pond: Fifteen bottles of muck mixed with teacher ingenuity can bring a pond to the schoolyard; *Science and Children*, Feb 1991, pp. 27-29

Friedrich, Hermann; (1973) *Marine Biology: An Introduction to its Problems and Results*; pp.474; University of Washington Press

Fryer, Lee; Simmons, Dick; (1997); *Food Power from the Sea: The Seaweed Story.* Mason/Charter N.Y.; 0-88405-383-0 or 0-88405-404-7 (paperback)

Fujita; S; (1979); Culture of Red Sea Bream, *Pagrus major*, and its Food; in *Cultivation of Fish Fry and its Live Food*; Styczynska-Jurewicz, E; Backiel, T; Jaspers, E; Persoone (eds.); European Mariculture Society, Special Publication 4: 183-197

Fusaro, B. A.; & Kenschaft, P. C.; (2003); *Environmental mathematics in the classroom*; The Mathematical Association of America

Garber, Steven D.; (1987); *the Urban Naturalist*; John Wiley & Sons

Gardner, Todd; Larval Rearing of the Masked Goby, Coryphopterus personatus; *Seascope*; Volume 16, Winter 1999

Gardner Todd, & Seahorse.org; 2002-2004; *Solving Problems in Seahorse Culture*; http://www.seahorse.org/library/articles/seahorseCulture.shtml

Ghost Shrimp General Biology - *Biology of Thalassinideans*; http://www.mov.vic.gov.au/crust/thalbiol.html

Gilley, J Wade; Dietz, Thomas; *the Virginia Survey: Results from a Statewide Survey on Higher Education and Environmental Problems*; 1989-Jan-30; Pages: 15; ERIC Number: ED328105

Giwojna, Peter; (1990); *A Step-by-Step Book about Seahorses*; pp.64; TFH Press

Glasser, William M.D.; (1990); *The Quality School: Managing Students without Coercion*; Harper and Row Publishers, Inc., 10 East 53rd Street, New York, NY 10022

Glasser, William M.D.; (1993); *The Quality School Teacher: A Companion Volume to the Quality School*; Special Markets Department, Harper-Collins Publishers, Inc., 10 East 53rd Street, New York, NY 10022. Harper Perennial

Goldfarb, Theodore.D; (1991) *Taking Sides: Clashing Views on Controversial Environmental Issues*; 4th Ed.; Dushkin Publishing Group; p.334

Goldsby, *Basic Biology*; (1976); Harper's College Press

Grass Shrimp; http://fwie.fw.vt.edu/WWW/macsis/lists/M070010.htm

Gratzek, J.B.; Loiselle, P.V.; Norton, Joanne; (1992); *Aquariology: Fish Breeding & Genetics*. 1-56465-106-1; Tetra Press Item # 16856

Greenberg, D; Sadofsky, M; & Lempka, J; (2005); *the Pursuit of Happiness: The Lives of Sudbury Valley Alumni*; Framingham, MA: Sudbury School Press

Greene, H. David; *Aquatic Activities for Youth*; New York Sea Grant; May 1980; pp. 44; Prepared by the Youth Coastal Education Program and the Marine Education Program; ED 230 420

Grossman, John; the Edible Schoolyard; *Sky Magazine*; April 2006

Grossman, Margaret R.; (1983); *Legal and Regulating Environment for Illinois Aquaculture*; IL Dept. of Energy and Natural Resources; Document Number 83/05

Guillard, R.R.L.; (1975); Culture of Phytoplankton for Feeding Marine Invertebrates; in: *Culture of Marine Invertebrate Animals*. W.L. Smith and M.H. Chanley (eds.), Plenum Publishing Corp, New York, NY pp. 29-60

Harris, Daniel C.; *Quantitative Chemical Analysis*, Fifth Edition; 1999; W.H. Freeman and Company

Harris, L. Kasimu; http://www.louisianaweekly.com/aquaponics-being-taught-in-vietnamese-community/; *Louisiana Weekly*; 19, December 2011

Hanes, F.; Audet, R.; Brown, C.; Buttner, J.; Leavitt, D.; and Soares, S.; (2002); *Explorations in Aquaculture: Volume 1; Marine Resources for the Classroom*; Aquatic Ecosystems, Inc. Apopka, FL

Hanes, F.; Audet, R.; Brown, C.; Buttner, J.; Leavitt, D.; and Soares, S.; (2002); *Explorations in Aquaculture: Volume 2; Freshwater Resources for the Classroom*; Aquatic Ecosystems, Inc. Apopka, FL

Haywood, Martyn; Wells, Sue; (1989); *the Manual of Marine Invertebrates*; pp. 208; Tetra Press

Helm, Mike; Tukel, George; *Restoring Cities From the Bottom Up; A Bi-Coastal View From The Street*; Earth Island Journal; pp. 5; http://www.earthisland.org/

Hellweg, Mike; *Culturing Live Foods: A Step-by-Step Guide for Culturing One's Own Live Foods for the Home Aquarium*: TFH Publications, Inc.; June 2008; 240 pp.; ISBN-13: 9780793806553

Hirayama, (1990); *A Physiological Approach to Problems of Mass Culture of the Rotifer;* NOAA Technical Report Number NMFS 85; US Department of Commerce; 73-79

Hoegh-Guldberg, O. et.al; Coral Reefs under Rapid Climate Change and Ocean Acidification; *Science*; 14 December 2007: 1737-1742

Hoff, Frank H.; Snell, Terry W.; *Plankton Culture Manual*; 3rd Edition; 1987; pp. 155; Florida Aqua Farms, Inc.

Hoff, Frank; *Conditioning, Spawning, and Rearing of Fish: with Special Emphasis on Marine Clownfish;* 1996; pp. 213; Aquaculture Consultants Inc.

Holt, Joan G.; Cecilia M. Riley; (1999); *Laboratory Spawning of Coral Reef Fishes: Effects of Temperature and Photoperiod*; United States-Japan Cooperative Program in Natural Resources (UJNR) Technical Report No 28

Horst, William; Mangroves, Is this a Fad?; *The Wave: Monthly Newsletter of the Eastern PA Reef Club*, Jan 1998, Issue #3, and continued in Feb 1998, Issue #4

Houde, ED; Scheker, RC; (1980); Feeding Marine Fish Larvae: Developmental and Functional Responses; *Environmental Biology of Fish*; 5:315-334

Howell, Mark; The Reef is dying: Officials Rethink Why; *Solares Hill*; December 10, 2004

Hueter, Joanne; *The Dwarf Seahorse (Hippocampus Zosterae);* http://home1.gte.net/seahors1/; pp. 6

Hueter, Robert E; Gilbert, Perry W. (1991); *the Sensory World of Sharks;* In: *Discovering Sharks: A Volume Honoring the Work of Stewart Springer;* ed. S. H. Gruber; pp. 48 – 55; New Jersey; American Littoral Society

Hughes, John, H.; Schleser, Robert A.; Tchobanoglous, George; A Rearing Tank for Lobster Larvae and other Aquatic Species, *The Progressive Fish-Culturist*, Volume 36, Issue 3, 1974, Pages 129 – 132

Hung, W. (2008); the 9-step problem design process for problem-based learning: Application of the 3C3R model. *Educational Research Review*, 4(2) 118-141

Hunt, John D. (Ed.); *Marine Organisms in Science Teaching*; 1980; Texas A&M University; TAMU SG-80-403

Hunter, JR; (1980); The Feeding Behavior and Ecology of Marine Fish Larvae; in *Fish Behavior and its Uses in the Capture and Culture of Fishes*; Bardach, JJ; Magnuson, RC; May, RC; Reinhart, JM (eds.); ICLARM Conference Proceedings; 5: 287-330

Illuminating Engineering Society of North America; 2000; IESNA Lighting Handbook: Reference & Application; 9th ed.; New York, NY; Edited by Mark S. Rea

Inform, Inc.; *Environmental Report Card for Schools*; http://www.cmc-ocean.org/mdio/reptcardprint.html

International Democratic Education Network; (2012); *International Democratic Schools*; International Democratic Education Network; http://www.idenetwork.org/democratic-schools.htm

International Society for Technology in Education (ISTE); (2010); *ISTE's U.S. Public Policy Principles and Federal & State Objectives*; http://www.iste.org/docs/pdfs/publicpolicybrief-255-10.pdf?sfvrsn=2; accessed June, 2012

Iwai, T; (1980); Sensory Anatomy and Feeding of Fish Larvae; in Fish Behavior and its Use in the Capture and Culture of Fishes; Bardach, JJ; Magnuson, RC; May, RC; Reinhart, JM (eds.); ICLARM Conference Proceedings; 5: 124-145

Jacobs. William P.; Caulerpa; Dec. 1994, Vol. 271, # 6, pp. 100-105; *Scientific American*;

Jaffe, David A.; Spectrum Analysis Tutorial, Part 1: The Discrete Fourier Transform; *Computer Music Journal*; Vol. 11; No.2, summer 1987

Jaffe, David A.; Spectrum Analysis Tutorial, Part 2: Properties and Applications of the Discrete Fourier Transform; *Computer Music Journal*; Vol. 11; No.3, fall 1987

Johnson, D.; Wardlow, G.; a Prototype Recirculating Aquaculture-Hydroponic System; *Journal of Agricultural Mechanization*; November, 1997

Jonklaas, Rodney; (1975); *Collecting Marine Tropicals*; TFH; 0-87666-119-3

Jones, John Lemuel; *Home Hydroponics . . . and How to Do It*; 1975; p. 156; rev. and edited by Cay and Paul Beardsley; Paradise Valley, Ariz.; Beardsley Pub. Co; OCLC ID No. 01551311; ISBN 0-517-53760-5

Kaye, Cathryn Berger; 2010; *Going Blue: A Teen Guide to Saving Our Oceans, Lakes, Rivers, & Wetlands*; Free Spirit Publishing, Inc.

Kerr, Geoff; Corfield; Association between the ghost shrimp Trypae australiensis Dana 1852 (Crustacea: Decapoda) and a small deposit-feeding bivalve Mysella vitrea Laserson 1956 (Mollusca: Leptonidae); *Marine and Freshwater Research*, 1998, 49, 801-6

Kheiralla, A.F.; Design and Development of a Low Cost Programmable Logic Controller Workbench for Education Purposes; *International Conference on Engineering Education – ICEE - 2007*

Kirkland, Jill; (2007); *No Student Left Indoors: Creating a Field Guide to Your Schoolyard*; Stillwater Publishing

Kitajima, et al (1979); Improvement of dietary Value of Red Sea Bream Larvae of Rotifers Branchionus plicatus, cultured with baker's yeast, *Saccharomyces cerevisiae*; *Bulletin Japanese Society Scientific Fisheries*; 45:469-471

Klinger-Bowen, R; Tamaru, C; et al; March 2011; *Testing your Aquaponic System Water: A Comparison of Commercial Water Chemistry Methods*; Center for Tropical and Subtropical Aquaculture; pp. 20

Knuckey, Richard M.; Semmens, Gale L.; Mayer, Robert J.; Rimmer, Michael A.; Development of an optimal microalgal diet for the culture of the calanoid copepod Acartia sinjiensis: Effect of algal species and feed concentration on copepod development; *Aquaculture*; 12 September 2005

Koeniger, John; Making a Splash in the Classroom; *Carolina Tips*, January 1997, Vol. 60, No. 1

Lawrence, Frances L.; Section II, Activities: Marine Biology Fishy Business, in Oceanography for Landlocked Classrooms; Madrazo, Jerry M.; Hounshell, Paul B. (Eds.); National Association of Biology Teachers; 1990

Lear, Linda: (1997) *Rachel Carson: Witness for Nature*; Henry Holt & Company; 0-8050-3427-7

Lear, Linda; (1998) *Lost Woods: The Discovered Writings of Rachel Carson*; Beacon Press;

Leber, K.; Lorenzen, K.; Main, K.; Moe, M.; Vaughn, D.; Capo, T.; Bardales, A.; Gillette, P.; 2009; *Developing Restoration Methods to Aid recovery of a Key Herbivore, Diadema antillarum, on Florida Coral Reefs: "Protect Our Reef" Grants Program 2009/2009 Final Report*; https://dspace.mote.org/dspace/bitstream/2075/316/1/POR%202007-29f.pdf

Lee, Jasper S.; *Producing Catfish: Aquaculture Curriculum Guide; Year Two Species Module*; 1995 Preliminary Edition; pp. 277; National Council for Agricultural Education

Lee, Richard S.; *Construction and Maintenance of Classroom Aquaria*; 21 pages; Marine Science Curriculum Aid No. 2; Alaska Univ., Anchorage; Alaska Sea Grant Program; 1976 - SGR-76-13; NOAA-04-6-158-44039; ED 162 915

Leedy, Paul D.; *Practical Research: Planning and Design*; Jeanne Ellis Ormrod; Upper Saddle River, N.J.: Merrill Prentice Hall, 1980 and 2010

Leewis, Rob; Wijgerde, Tim; Laterveer, Michael, Osinga, Ronald; (2009); *Working with aquarium corals: A Book of Protocols for the Breeding and Husbandry of Scleractinian Corals*; The European Association for Zoos and Aquaria http://www.coralzoo.org/home/

Lennard, Wilson A.; Leonard, Brian V. (2006), A comparison of three different hydroponic sub-systems (gravel bed, floating and nutrient film technique) in an Aquaponic test system, *Aquacult Int* (14): 539–550

Leslie, R. P. & Conway, K. M.; (1996); *the lighting pattern book for homes*; Lighting Research Center; Rensselaer Polytechnic Institute

Lethbridge College; (2013); http://www.aquaponicsgardner.com/aquaponic-systems.php; Accessed April 29, 2013;

Levine, Joseph S; (1991); *the complete fishkeeper: Everything aquarium fishes need to stay happy, healthy, and alive*; HarperCollins Publishers

Losordo, TM; Piedrahita, RH; Ebeling JM; (1988) an automated water quality data acquisition system for use in aquaculture ponds; *Aquacultural Engineering*

Loveless, T; (2012); *The 2012 Brown Center Report on American Education: How Well Are American Students Learning*? Washington DC: Brookings Institute

Lubzens, E; Marko, A; Tietz, A; (1984); Lipid Synthesis in the Rotifer *Brachionus plicatus*; *European Mariculture Society, Special Publication*; 8:201-210

Lundegaard G; (1985); *Keeping Marine Fish*; Blandford Press, Dorset, England; pp. 94

MacLeish, William H; (1989); *the Gulf Stream: Encounters with the Blue God*; 0-395-40621-8

Marchant, R; Hehir, G.; A Method for Quantifying Hand-Net Samples of Stream Invertebrates; *Marine and Freshwater Research*; 1999, 50, 179-82

Marini, Frank; *My Notes and Observations on Raising and Breeding of the Banggai Cardinalfish;* 1998; the Breeders Registry

Marten, Gerald; Environmental Tipping Points: A New Paradigm for Restoring Ecological Security; *Journal of Policy Studies* (Japan), No. 20 (July 2005), pp. 75-87

Marten, Gerald G.; (2001); *Human Ecology: Basic Concepts for Sustainable Development*. Earthscan Publications, London

Marten, Gerald, Matthews, Catherine; Ecotipping Points: Sharing Environmental Success Stories with Students; *the Science Teacher*, Vol. 76, No. 7, October 2009

Marten, Gerald; *EcoTipping Points: Strategies for Shifting from Decline to Restoration*; Jan. 2011; http://www.yesmagazine.org/planet/ecotipping-points-strategies-for-shifting-from-decline-to-restoration;

Marten, G., & Points, T. Marine Sanctuary: Restoring a Coral-Reef Fishery (Apo Island, Philippines) - How Success Works - http://www.ecotippingpoints.org/education/how-success-works-apo/index.html

Maryland Sea Grant Schools Network News; http://www.mdsg.umd.edu/extensionconf/program.html ; *Bob Foor-Hogue South Carroll High School: SciTech: Center of Marine Biotechnology Sea Grant Extension Program*; Aquaculture in Action Workshop - July 6 - 11, 1998

Mathews, Jay; 1988; *Escalante: The Best Teacher in America*; Henry Holt and Company/ New York

Maypa, A.P; G. R. Russ; A. C. Alcala; and H. P. Calumpong; 2002; Long-term trends in yield and catch rates of the coral reef fishery at Apo Island, central Philippines. *Marine Freshwater Research* 53:207-213.

McKenny, C. L. Jr.; Weber, D. M.; (1998); Altered Growth and Metabolism of an Estuarine Shrimp (Palaemonetes pugio) During and After Metamorphosis onto Fenvalerate-Laden Sediment; *Arch Contam Toxicol*; 35(3): 464-471; http://www.epa.gov/gbwebdev/ged/publica/c0146.htm

McMurtry; Nelson & Sanders, (1988); Aqua-vegeculture Systems; *International Ag-Sieve*, 1(3), article 7

McVey, Jim; (2000); *the NOAA/DOC Aquaculture Initiative and Policy and Its Implication for Recirculation Systems*; National Sea Grant Program; http://nsgl.gso.uri.edu/vsgcp/vsgcpc00001/2000/COMPLETE_document.pdf; accessed May 21, 2013

Mechling, Kenneth R.; Oliver, Donna L.; (1983); *What Research Says about Elementary School Science*; Washington, DC; National Science Teachers Association.

Meinesz, Alexandre; (1999); *Killer Algae: The True Tale of a Biological Invasion*; pp. 360; 0-226-51922-8

Meyerson, A. Lee; (1988); *Sea Water: A Delicate Balance*; Earth Presses Book - Easlow Pub; 0-89490-157-5

Michener, J. A. (1947); Tales of the South Pacific

Milgram, James R; and Wu, Hung-Hsi; (2008); *The Key Topics in a Successful Math Curriculum: Intervention Program*; http://math.berkeley.edu/~wu/six-topics1.pdf

Miller, Wade W.; (1995); National Council for Agricultural Education;

- *Tilapia: Aquaculture Curriculum Guide; Year Two Species Module*; 1995; pp. 86
- *Baitfish: Aquaculture Curriculum Guide; Year Two Species Module*; 1995; pp. 25
- *Striped Bass: Aquaculture Curriculum Guide; Year Two Species Module*; 1995; pp.85
- *Yellow Perch: Aquaculture Curriculum Guide; Year Two Species Module*; 1995; pp. 100
- *Shellfish Culture: Aquaculture Curriculum Guide; Year Two Species Module*; 1995; pp. 99
- *Crawfish farming: Aquaculture Curriculum Guide; Year Two Species Module*; 1995; pp. 122
- *Saltwater Shrimp: Aquaculture Curriculum Guide; Species Specific Module*; 1995; pp. 95
- *Salmon: Aquaculture Curriculum Guide; Year Two Species Module*; 1995; pp. 154
- *Red Drum: Aquaculture Curriculum Guide; Species Specific Module*; 1995; pp. 120
- *Walleye: Aquaculture Curriculum Guide; Year Two Species Module*; 1995; pp. 86
- *Trout: Aquaculture Curriculum Guide; Year Two Species Module*; 1995; pp. 109
- *Sturgeon: Aquaculture Curriculum Guide; Year Two Species Module*; 1995; pp. 116
- *Carp: Aquaculture Curriculum Guide; Year Two Species Module*; 1995; pp. 166

Mills, Dick; (1991) *You and Your Aquarium: A Complete Guide to Collecting and Keeping Aquarium Fishes*; pp. 288; Knopf

Minkoff, G; Lubzens, E; Meragelman, E; (1985) Improving Asexual Reproduction Rates in a Rotifer (*Brachionus plicatus*) by Salinity Manipulations; *Israel Journal Zoology*; 33: 195-203

Mitchell, Mark K.; Stapp, William B.; (1997) *Field Manual for Water Quality Monitoring: An Environmental Education Program for Schools;* 11th Edition; Kendall Hunt Publishing Co.; 0-7872-3730-2

Moe, Martin A. Jr.; (1992a) *The Marine Aquarium Handbook: Beginner to Breeder*; pp. 320; Green Turtle Press

Moe, Martin A. Jr.; (1992b) *The Marine Aquarium Reference: Systems and Invertebrates*; pp. 512; Green Turtle Press

Moe, Martin A. Jr.; (2014a) *The (New) Marine Aquarium Reference Volume I: The Chemical Environment*; Ebook edition, Volume I ISBN 978-0-939960-16-3; pp. 225 Green Turtle Publications

Moe, Martin A. Jr.; (2014b) *The (New) Marine Aquarium Reference Volume II: The Physical and Biological Environment*; Ebook edition ISBN 978-0-939960-17-0; pp. 103; Green Turtle Publications

Moe, Martin A. Jr.; (1997) *Breeding the Orchid Dottyback Pseudochromis fridmani: An Aquarist's Journal*; pp.285; Green Turtle Press; 0-939960-09-5

Moore, F. R.; an Introduction to the Mathematics of Digital Signal Processing; pp. 1-68; in Strawn, John (Editor) *Digital Audio Signal Processing: An Anthology*; 1985; pp. 283; William Kauffman, Inc.; ISBN 0-86576-082-9

Moran, Brandy; Goudy, Clifford; and Hankin, Alan; *Marine Aquaculture: Raising Saltwater Fish in Your Classroom*; 2001; 94 pp.; MITSG 01-13 Activity One

Moretti, A; Pedini, Fernandez-Criado; Cittolin, G; Guidastri, R; *Manual on hatchery production of seabass and gilthead seabream, Volume 1*; Rome, FAO, 1999; 194 pages

Moretti, A.; Fernandez-Criado, Mario-P; Vetillart, René ; (2005); *Manual on Hatchery Production of Sea bass and Gilthead Sea bream*; Food and Agriculture Organization of the United Nations; Rome; PP. 166

Morgan, Ann Haven; (1930); *Field Book of Ponds and Streams: An Introduction to the Life of Freshwater*; G.P. Putnam's Sons, New York, London; Mt Holyoke College.

Moss, Graeme A.; Tong, Lennard J.; Illingworth, John; Effects of Light Intensity and Food Density on the Growth and Survival of Early-Stage Phyllosoma larvae of the Rock Lobster *Jasus edwardsii*; *Marine and Freshwater Research*; 1999, 50, 129-34.

Murray, John A. editor; (1999); *The SeaCoast Reader*; pp. 294; Lyons Press; 1-55821-782-7

NAAEE; 1990; *Computer Aided Environmental Education: Monographs in Environmental Education and Environmental Studies;* Vol. VII; PO Box 400; Troy OH; 45373

NASA Mission Geology: Geography (What's up with Earth's Water Resources**?):** http://er.jsc.nasa.gov/seh/Mission_Geography/9-12/Module_1/III-1-1.pdf (accessed December 2013)

National Aquarium, Baltimore; *Activity 18: What's in Our Water: Do the amounts of nitrate in natural bodies of water and drinking water vary? Students test and compare water samples.* Also @ same site: *Marsh Mania: What Lives Here*?

National Governors Association Center for Best Practices (NGA Center) and the Council of Chief State School Officers (CCSSO), 2011; http://www.corestandards.org

NOAA Marine Aquaculture Policy; 2011; http://www.nmfs.noaa.gov/aquaculture/docs/policy/noaa_aquaculture_policy_2011.pdf

Neill, A.S.; (1960); *Summerhill: A Radical Approach to Child Rearing*; Oxford, UK: Hart Publishing

Nelson, R. L. (2007); 10 Systems around the World; *Aquaponics Journal*, 46(3), 8

Newsman, Michael E.; *Plant Aquaculture Module: Aquaculture Curriculum Guide*; 1995 Preliminary Edition; pp. 35; National Council for Agricultural Education

Nicol, Ernest; How to Do It: Hydroponics and Aquaculture in the High School Classroom. *The American Biology Teacher*; 52 (1990): 182-4.

Niesen; (2000); *the Marine Biology Coloring Book*; Coloring Concepts, Inc. pp. 290

North Carolina Agricultural Extension Service - Natural Resources Work Group; *Natural Resources in North Carolina: In Inventory and Conservation Issues*

O'Hara, Kathryn; National Marine Debris Data Base: Findings on Beach Debris Reported by Citizens; in R.S. Shomura and M. L. Godfrey (editors); Proceedings of the Second International Conference on Marine Debris, 2 7 April 1989, Honolulu, Hawaii; U. S. Dept. Commerce, NOAA Tech. Memo. NMFS, NOAA-TM-NMFS-SWFSC-154; 1990

Oregon State University (2012, August 7); New pathway for invasive species - science teachers; *ScienceDaily*; Retrieved February 22, 2013

Osborne, K; 1983; Full Spectrum Lighting in the Planted Aquarium; *Freshwater and Marine Aquariums*; Vol. 6; Number 1; p. 8

Osinga, Ronald; Janse, Max; September, 2009; *Fragmentation;* CORALZOO

Outwater, Alice; *Water: A Natural History*; 1996; Basic Books; pp. 212; 0-465-03779-8

Palko, Barbara Jayne; ED 247-108; SE 044-704; *A Balanced Marine Aquarium* [and] *The Biology of Marine Aquarium Fishes Collected in Monroe County, Florida.* A Two-Paper NOAA Memorandum; Jan 1981; pp. 90;

Patton, A; & Robin, J; (2012); *Work that matters: The teacher's guide to project-based learning*; London, England: Paul Hamlin Foundation, Learning Futures.

Peace Corps, Washington D. C. Information Collection and Exchange Division; *Peace Corps Aquaculture Training Manual Training Manual* T0057; ED 327 732; CE 056 826; 1990; pp324

Pennington, PL.; DeLorenzo, M.E.; Key, P.B.; Wirth, E.F.; Fulton, M.H.; Scott, G.L.; 2007; The Design, Construction, Operation, and Maintenance of the Replicated Modular Estuarine Mesocosm; NOAA Technical Memorandum - NOS NCCOS 62. 77pp.

Perrin, Richard. 1993. Coral farming in Detroit. *SeaScope* 10:2

Platt, Rutherford H; Rowntree, Rowan A; Muick, Pamela C; *the Ecological City: Preserving and Restoring Biodiversity*; 1994; pp. 291; University of Massachusetts Press.

Popper, K. R; (1959); *the Logic of Scientific Discovery*; New York, Harper & Row

Posner, R; (2009); *Lives of Passion, School of Hope*; Boulder, CO: Sentient Publications

Prodigo, Denise M; Doin' the Sanibel Shuffle: Awareness Can Help Prevent Stingray Stings; 1992; pp. 1; *Fathom*

Quesada, Laura; (2011); *Automatic Clownfish Hatchery*; http://digitalcommons.calpoly.edu/

Quinn, H.; Schweingruber, H.; & Keller, T; (Eds.); (2011*); A framework for K-12 science education: Practices, crosscutting concepts, and core ideas*; National Academies Press

Rakocy, James E.; Masser, Michael P.; Losordo, Thomas M. (2006**),** *Recirculating aquaculture tank production systems: Aquaponics — integrating fish and plant culture*, Southern Region Aquaculture Center, http://www.aces.edu/dept/fisheries/aquaculture/documents/309884-SRAC454.pdf

Rakocy; Shultz; Bailey & Thoman, (2004), M.A. Nichols, ed., Aquaponic production of tilapia and basil: Comparing a batch and staggered cropping system, *Acta Hort* (ISHS) (648)

Ramel, Gordon; *The Picturesque World of Fish*; http://www.earthlife.net/fish/intro.html; accessed 1/29/2013

Rand, W., & Wilensky, U. (June, 2006). *Verification and Validation through Replication: A Case Study Using Axelrod and Hammond's Ethnocentrism Model*; Conference of the North American Association for Computational Social and Organizational Sciences; South Bend, IN

Raymundo, L. J; and A. P. Maypa; (2003); Chapter 14; *Apo Island marine sanctuary, Dauin, Negros Oriental*; Pages 61-65 in *Philippine coral reefs through time*. The Marine Science Institute, University of the Philippines, Diliman, Philippines.

Raymundo, L. J; and A. T. White; (2005); 50 years of scientific contributions of the Apo Island experience: a review; *Silliman Journal* (50th Anniversary Issue), Silliman University, Dumaguete, Philippines.

Reboton, C; and H. P. Calumpong; (2003); Coral damage caused by divers/snorkelers in Apo Island marine sanctuary, Dauin, Negros Oriental, Philippines. *Philippine Scientist* 40:177-190.

Redmond, S.; Green, L.; Yarish, C.; Kim, J.; Neefus, C.; (2014); *New England Seaweed Culture Handbook-Nursery Systems*; Connecticut Sea Grant CTSG-14-01; 92 pp.; PDF file; URL: http://seagrant.uconn.edu/publications/aquaculture/handbook.pdf

Reebs, Stephan; (2001); *Fish Behavior in the Aquarium and in the Wild*; Cornell University; ISBN 0-8014-3915-9 (cloth) or 0-8014-8772-2 (paperback)

Renner, John W.; et.al.; (1973); An Evaluation of the Science Curriculum Improvement Study; *School Science and Mathematics*; Vol. 73; No. 4: 291-318

Reverberi, G.; (editor) (1971); *Experimental Embryology of Marine and Freshwater Invertebrates*; 1971; North Holland Publishing Co; Chapter 12 Echinoids by Czihak, G.

Riddle, Dana; *the Captive Reef: A Concise Guide to Reef Aquaria in the Home*; 1995; Energy Savers Unlimited, Inc.; pp. 297

Riley, Cecilia M.; Captive spawning and rearing of the Peppermint Shrimp (Lysmata wurdemanni); *Seascope*, Volume 10; summer, 1993

Rippingale, R.J; Payne, M.F; (2001); *Intensive Cultivation of a Calanoid Copepod - Gladioferens imparipes: A Guide to Procedures*; ISBN 1 74067 070 1;

Rivers, J. L. (2011). Common core standards and assessments; http://lwvthurston.org/files/Standards_Assessments.pdf

Roast, SD; Thompson, RS; Widdows, J; Jones, MB; Mysids and Environmental Monitoring: A Case for Their Use in Estuaries; *Marine and Freshwater Research*, 1998, 49, 827 - 32

Rogers-Bennett, Laura; Fastenau, Henry C.; et.al; *Outplanting Study: Culturing Sea Urchins for Experimental Outplanting in Northern California*; July 18, 1994; Contract FG-2230-MR; http://seaurchin.org/studies/outplant.html

Rogosa, Eli; (2010); Organic Aquaculture" Heritage Wheat Conservancy website http://www.growseed.org/aquaponics.html

Rogosa, E. (2010). Organic aquaponics; A Tour of Will Allen's Growing Power Facility in Milwaukee, Wisconsin; http://www.growseed.org/growingpower.html

Romero, Filemon G.; *CREST: coral reef education for students and teachers: manual*; International Marinelife Alliance; pp.81; 971-91787-0-1 (and the accompanying student manual: 971-91787-6-0); Accessed 6/12/2003

Rounds, A.A.; Wilde, F.D.; and Ritz, G.F.; (2013) Dissolved oxygen (ver. 3.0) U.S. Geological Survey Techniques of Water Resources Investigations, book 9, chap, A6; sec 6.2, http://water.usgs.gov/owq/FieldManual/Chapter6/6.2_v3.0.pdf

Rowe, Mary Budd; (1975); Help is Denied to Those in Need; *Science and Children*; Vol. 12; No. 6; pp. 23-25.

Royte, Elizabeth (July 5, 2009); Street Farmer; *the New York Times Company*; http://www.nytimes.com/2009/07/05/magazine/05allent.html?sq=aquaponics&st=cse&adxnnl=1&scp=10&pagewanted=2&adxnnlx=1299564045-NgXx2LMa3/ee2z6MNr9YdA

Rubenstein, G; (2008, December 3); *Real World, San Diego: Hands-On Learning at High Tech High*; Edutopia; http://www.edutopia.org/collaboration-age-technology-high-tech

Russ, G. R and A. C. Alcala; (1996); Do marine reserves export adult fish biomass? Evidence from Apo Island, central Philippines; *Marine Ecology Progress Series* 132:1-9

Russ, G. R. and A. C. Alcala; (1998); Natural fishing experiments in marine reserves 1983-1993: community and tropic responses. *Coral Reefs* 17:383-397

Russ, G. R; and A. C. Alcala; 1999; Management histories of Sumilon and Apo marine reserves, Philippines, and their influence on national marine resource policy; *Coral Reefs* 18:307-319.

Russ, G. R; and A. C. Alcala; 2004; Marine reserves: long-term protection is required for full recovery of predatory fish populations. *Oecologia* 138:622-627.

Russ, G. R., A. C. Alcala, A.P. Maypa, H. P. Calumpong, and A. T. White; (2004); Marine reserve benefits local fisheries. *Ecological Applications* 14:597-606

Russell, M. (2000); a tank-system design for the hatchery production of sea-urchin larvae; In *Proceedings of the Workshop on the Co-ordination of Green Sea Urchin Research in Atlantic Canada Université de Moncton,* http://www86. Homepage: Villanova.edu/michael.russell/PDF_Links/Moncton_workshop_2000; PDF (accessed on 25 August 2012)

Rust, Thomas G.; *A Guide to Biology Lab*; (1983); S.W. Educational Enterprises

Santos, Elsa; Baylina, Núria; Fragmentation of a Branching Coral (Type I), e.g. Acropora sp., Stylophora pistillata; Pocillopora, damicornis; in Osinga, Ronald and Janse, Max (eds.); September, 2009; Fragmentation; CORALZOO

Santos, Elsa; Baylina, Núria; Fragmentation of a Branching Coral (Type II) e.g.; Euphyllia Divisa; in Osinga, Ronald and Janse, Max (eds.); September, 2009; Fragmentation; CORALZOO

Santos, Elsa; Baylina, Núria; Fragmentation of Plate Shape Coral, like Montipora Sp.; in Osinga, Ronald and Janse, Max (eds.); September, 2009; Fragmentation; CORALZOO

Sasala, Thomas; 1995; *The Lighting Comparison to End All Lighting Comparisons - (AKA: A Lighting Cost Guide)* Contributed by Thomas Sasala; http://www.thekrib.com/Lights/cost-comparison.html

Sato, Gordon H.; Ghezae, and Negassi, Samuel; 1998; The Manzanar Project: Towards a Solution to Poverty, Hunger, Environmental Pollution, and Global Warming through Sea Water Aquaculture and Silviculture in Desserts; http://mangroves.org/sato.htm; *In Vitro Cellular and Developmental Biology* - Animal (July-August issue)

Schauer, Jerry L.; *Simulated Laminar Flow Leapfrog Fountain*; 1997; http://www.hiddenmickeys.org/Imagineering/LeapFrog.html

Schutter, Miriam; Shafir, Shai; Fragmenting Branching Corals; in Osinga, Ronald and Janse, Max (eds.); September, 2009; Fragmentation; CORALZOO

Schwartz, David M.; It Takes More than Repellent to Make Flying Pests Bug Off; *Smithsonian Magazine;*

Scott, Thomas M.; Rust, Michael B.; (1996); *A Computer Automated Cold-Water Recirculating System for Aquaculture Research*; Conference Presentation at Virginia Tech, University July 19-21

Seely, Jeffery K.; How to Do It: Environmental Sampling with an Inexpensive Electric Pump; *the American Biology Teacher*, Volume 54, No. 1, January 1992; pp. 37-39

Seidel, S., Walters, J., Kirby, E., Olff, N., Powell, K., Scripp, L., & Veenema, S. (1997). Portfolio practices: Thinking through the assessment of children's work. Washington, DC: National Education Association Publishing Library

Shafir, S., J. van Rijn, and B. Rinkevich, 2006; Steps in the construction of underwater coral nursery, an essential component in reef restoration acts. *Marine Biology* 149: 679-687.

ShaHani, A.K.; Parsons, P.S.; and Meacock, S.E.; *Animals in A Pond: Biology and Statistics in Mutual Support*; Daphnia are studied; http://www.maths.nott.ac.uk/rsscse/TS/bts/shahami/text.html

Shakerin, S. A. I. D. (2004). Microcontrolled water fountain: A multidisciplinary project. *International Journal of Engineering Education*, *20*(4), 654-659.

Shakerin, S.; (2005); Water fountains with special effects; *American Scientist*, *93*(5), 444

Shawl, A.; M. Davis; (2004); Captive breeding behavior of four Strombidae; Conch; *Journal of Shellfish Research*; 23(1):157-164

Shawl, A.; M. Davis; (2005); Harbor Branch Oceanographic Institution Develops Recirculation Systems for Queen Conch: Captive Breeding to Juvenile Grow-out; *Recirc Today*

Shawl, Amber; Davis, Megan; Effects of Dietary Calcium and Substrate on Growth and Survival of Juvenile Queen Conch (Strombus gigas) Cultured for Stock Enhancement; 2006; 57th Gulf and Caribbean Fisheries Institute; http://procs.gcfi.org/pdf/gcfi_57-70.pdf;

Shrimp: The Real Cost; http://www.sierraclub.ca/national/shrimp/cost.html

Shymansky, James; Kyle, William Jr.; Alport, Jennifer; (1982); How Effective Were the Hands-On Science Programs of Yesterday? *Science and Children*; Vol. 20; No. 3; 1415

Smith, DeBoyd L; (1977); *A Guide to Marine Coastal Plankton and Marine Invertebrate Larvae*; Kendall/Hunt Publishing Co

Smith, John Wesley; the Genius of Simplicity: Integrated Aquaculture/Hydroponic System - *The Growing Edge*; Winter 1993/1994; pp. 6; S & S Aqua Farm, Rt. 1, Box 747; West Plains, MO 65775

Smith, Walter L.; Chanley, Matoria H.; (1972); *Culture of Marine Invertebrate Animals*; Plenum Press; from a Conference on Culture of Marine Invertebrate Animals, Greenport, N.Y

Sneyink, Vernon L.; & Jenkins, David; (1980); *Water Chemistry*; John Wiley and Sons; 0-471-05196-9

Soderberg, Richard W.; (1995); *Flowing Water Fish Culture*; Mansfield University, Pennsylvania; pp.147; ISBN 1-56670-081-7

Spede, Jani Farlow; *UF Scientist Conducts Taste Test for Manatees*; pp. 2; http://gnv.ifas.ufl.edu/seaweb/homepage/v63p12.htm

Spence, Lundie; *Science Education Critical to Our Children's Future*; Sea Grant Editorial; Opinions/Editorials; http://www.mdsg.umd.edu/seagrantmediacenter/opscience.html

Spotte, Stephen; (1973); *Marine Aquarium Keeping: The Science, Animals, and Art*; J. Wiley & Sons; 0-471-81759-7 (cloth), or 0-471-82591-3 (paperback)

Spotte, Stephen; (1995); *Reef Notes*; Ricordea Publishing; pp. 185

Spring, Ashley; Davis, Megan; *Recommendations for Culturing Juvenile Queen Conch, Strombus gigas, for Restocking and Commercial Purposes*; 2005; 56th Gulf and Caribbean Fisheries Institute; http://procs.gcfi.org/pdf/gcfi_56-71.pdf

Staff (29 April 2012) Fish farming in a high-rise world; *BBC News US and Canada*, retrieved 30 April 2012

Stake, Robert E.; Easley, Jack; (1978); Case Studies in Science Education; Vol. 1; *The Case Reports*; Washington, DC: US Gov't Printing Office

Stavis, Barrie; (1966); *Lamp at Midnight: The Great Drama of Galileo and the Fight for Truth in the Dark Days of the Inquisition*; Bantam Books Inc.

Steinbeck, J.; (1945); *Cannery row*; Penguin

Steinbeck, J.; & Ricketts, E. F.; (1951); *the log from the Sea of Cortez*; New York: Viking Press

Stone, Irving; (1980); *the Origin: A Biographical Novel of Charles Darwin*; pp. 743; Doubleday & Co.; 0-385-12064-8

Storey, Nathaniel, R.; (2012) Vertical Aquaponic Crop Production Towers and Associated Produce Sales and Distribution Models: Design, Development, and Analysis; Ph.D.; Understanding Biological Surface Area in Aquaponics (Table 2.01)

Strathmann, Megumi F.; (1987); *Reproduction and Development of Marine Invertebrates of the Northern Pacific Coast: Data and Methods for the Study of Eggs, Embryos, and Larvae*; ISBN 0295965231

Strawn, John; (1985); *Digital Audio Signal Processing: An Anthology*; pp. 283; William Kauffman, Inc.

Strong, Maurice; (2001), introduction to: Marten, Gerald G. 2001; *Human Ecology: Basic Concepts for Sustainable Development*; Earthscan Publications, London, pp. 256

Studio Schools Trust; (2010); *Rooting Learning in the Real World: An Introduction to Key Stage 4 Projects*; http://studioschooltrust.org/sites/default/files/Studio%20School%20Curriculum%20Introduction.pdf

Sutton, Suzy; *Reaching Beyond*;

Suutari, Amanda; USA - California (Berkeley) - *Edible Schoolyards: Connecting Kids to the Earth and Encouraging Healthy Eating Habits*; EcoTipping Points Project. Retrieved on March 18, 2013; from: http://www.ecotippingpoints.org

Swenson, William; Barman, Charles; Koch, Rudy; University of Wisconsin - Superior; *Aquaria on a Classroom Budget*; March 1976; pp. 35-36

Szyper, JP; (1989); Nutritional Depletion of the Aquaculture Feed Organisms *Euterpina acutfrons*, *Artemia* sp. and *Branchionus plicatus* During Starvation; *J World Aquaculture Society*; 20: 162-169

Tamaru, Clyde; Rivera, Malia; Haverkort, Roxanne; Gorospe, Kevin; (2009); *Sea Urchin Fertilization Lab*; Hawai'i Institute of Marine Biology Education Program; http://www2.hawaii.edu/~himbed/forms/HIMB-Sea-Urchin-Fertilization-Lab.pdf

Tamaru, C; Ako, H. et al; *Enrichment of Artemia for Use in Freshwater Ornamental Fish Production*; Center for Tropical and Subtropical Aquaculture; pp. 21

Taylor, Alma E.F.; *Illumination Fundamentals*; Rensselear Polytechnic Institute (2000); Optical Research Associates

Taylor, Scott D.; the Ubiquitous Mosquito; winter 1988; *Florida Naturalist*

Teal, John and Mildred; (1975); *the Sargasso Sea*; Atlantic Monthly Press; 0-316-84351-3

The Independent; 'Backyard Aquaponics System' An outstanding research by BAU teacher; *The Independent*; January 26, 2012

Their, Herbert; (1986); Heads-On Elementary Science: The Challenge of the Nineties. *Monograph and Occasional Paper Series #1*; Council for Elementary Science International

Thurman, H,V.; Webber, H.H.; (1984); *Marine Biology*; pp. 436; Charles E. Merril Publishing Company; ISBN 0-675-20139-X

Thurston, R.V.; Russo, R.C.; Emerson, K.; (1979, August); Aqueous Ammonia Equilibrium – Tabulation of Percent Un-Ionized Ammonia; USEPA Environmental Research Lab – Fisheries Biology Laboratory; Grant R800861; EPA 600/3-79-091; http://nepis.epa.gov/Adobe/PDF/30000I7U.PDF

Toonen, Rob; Home Breeders FAQ for Marine Invertebrates; (Jan 1996); *Journal of MaquaCulture*

Ukens, Leon; Hands-On Education: What the Research Says; (preface) in Chase, Valerie; *Living in Water Curriculum*; 1987; National Aquarium in Baltimore; pp. 302

United Nations Food and Agriculture Organization, (2009); FISHSTAT

University of Maryland Sea Grant College; *Decision Making/ the Chesapeake Bay*; 1985; pp. 100; University of Maryland Sea Grant College; ED 306 082; UM-SG-ES-85-01

U.S. Commission on Ocean Policy; (2004); http://www.oceancommission.gov/documents/full_color_rpt/welcome.html

USD; (1982), revised 1997; Agri. Handbook # 590; *Ponds: Planning, Design, Construction.*

U.S. Department of Commerce Aquaculture Policy; (2011); http://www.nmfs.noaa.gov/aquaculture/docs/policy/doc_aquaculture_policy_2011.pdf

U.S. Department of the Interior/U.S. Geological Society; *Why Is the Ocean Salty?* pp. 15; 1991-302-674

U.S. Fish and Wildlife Service; (1998); *Adopt a Salmon Family: An Interdisciplinary Watershed Education Program*; http://www.fws.gov/northeast/cnefro/pdf/intro.pdf

U.S. Fish and Wildlife Service; (1998); *From Egg to Stream: A Guide to Building and Maintaining Classroom Incubators for Atlantic salmon*; http://www.fws.gov/northeast/cnefro/pdf/incubator.pdf

Vendetti, Michael; How to Make "Egg" Plugs; Reef Aquarium Framing News Volume 15, page 3; Geothermal Aquaculture Research Foundation; Accessed February 1, 2000

Viladomat, Lorena; Jones, Philip; (2011); Development of aquaponic systems for space and water efficient food production: pilot project final report; http://www.byspokes.org/wp-content/uploads/documents/basma_report.pdf; Accessed April 2013

Walch, John; Reproduction of Yellowhead Jawfish in Captivity; *SeaScope; Volume 11; 1994*

Walker, Cyril; Walker, David; *Fossils*; (1992); pp. 320; Dorling Kindersley; ISBN 1-56458-071-7

Walker, Susan S.; *Aquaculture: Teacher Edition; Mid-America Vocational Curriculum Consortium*, Stillwater, Oklahoma; 1990; pp. 876; ED 327 657; CE 056 667

Walton, Susan; (1980); The Reef's Tale: Smithsonian transplants coral system from tropics to tank; *Bioscience*, 30(12): 805-808

Walton, Susan; McLamb, L. W.; Bibliography of Marine Education Computer Software; *Journal of Computers in Mathematics and Science Education;* Fall, 1985; pp. 42-43

Watanabe, T; Kitajima, C; Fujita, S; (1983); Nutritional Value of Live Food Organisms Used in Japan for Mass Culture of Fish: A Review; *Aquaculture*; 34:115-143

Weast, R; (1987); Editor; *Handbook of Chemistry and Physics*; CRC Press, Inc. Boca Raton, FL, USA

Wellman, Ruth T.; (1978); Science: A Basic for Language and Reading Development in *What Research Says to Science Teachers;* Vol. 1; Washington, DC; National Science Teachers Association

Westerd, Leon; Janse, Max; *Fragmentation of Massive Corals: Favia Sp.; Lobophyllia Sp.*; in Osinga, Ronald and Janse, Max (eds.); September, 2009; Fragmentation; CORALZOO

Whitern, Wilfred A; (1980); *Guppies*; pp. 95; TFH Press

Wijgerde, Tim; *Coral reproduction - part I: Biology*; 2009 coralscience.org; http://www.coralscience.org/main/articles/reproduction-10/coral-reproduction-part-i

Wijgerde, Tim; *Coral reproduction - part II: Challenges and future perspectives;* 2009 coralscience.org; http://www.coralscience.org/articles/PDF/Coral%20reproduction%20-%20part%20II.pdf

Williamson, Robert; Smoak, Ellen; (1998); *Curious About Water;* NC A&T State University – Cooperative Extension

Wise, Kevin C.; Okey, James; 1983; A Meta-Analysis of the Effects of Various Science Teaching Strategies on Achievement; *Journal of Research in Science Teaching*; Vol. 20; pp. 419-435

Wray, Kris; *Coral Fragging 101; Getting Started with Coral propagation*; 2013 Blue Zoo Aquatics; http://www.bluezooaquatics.com/resources.asp?show=397; accessed January 2013

Yamashita, Masakane; Embryonic Development of the Brittle star Amphipholis kochii in Laboratory Culture; *The Biological Bulletin*; August 1985; v169, 131-142

Yao, Dr. Guo Jun; *Development of Commercial Breeding and Production Technology of the Chinese Hairy Crab Eriocheir sinensis* The School of Biological Sciences - Ecosystems Lab; http://www.science.nus.sg/~sbs/Labs/ecos-lab/index.html

Yaquina Bay: Where the River Meets the Sea, What is an Estuary? Anatomy of an Estuary, Biological Importance of an Estuary, Importance of Yaquina Bay Estuary to Humans, Human Disturbances, Why Protect? http://www.hmsc.orst.edu/education/estuary/index.html

Zapp, J; 1992; Cultivos sin terra; *Hidroponia Popular*; Proyecto regional para la superacion de la pobreza; PNUB-UNIFEM, Colombia

Závodská & Dolly (2009); *a comparison of Small Scale Farming in Barbados, Dominica, and Trinidad and Tobago*; San Juan: Association for International Agricultural and Extension Education (AIAEE).

Zhao, Yong; (2011); *Teach Children to Invent Jobs, not find them; students as global entrepreneurs*; http://www.youtube.com/watch?v=NOXAJzqm2Rw; accessed Feb. 22, 2013

Zhao, Yong; (2012); *World Class Learners: Educating Creative and Entrepreneurial Students*; Corwin SAGE Publications;

Index

Airborne Visible Infrared Imaging Spectrometer....272, 276
algae *24, 27, 34, 41, 64, 65, 67, 68, 69, 70, 97, 103, 117, 149, 150, 153, 154, 155, 172, 191, 193, 194, 195, 202, 204, 206, 220, 225, 247, 274, 317, 319*
Allen Bradley..*200, 202*
An Ocean Blueprint for the 21st Century.................... 5
Apo Island..*243, 250, 261, 277, 278, 279, 281, 288, 290, 317, 321, 326, 329, 330*
Aquaculture......*10, 46, 47, 127, 128, 133, 135, 136, 169, 172, 173, 174, 294, 306, 319, 320, 321, 322, 323, 324, 325, 326, 327, 328, 329, 330, 331, 332, 333*
aquaponics... *1, 4, 23, 24, 26, 42, 50, 53, 60, 76, 77, 88, 91, 105, 106, 121, 123, 124, 126, 127, 130, 131, 132, 133, 134, 135, 136, 137, 141, 142, 143, 144, 165, 186, 191, 225, 283, 285, 299, 304, 320, 322, 324, 329, 330*
Aquatic Ecosystem Conference.......................*242, 310*
aquatic ecosystem design............ *11, 14, 24, 60, 249*
Assessment.. *2, 3, 13, 14, 17, 19, 54, 55, 56, 87, 88, 90, 121, 122, 123, 125, 142, 143, 145, 166, 167, 168, 187, 188, 190, 203, 204, 205, 220, 224, 246, 248, 258, 260, 283, 285, 286, 288, 289, 293, 295, 307, 308, 310, 311, 312, 313, 314, 315, 316, 317, 336*
Automatic Clownfish Hatchery *306, 329*
Bibliography...319
Bioethics Research Form 129
black worm. 126, 128, 141, 165, 169, 172, 186, 188
brine shrimp.......126, 128, 141, 153, 165, 169, 172, 186, 206
broodstock and larval rearing system *1, 50, 53, 83, 84, 113, 117, 126, 169, 188, 189*
Brylske...155, 156, 167, 320
Buchheim *95, 96, 148, 175, 320*
Budget Sheet (Chronologically Itemized Project Cost Assessment)..22
Business..*13, 243*
carpentry 4, 53, 76, 80, 83
Chesapeake Bay *249, 250, 261, 262, 263, 264, 265, 266, 267, 268, 270, 271, 272, 273, 275, 276, 277, 291, 332*
community service..................................*11, 156, 283*
computer automation.........*10, 11, 12, 191, 199, 299*
Copepod 202, 329
coral frag *146*
coral fragging.......*123, 126, 127, 141, 161, 165, 186*
coral propagation system 23, 41, 50, 80, 81, 82, 107, 112, 113, 121, 123, 127, 141, 146, 147, 160, 161, 165, 166, 167, 186
Coral Shores High School...............................*155, 167*
CRI...*66, 71, 89*
critique. *3, 21, 46, 50, 53, 55, 76, 80, 83, 88, 89, 91, 105, 112, 117, 121, 143, 144, 167, 189, 191, 199, 226, 231, 234, 247, 250, 286, 300, 309, 310*
critiquing..................4, 24, 76, 80, 83, 191, 225, 226
Crustaceans - Crab Life Cycle
– text and drawing.. *36*
Daily Journal 54, 87, 122, 142, 166, 187, 203, 223, 246, 258
data collection *5, 10, 42, 43, 127, 146, 169, 206, 226, 237, 247, 283, 285, 290, 299, 303*
design.*1, 4, 7, 10, 11, 12, 13, 14, 15, 21, 23, 24, 39, 40, 41, 42, 53, 54, 59, 60, 64, 70, 74, 75, 76, 79, 80, 81, 82, 83, 84, 85, 86, 87, 88, 91, 96, 100, 103, 105, 112, 117, 119, 121, 122, 123, 124, 126, 127, 137, 141, 142, 143, 144, 146, 166, 167, 169, 187, 189, 191, 192, 193, 199, 202, 203, 206, 209, 214, 223, 225, 228, 229, 237, 242, 244, 246, 247, 250, 251, 258, 261, 282, 283, 284, 286, 290, 294, 296, 299, 300, 301, 302, 303, 304, 305, 306, 309, 311, 325, 330, 336*
ecosystem frame 4, 24, 53, 76, 80, 83, 89
electrical..... 4, 11, 41, 42, 53, 60, 69, 70, 74, 75, 94, 95, 137, 161, 163, 164, 184, 191, 199, 209, 301
electrolysis 92, 93, 94, 124
engineering *1, 12, 13, 53, 54, 86, 204, 247, 250, 296, 301, 336*
Explorations in Aquaculture....................................*324*
Field Trip Feedback Form.. 21
field trip five 192
field trip four 191
field trip one... 23
Field Trip Seven 234
Field Trip Six...231
field trip three 126
field trips *4, 23, 24, 126, 225, 226*
Framework for K–12 Science Education: Practices, Crosscutting Concepts, and Core Ideas *7*
fund-raising events *18, 128*
Hoff and Snell *195, 238*
Illuminating Engineering Society of North America .. 64
illumination *66, 67, 71, 75, 78, 81, 82, 83, 153, 214, 215, 216, 276*
Index...*334*
International Society for Technology in Education... *10, 325*
Introductory Aquatic Ecosystem Inhabitant Form..... 57
Invertebrates.38, 322, 324, 326, 327, 329, 331, 332
laminar flow fountain............*250, 296, 297, 301, 307*
LANDSAT 7 ....268, 272, 277
Larval Rearing System........ 13, 83, 84, 117, 118, 185
lenses....*60, 61, 183, 206, 207, 208, 209, 210, 211, 212, 213, 214, 216, 220*
light intensity.. 24, 50, 51, 65, 66, 67, 68, 69, 70, 73, 74, 75, 89, 238
lighting system components 4, 60, 105, 112
Marine Flowering Plants
– text and drawing.. *34*
mathematics.. 1, 8, 12, 15, 16, 17, 56, 90, 125, 145, 168, 190, 205, 224, 248, 260, 289, 295, 299, 308
Michener *327*
Microcontroller......*See* Programmable Logic Controller

microscope .. 11, 143, 191, 206, 209, 210, 211, 212, 213, 214, 215, 216, 217, 218, 220
Moe 38, 60, 64, 68, 71, 104, 326, 327
National Aquaculture Act of 1980 .... *128*
NEPTUNE *303*
Niesen *25, 328*
NOAA ... *60, 128, 174, 272, 292, 321, 323, 324, 326, 327, 328*
Non-Human Vertebrate Endorsement Form *129*
optics *192, 206, 211*
oysters 41, 262, 264, 265, 272
Pair History Data Sheet 188
plankton culture and collection system *10, 11, 191, 199*
plumbing 13, 24, 85, 91, 105, 112, 117, 136, 298, 301, 311
portfolios 3, 155, 300, 309, 310
Programmable Logic Controller 325
Project Eight *206*
Project Eleven *261*
Project Five *146*
Project Four *130*
Project Fourteen *310*
Project Nine *227*
Project One 25
Project Seven *193*
Project Six *169*
Project Ten *251*
Project Thirteen *296*
Project Three *91*
Project Twelve *290*
Project Two *60*
Research Project 225, 227, 237
Research-Ability Potential Topics Form 59
Rockwell Automation *10, 199*
rotifers 117, 170, 191, 193, 195, 201, 204, 220, 225, 238, 323
school garden 11, 138
Sea Perch *303*
Sea-Viewing Wide Field-of-view Sensor 272
Smithsonian 152, 153, 154, 330, 333
STEM *1, 2, 3, 4, 7, 10, 12, 14, 15, 16, 17, 18, 53, 54, 56, 87, 88, 90, 122, 125, 142, 145, 166, 168, 187, 190, 191, 203, 205, 220, 224, 243, 244, 246, 248, 249, 258, 260, 288, 289, 290, 293, 295, 307, 308, 309, 310, 336*
Summary of Decisions about 3 Classroom Ecosystems Form 86
technological 5, 10, 11, 53, 128, 136, 173, 174, 309, 336
Tides – text and drawing 28
Types of Coral Reef – text and drawing *32*
vicious cycle *277, 278*
Victoria Experimental Network Under the Sea *303*
virtuous cycle *277*
Water Chemistry 16, 325, 331
water quality 24, 42, 67, 69, 92, 97, 98, 99, 100, 101, 103, 105, 112, 117, 119, 121, 123, 124, 130, 144, 147, 148, 162, 167, 174, 189, 191, 204, 220, 261, 263, 268, 270, 273, 275, 277, 299, 303, 326
water quality parameters *98, 144*
wavelength. *24, 64, 65, 67, 68, 70, 71, 88, 210, 211, 212*

Pre and Post Assessment

How often do you use science, technology, engineering, and/or math STEM knowledge or skills? Use the number one to mean almost never and the number ten to mean almost always. Circle your choice from 1 to 10.

In-School? —1—2—3—4—5—6—7—8—9—10—

Out of school? —1—2—3—4—5—6—7—8—9—10—

What kinds of science things do you usually do (and expect to do) in school?

What are the five most common technological things you do in school?

1.

2.

3.

4.

5.

What kinds of things do you usually design and assemble in school?

List five math ideas you use most often in school and explain how you use each

1.

2.

3.

4.

5.

Image Credits

p. 29 Tides; p. 31 Photic Zone; p. 33 Types of Coral Reef; p. 35 Marine Flowering Plants; p. 37 Crustaceans – Crab Life Cycle (Permission by Niesen)
p. 48 http://commons.wikimedia.org/wiki/File:Tautoga_onitis_(line_art).jpg
p. 48 http://commons.wikimedia.org/wiki/File:Centropristis_striata.png
p. 48 http://commons.wikimedia.org/wiki/Category:Paralichthys_dentatus
p. 48 http://commons.wikimedia.org/wiki/File:Gadus_morhua.png
p. 48 http://commons.wikimedia.org/wiki/File:Melanogrammus_aeglefinus.png
p. 48 http://commons.wikimedia.org/wiki/File:BlackGrouper.jpg
p. 49 http://commons.wikimedia.org/wiki/File:Lutjanus_campechanus.png
p. 49 http://commons.wikimedia.org/wiki/File:Sciaenops_ocellatus.jpg
p. 49 http://commons.wikimedia.org/wiki/Sphyraena_barracuda#mediaviewer/File:Sphyraena_barracuda.jpg
p. 49 http://commons.wikimedia.org/wiki/File:Carcharodon_carcharias01.jpg
p. 49 http://commons.wikimedia.org/wiki/Thunnus_thynnus#mediaviewer/File:Bluefin-big.jpg
p. 49 http://commons.wikimedia.org/wiki/Xiphias_gladius#mediaviewer/File:Xiphias_gladius1.jpg
p. 49 http://commons.wikimedia.org/wiki/Category:Dasyatis_americana#mediaviewer/File:SStringray.jpg
p. 64 Light Wavelength and Color - http://infraredtv.com/IR_Explained.html (permission granted)
p. 66 Depiction of "Lux" as a unit of light measurement; Ryer, A;(1997); Light Measurement Handbook, permission granted by International Light Technologies, Inc.
p. 67 Transmission of Light in Water (Buchheim, 1998 permission granted on website)
p. 69 fluorescent bulb Leslie & Conway, (1993); permission granted from Rensselaer Polytechnic University
p. 113 illustration of seven unique morphological characteristics of coral from a Shedd Aquarium PowerPoint by George Parsons (2012)
p. 146 http://upload.wikimedia.org/wikipedia/commons/a/a1/Reef0814_-_Flickr_-_NOAA_Photo_Library.jpg (Public Domain)
p. 147 illustration depicting Anthozoan diversification from a Shedd Aquarium PowerPoint by George Parsons (2012)
p. 148 colorful coral polyps; flowergarden.noaa.gov (Public Domain)
p. 148 coral polyp anatomy - http://upload.wikimedia.org/wikipedia/commons/4/49/Coral_polyp.jpg (Public Domain)
p. 149 http://commons.wikimedia.org/wiki/Category:Corals_of_the_Philippines#mediaviewer/File:Reef4313_-_Flickr_-_NOAA_Photo_Library.jpg (Public Domain)
p. 149 http://commons.wikimedia.org/wiki/File:Elkhorn_coral.jpg (Public Domain)
p. 149 http://en.wikipedia.org/wiki/Pacific_Proving_Grounds#mediaviewer/File:Operation_Crossroads_Baker_Edit.jpg (Public Domain)
p. 150 coral image before and after; http://www1.ncdc.noaa.gov/pub/data/cmb/images/content/coral-bleaching.gif (Public Domain)
p. 151 http://upload.wikimedia.org/wikipedia/commons/4/4b/Projoverview_coral_figure_2_750%28John_Reed%29.jpg Public Domain)
p. 158 http://www.lib.noaa.gov/retiredsites/docaqua/imageschool/liverock2.jpg (Public Domain)
p. 161 http://www.noaa.gov/features/04_resources/corals.html (credit coral restoration foundation) (Public Domain)
p. 169 image of black worm from Drewes work (permission granted on website)
p. 181 fish vascular system; p. 182 gill counter exchange (Ramel G.; http://www.earthlife.net/fish/intro.html; granted permission to use the two images)
p. 199 image of a Rockwell Automation Micro 830 programmable logic controller (permission to use image granted by Rockwell Automation)
p. 209 image of microscope students may design in class by Mike Graff with kind permission
p. 272 (Public Domain Images via NOAA Chesapeake Bay activity) location of Chesapeake Bay and rivers flowing into it; p. 273, Historical changes of river flow into the Chesapeake Bay; p. 276 SeaWiFS data from September 16, 1997; p. 276 contrasting LANDSAT images of the Chesapeake Bay; p. 277 LANDSAT 7 image showing land cover types in Chesapeake Bay area
p. 281 Apo Island Marine Sanctuary Negative Tip Feedback Diagram; p. 282 Apo Island Marine Sanctuary Positive Tip Feedback Diagram (Permission of Gerry Marten)

Drawings / Photos on pages listed below are by Robert Jakus
p.39 Classroom Aquatic Ecosystem Model Number One**; p.40** Classroom Aquatic Ecosystem Model Number Two**; p.41** Classroom Aquatic Ecosystem Model Number Three; **p.42** Classroom Aquatic Ecosystem Model Number Four; **p.62** Goggles Template; **p.73** Metal Halide light bulb; **p.74** wiring diagram ballast to tubes; **p.76** greenhouse light; **p.77** schematic for aquaponic light frame. **p.79** top down view of layout of five pipes in aquaponics system; **p.79** aquaponics pool with grow frames in background; **p.80** Students Assembling Coral Propagation System on Frame; **p.81** details (2) of coral propagation frame (tongue and groove); **p.82** coral frags in an aquarium; **p.83** illustration of brood stock larval rearing system front view; **p.85** newly hatched clownfish; **p.95** water molecule; **p.95** hydrogen bonds in water; **p.95** the water cycle; **p.105** aquaponics reservoir with water; **p.105** PVC pipes return water to reservoir; **p.106** pipes set on aquaponics frame; **p.107** filling aquaponics system with water; **p.109** students conducting water tests; **p.112** light reflecting on water over coral frags; **p.113** painted coral propagation system reservoirs; **p.115** students collecting water quality data; **p.117** illustrated profile of broodstock larval rearing system; **p.118** assembled working prototype of broodstock larval rearing system; **p.118** view of plumbing in top reservoir of broodstock larval rearing system; **p.119** broodstock larval rearing plumbing view beneath top shelf; **p.120** students recording water quality data; **p.123** water quality test results; **p.124** a water quality test; **p.130** fish being transferred to the aquaponics pool; **p.130** students germinating seeds in an inert soilless medium; **p.131** leafy veggies growing in aquaponic pipe; **p.136** plants after three months in the system; **p.137** students drilling holes in PVC pipe to hold germinated seedlings; **p.137** side view of plants growing in aquaponic system; **p.138** an aquaponically grown plant's root system; **p.138** perch in the aquaponics system waiting to be fed; **p.146** student assembled coral propagation system in a classroom; **p.155** fragged soft coral in classroom system; **p.169** nest of eggs collected from classroom clownfish; **p.169** juvenile clownfish in classroom growout tank; **p.170** illustration of black worm culture system; **p.171** microscope photo of newly hatched brine shrimp in classroom; **p.177** shark epidermal layers; **p.177** variation in shark teeth; **p.180** fish external anatomy; **p.180** forces at work as a fish swims; **p.180** illustration of the water resistance to a fishes swimming called drag; **p.180** fish tail illustrations; **p.180** illustration of pitch, yaw, and roll; **p.180** fish body shapes; **p.183** gill ventilation; **p.185** salmon hatchlings; **p.193** illustration of a rotifer; **p.193** container of unicellular algae growing in the classroom; **p.195** algae culture system during a field trip; **p.197** student cultured algae next to aquaponics system algae culture system; **p.200** Micro 830 mounted on a rail; **p.200** Micro 830 showing all wiring to terminals; **p.200** Micro 830 wiring diagram; **p.201** illustration of a computer automated copepod (plankton) culture system; **p.206** microscope image of a peppermint shrimp larvae (taken in class); **p.207** magnifying glass; **p.207** focal distance activity; **p.210** microscope photo of larval peppermint shrimp eye (taken in class); **p.216** exploded illustration of microscope mechanical stage (students may build in class); **p.216** microscope photo of abdomen of egg bearing peppermint shrimp taken in class; **p.218** hemocytometer; **p.219** illustration of a grid on a hemocytometer slide; **p.247** photo of an adult peppermint in a classroom aquarium; **p.249** simple model of spherical kreisel; **p.249** operational student fabricated spherical kreisel; **p.251** model juvenile growout system #1; **p.251** model juvenile growout system #2; **p.251** model juvenile growout system #3; **p.256** unpainted, student fabricated (with fiberglass) exterior of kreisel (for raising aquatic larvae); **p.256** model upweller in water; **p.256** model upweller upside down view (during a field trip); **p.296** Laminar Fountains at Navy Pier - Chicago, IL; **p.297** Illustration of positioning of holes in top and bottom of laminar fountain; **p.298** Six images of a prototype laminar fountain - RGB LED wired-up and running via an Arduino MEGA, (3D-printed CAD-designed "cutter mechanism" - not shown); **p.301**Audio/Visual Underwater Recording Apparatus (designed to study dolphin bioacoustics); **p.303** Sea Perch in a pool; **p.305** Calcium Reactor Design

Beyond the Classroom Aquarium Connections with Next Generation Science Standards Matrix - by Project
A = Project 1, B = Project 2, C = Project 3, …, L = Project 12, M = Project 13, N = Project 14

		Physical Sciences				Life Sciences				Earth & Space Sciences			Engineering, Technology & Applications of Science	
	Scientific & Engineering Practices (1 – 8)	PS 1 Matter & its Inter-actions	PS 2 Motion & stability: Forces & interactio ns	PS 3 Energy	PS 4 Waves & their applications in technologies for information transfer	LS 1 From molecules to organisms: Structures & processes	LS 2 Ecosystems: Interactions, energy, & dynamics	LS 3 Heredity: Inheritance & variation of traits	LS 4 Biological evolution: Unity & diversity	EES 1 Earth' s place in the universe	EES 2 Earth' s systems	EES 3 Earth & human activity	ETS 1 Engineering design	ETS 2 Links among engineering, technology, science, & society
1	. Asking questions & defining problems	Projects I, L, M, and N	Projects I, L, M, and N	Projects I, L, M, and N	Projects G, I, M, and N	Projects A - N	Projects A - N	Projects F and I	Projects I and N	Not Applicable	Projects A - N	Projects K - N	Projects A – J, M, N	Projects A - N
2	Developing & using models	Projects I, L, M, and N	Projects I, L, M, and N	Projects I, L, M, and N	Projects G, I, M, and N	Projects A - N	Projects A - N	Projects F and	Projects I and N	Not Applicable	Projects A - N	Projects K - N	Projects A – J, M, N	Projects A - N
3	Planning & carrying out investigations	Projects I, L, M, and	Projects I, L, M, and	Projects I, L, M,	Projects G, I, M, and N	Projects A - N	Projects A - N	Projects F and	Projects I and N	Not Applicable	Projects A - N	Projects K - N	Projects A – J, M, N	Projects A - N
4	Analyzing & interpreting data	Projects I, L, M, and N	Projects I, L, M, and N	Projects I, L, M, and N	Projects G, I, M, and N	Projects A - N	Projects A - N	Projects F and	Projects I and N	Not Applicable	Projects A - N	Projects K - N	Projects A – J, M, N	Projects A - N
5	Using mathematics & computational thinking	Projects I, L, M, and N	Projects I, L, M, and N	Projects I, L, M, and N	Projects G, I, M, and N	Projects A - N	Projects A - N	Projects F and	Projects I and N	Not Applicable	Projects A - N	Projects K - N	Projects A – J, M, N	Projects A - N
6	Constructing explanations & designing solutions	Projects I, L, M, and N	Projects I, L, M, and N	Projects I, L, M, and N	Projects G, I, M, and N	Projects A - N	Projects A - N	Projects F and	Projects I and N	Not Applicable	Projects A - N	Projects K - N	Projects A – J, M, N	Projects A - N
7	Engaging in argument from evidence	Projects I, L, M, and N	Projects I, L, M, and N	Projects I, L, M, and N	Projects G, I, M, and N	Projects A - N	Projects A - N	Projects F and	Projects I and N	Not Applicable	Projects A - N	Projects K - N	Projects A – J, M, N	Projects A - N
8	Obtaining, evaluating, and communicating information	Projects I, L, M, and N	Projects I, L, M, and N	Projects I, L, M, and N	Projects G, I, M, and N	Projects A - N	Projects A - N	Projects F and	Projects I and N	Not Applicable	Projects A - N	Projects K - N	Projects A – J, M, N	Projects A - N
	Crosscutting Concepts (1 – 7)	**PS 1**	**PS 2**	**PS 3**	**PS 4**	**LS 1**	**LS 2**	**LS 3**	**LS 4**	**EES 1**	**EES 2**	**EES 3**	**ETS 1**	**ETS 2**
1	. Patterns	Projects I, L, M, and	Projects I, L, M, and	Projects I, L, M,	Projects G, I, M, and N	Projects A - N	Projects A - N	Projects F and	Projects I and N	Not Applicable	Projects A - N	Projects K - N	Projects A – J, M, N	Projects A - N
2	Cause & effect: Mechanism & explanation	Projects I, L, M, and N	Projects I, L, M, and N	Projects I, L, M, and N	Projects G, I, M, and N	Projects A - N	Projects A - N	Projects F and	Projects I and N	Not Applicable	Projects A - N	Projects K - N	Projects A – J, M, N	Projects A - N
3	Scale, proportion, & quantity	Projects I, L, M, and	Projects I, L, M, and	Projects I, L, M,	Projects G, I, M, and N	Projects A - N	Projects A - N	Projects F and	Projects I and N	Not Applicable	Projects A - N	Projects K - N	Projects A – J, M, N	Projects A - N
4	Systems & system models	Projects I, L, M, and	Projects I, L, M, and	Projects I, L, M,	Projects G, I, M, and N	Projects A - N	Projects A - N	Projects F and	Projects I and N	Not Applicable	Projects A - N	Projects K - N	Projects A – J, M, N	Projects A - N
5	Energy & matter: Flows, cycles, & conservation	Projects I, L, M, and N	Projects I, L, M, and N	Projects I, L, M, and N	Projects G, I, M, and N	Projects A - N	Projects A - N	Projects F and	Projects I and N	Not Applicable	Projects A - N	Projects K - N	Projects A – J, M, N	Projects A - N
6	Structure & function	Projects I, L, M, and	Projects I, L, M, and	Projects I, L, M,	Projects G, I, M, and N	Projects A - N	Projects A - N	Projects F and	Projects I and N	Not Applicable	Projects A - N	Projects K - N	Projects A – J, M, N	Projects A - N
7	Stability& change	Projects I, L, M, and	Projects I, L, M, and	Projects I, L, M,	Projects G, I, M, and N	Projects A - N	Projects A - N	Projects F and	Projects I and N	Not Applicable	Projects A - N	Projects K - N	Projects A – J, M, N	Projects A - N

About the Author – Robert M. Jakus - www.robertjakus.wordpress.com

Mr. Jakus received his B.A., in education from the University of Illinois (with K-9 certification); his M.S. in Public Health (also from the University of Illinois); and attended DePaul University from which he received his M.A. in math education (he also obtained endorsement to teach math for grades 6 – 12).

Robert Jakus has been a math teacher for thirty-five years. He has taught math at middle school, high school and community college levels and even worked for "Teach for America" as a math education professor for one semester. He has also worked as a middle school science teacher for four years. Mr. Jakus loves helping students learn math (and science), and believes there is great value in helping students find ways to apply daily life experiences to their efforts to learn math. His vision for developing and using *Beyond the Classroom Aquarium* projects as a platform from which to teach math via aquatic ecosystem themed applications has existed since 1995. As a result, many of his students have experienced *Beyond the Classroom Aquarium* project prototypes as he worked to incorporate and integrate additional aquatic ecosystem themed projects into his middle school (math/science) curriculum.

As he developed an outline and began reading literature related to this project, he soon realized a need to gain additional knowledge and professional credibility and subsequently participated in a series of professional development workshops related to it between 1997 and 2012. Some of the related workshops and professional development activities he participated in are listed below.

- July 2014, National Marine Educators Association Conference, Annapolis, MD; presented - A Brood-Stock/Larval Rearing System Your Students Will Love (featuring a prototype system for session participants to view and help assemble)
- July 2011, National Marine Educators Association Conference, Boston, MA: presented - Student-Developed Coral Propagation and Brood-Stock/Larval-Rearing Aquatic Ecosystems (featuring a prototype coral propagation system for session participants to view and help assemble)
- July 2010, National Marine Educators Association Conference Gatlinburg, TN: presented - 16 Aquatic Ecosystem Projects: A Secondary STEM Curriculum
- Winter 2010, Student Development of a 500-Gallon Aquaponics System; National Marine Educators Association - Expanding Audiences Committee - Newsletter
- July 2009, National Marine Educators Association Conference in Monterey, CA: presented - Aquatic Ecosystem Math: A Series of Projects to Integrate Math Instruction
- July 2008, National Marine Educator's Association Conference in Savannah, Georgia: presented - Peppermint Shrimp in the Classroom: A Model for Inquiry-Based Math/Science Activities
- OCEANS '07, in Vancouver, BC; presented - Mathematics of Aquatic Ecosystems – A Curriculum: Integration of Math Instruction Using a Sequence of 16 Aquatic Ecosystem Activities: published by the IEEE (2007)
- 1999 Summer; Marine Quest Counselor; University of North Carolina, Wilmington
- 1998, Developed and presented Logical Puzzle Contests and Mental Calculation Curriculum for the School of Continuing Education (Arts & Sciences Summer Camp), University of North Carolina – Greensboro

Past and Present Organizational Affiliations and Professional Development

- Member of National Marine Educators Association (2017 - 2008); member of expanding audiences committee
- Shedd Aquarium Summer Class for Teachers: Biology in the Bahamas (2012, Shedd Aquarium and aboard the Coral Reef II R/V in the waters of Bimini for one week)
- Horn Point Marine Laboratory, Cambridge, MD (2010 - Centers for Ocean Science Education Excellence – 1 week workshop)
- Marine-Lab Summer Class for Teachers (2007 - Key Largo, FL)
- Florida State University (2003 - Using Inquiry in the Classroom – Five Week Summer Class for Teachers – Panama City, FL)
- Mote Laboratory Two Week Summer Class for Educators (1997: Sarasota and Pigeon Key, FL)
- Past Member of the Marine Technology Society, National Council For the Teachers of Mathematics; and the Association for Education of the Gifted

Mr. Jakus has attempted to present the activities in this book in a style that encourages individual secondary school educators, from grades six through twelve (and in some cases beyond) to modify any activity within it so that it may be of the greatest value to their own students. Mr. Jakus hopes to see this curriculum piloted and reviewed more fully as educators learn about its' availability. He has even developed a weeklong professional development workshop (featuring the project sequence in this book) for educators who would like to implement this curriculum for their students. As with the entire curriculum, the professional development workshop is designed so that secondary school educators may gain further insights into these projects and ways they might modify (or "differentiate") the curriculum to best meet the skills and needs of their own students.

Made in the USA
Monee, IL
12 November 2026

68342115R00207